W9-ARI-754

DISCARDED
JENKS LRC
GORDON COLLEGE

Neuropharmacology of
Cyclic Nucleotides

Proceedings of the symposium on Neuropharmacology of Cyclic Nucleotides presented at the fall meeting of the American Society for Pharmacology and Experimental Therapeutics, Ohio State University, Columbus, Ohio, August 22, 1977

Neuropharmacology of Cyclic Nucleotides

Role of Cyclic AMP in Affective Disorders, Epilepsy and Modified Behavioral States

Edited by
Gene C. Palmer, Ph.D.

Urban & Schwarzenberg • Baltimore-Munich 1979

Urban & Schwarzenberg, Inc.
7 E. Redwood Street
Baltimore, Maryland 21202
U.S.A.

Urban & Schwarzenberg
Pettenkoferstrasse 18
D-8000 München 2
GERMANY

Library of Congress Cataloging in Publication Data

Symposium on Neuropharmacology of Cyclic Nucleotides, Ohio State University, 1977.
Neuropharmacology of cyclic nucleotides.

Presented at the fall meeting of the American Society for Pharmacology and Experimental Therapeutics, Ohio State University, Columbus, August 22, 1977.
Includes index.
1. Cyclic adenylic acid—Congresses. 2. Neuropharmacology—Congresses. 3. Cyclic nucleotides—Congresses. I. Palmer, Gene C II. American Society for Pharmacology and Experimental Therapeutics. III. Ohio. State University, Columbus. IV. Title. [DNLM: 1. Adenosine cyclic-3′,5′—Monophosphate—Physiology—Congresses. 2. Nucleotides, Cyclic—Pharmacodynamics—Congresses. 3. Central nervous system—Drug effects—Congresses. 4. Psychopharmacology—Congresses. QV185 N494 1977]
QP625.A35S97 1979 615′.78 78-23490

ISBN 0-8067-1521-9 (Baltimore)
ISBN 3-541-71521-9 (München)

Printed in the United States of America

This book is dedicated to the three people who have had the greatest impact with regard to my work in science: the late Dr. William Downs, Jr. formerly of Yale and Tennessee Tech Universities; Dr. G. Rodman Davenport of Vanderbilt University; and my wife, Jo.

Contents

Perspectives on Dopamine-sensitive Adenylate Cyclase in the Brain 1
Michael J. Schmidt

Actions of Neuroleptic Agents on Central Cyclic Nucleotide Systems 53
Gene C. Palmer and Albert A. Manian

Cyclic AMP and Adenylate Cyclase in Psychiatric Illness 112
G. N. Pandey and John M. Davis

Sensitivity of Noradrenergically Mediated Cyclic AMP Responses in the Brain 152
H. Ryan Wagner, Gene C. Palmer, and James N. Davis

Neuropharmacological Control of Pineal Gland Cyclic Nucleotide Systems 173
Samuel J. Strada and Michael W. Martin

Genetic Determination of Cyclic AMP Level in the Brain: Some Behavioral Implications 197
Albert Sattin

Effects of Seizures and Anticonvulsant Drugs on Cyclic Nucleotide Regulation in the CNS 211
James A. Ferrendelli, Robert A. Gross, Dorothy A. Kinscherf, and Eugene H. Rubin

Cyclic Nucleotide Levels in Brain During Ischemia and Recirculation 228
W. David Lust and Janet V. Passonneau

Microwave Inactivation as a Tool for Studying the Neuropharmacology of Cyclic Nucleotides 253
David J. Jones and William B. Stavinoha

Index 283

Contributors

James N. Davis, M.D.
Department of Medicine and Pharmacology
Duke University School of Medicine
Durham, North Carolina 27705

John M. Davis, M.D.
Illinois State Psychiatric Institute
Chicago, Illinois 60612

James A. Ferrendelli, M.D.
Department of Pharmacology and Neurology
Washington University School of Medicine
St. Louis, Missouri 63110

Robert A. Gross, B.A.
Department of Pharmacology and Neurology
Washington University School of Medicine
St. Louis, Missouri 63110

David J. Jones, Ph.D.
Department of Anesthesiology and Pharmacology
The University of Texas Health Science Center
San Antonio, Texas 78284

Dorothy A. Kinscherf, B.A.
Department of Pharmacology and Neurology
Washington University School of Medicine
St. Louis, Missouri 63110

W. David Lust, Ph.D.
Laboratory of Neurochemistry
National Institute of Neurological and Communicative Disorders and Stroke
National Institutes of Health
Bethesda, Maryland 20014

Albert A. Manian, Ph.D.
Psychopharmacology Research Branch
National Institute of Mental Health
Rockville, Maryland 20857

Michael W. Martin, B.A.
Department of Pharmacology
The University of Texas Medical School at Houston
Houston, Texas 77025

Gene C. Palmer, Ph.D.
Department of Pharmacology
University of South Alabama College of Medicine
Mobile, Alabama 36688

G. N. Pandey, Ph.D.
Illinois State Psychiatric Institute
Chicago, Illinois 60612

Janet V. Passonneau, Ph.D.
Laboratory of Neurochemistry
National Institute of Neurological and Communicative Disorders and Stroke
National Institutes of Health
Bethesda, Maryland 20014

Eugene H. Rubin, M.D., Ph.D.
Departments of Pharmacology and Neurology
Washington University School of Medicine
St. Louis, Missouri 63110

Albert Sattin, M.D.
Department of Psychiatry
Institute of Psychiatric Research
Indiana University
Indianapolis, Indiana 46260

Michael J. Schmidt, Ph.D.
Lilly Research Laboratories
Eli Lilly and Company
Indianapolis, Indiana 46206

Samuel J. Strada, Ph.D.
Department of Pharmacology
The University of Texas Medical School at Houston
Houston, Texas 77025

W. B. Stavinoha, Ph.D.
Department of Pharmacology
The University of Texas Health Science Center
San Antonio, Texas 78284

H. Ryan Wagner, Ph.D.
Department of Medicine and Pharmacology
Duke University School of Medicine
Durham, North Carolina 27705

Introduction

A vast amount of experimental evidence has accumulated within the past decade in an attempt to determine the role of cyclic nucleotides (cyclic AMP and cyclic GMP) and the enzymes responsible for their formation (adenylate and guanylate cyclases) and degradation (the phosphodiesterases) within the central nervous system. Most of these studies have been carried out by biochemists and pharmacologists. The latter have been particularly interested because of evidence linking adenylate cyclase to the adrenergic receptor. In this vein, psychoactive drugs which affect mood and behavior also appear to influence adenylate cyclase in a manner analogous to their actions on central catecholamine systems. Despite the present quantity of experimental findings, no symposium until now has been held for the sole purpose of discussing the neuropharmacology of cyclic nucleotides.

The history of cyclic nucleotide actions in central tissues began with the observation by Sutherland and Rall and their colleagues (see *Klainer et al.*, J Biol Chem 237: 1239, 1962) that catecholamines when added to cell-free particulate fractions of brain would result in an activation of adenylate cyclase. *DeRobertis et al.* (J Biol Chem 242:3487, 1967) then demonstrated the highest levels of adenylate cyclase and phosphodiesterase in subfractions of brain containing nerve endings—synaptosomes. At this time brain investigations were delayed because the tissue appeared to be too labile to demonstrate consistently significant effects of neurohumoral agents. For this reason cyclic nucleotide research in the brain lagged behind the more rapid advances made with certain other tissues, notably hepatic and adipose tissue. However, during this early period, initial steps were undertaken to characterize brain phosphodiesterase (*Cheung,* Biochem 6:1079, 1967) and guanylate cyclase (*Goldberg et al.*, J Biol Chem 244:4458, 1969), and to suggest a role for cyclic AMP in central glycogen metabolism (*Breckenridge* and *Norman,* 1965). Another important investigation was that of *Weiss* and *Costa* (Science 156:1750, 1967), who reported that homogenates of pineal gland contained an adenylate cyclase that was potently activated by β-adrenergic agonists. The pineal receives exclusively efferent noradrenergic nerve fibers and, when denervated, yields adenylate cyclase which is hypersensitive to stimulation by norepinephrine.

These findings strongly suggested a role for cyclic AMP in both the molecular mechanisms of neurotransmission and the phenomenon of denervation supersensitivity. The problem with central tissues remained, however, because the pineal consists of glandular cells and is non-neural. The most salient breakthrough came when *Kakiuchi* and *Rall* (Mol Pharmacol 4:379, 1968) found with incubated tissue slices (intact cellular preparations of brain) a profound elevation in cyclic AMP in response to added histamine or norepinephrine. Moreover, preliminary studies with the appropriate blocking agents revealed that these responses possessed specific receptor characteristics. Pharmacologists now had a potent research technique to evaluate the effects of psychotropic drugs in relation to neurohumorally mediated receptor responses in a wide variety of brain regions. At this time, many of the participants in the present symposium, along with other workers (notably Drs. Perkins, Daly, Paul, Gilman, Bloom, Siggins, Hoffer, Chasin, Abdulla and Greengard), began to make major contributions to a rapidly expanding field.

The major purpose of this symposium was to bring together new data from a number of productive laboratories that had been working for quite some time in the brain cyclic nucleotide field. The following chapters provide insights into several different aspects of the role of cyclic nucleotides in the brain. These insights were derived from both clinical and basic science approaches, and a number of important points including methodology are emphasized.

Dr. Schmidt tackles the difficult task of sorting out the conflicting data associated with the identification of the dopamine receptor and its relation to the dopamine-sensitive adenylate cyclase system. These findings may help to establish a correlation between drug effects on adenylate cyclase and central disease states such as schizophrenia and parkinsonism. In addition to these present observations, Dr. Schmidt conducted the first studies of cyclic nucleotides in the developing brain.

The chapter on neuroleptic actions on cyclic nucleotide systems brings together the vast amount of data indicating that cyclic AMP is involved in a major way in the molecular action of these compounds in the treatment of schizophrenia. The work with Dr. Manian began in 1968 and has included three major aspects of phenothiazine actions: 1) the specificity of the response, indicating that only neurohumoral stimulation of cyclic AMP formation appears to be influenced by neuroleptic agents; 2) the role of metabolites, which may contribute to the overall therapeutic and adverse effects of the parent drugs; and 3) the possibility that long-term treatment leads to either a refractoriness to neuroleptic action or a supersensitivity of the neurohumoral receptors.

The extreme difficulty in relating animal studies on the neuropharmacology of cyclic nucleotides to conditions and treatment of mental patients is emphasized by Drs. Pandey and Davis. A rather important multifaceted approach with human subjects was conducted using plasma, platelets, leukocytes, postmortem material and cerebrospinal fluid. In some cases indirect correlations can be made between affective disorders and cyclic nucleotide systems.

Since receptors for neurohumoral agents that are related to adenylate cyclase may undergo physiological alterations, Drs. Wagner and Davis have investigated the possibility of correlating direct receptor ligand binding of adrenergic agents to conditions of super- and subsensitivity of adenylate cyclase. Direct measurements of changes in affinity, intrinsic activity and receptor number are necessary prerequisites before conclusions can be drawn about the relationship between alterations in adenylate cyclase activity and behavioral modifications.

The pineal gland continues to serve as an extremely useful model of exclusively postsynaptic adrenergic receptors. Dr. Strada and his colleague have evaluated the functional alterations in adenylate cyclase activity that result from neuropharmacological, environmental, hormonal and physical (interruption of neural pathways) manipulations of this organ. Studies on a simple, basic system such as this are necessary prerequisites for future speculations as to how such manipulations affect key regions within the brain itself. In addition to his work with the pineal gland, Dr. Strada has made major contributions to the characterization of and the ontogeny of central phosphodiesterases.

Dr. Sattin conducted the initial experiments showing a stimulatory action of adenosine on hypothetical adenosine receptors that appear to influence adenylate cyclase in brain tissue. His most recent work is attempting to unravel the complex genetic relationships between behavioral activity and cyclic AMP-mediated responses in the brain. These studies may eventually lead to a genetic-biochemical link in normal/abnormal mental situations.

Dr. Ferrendelli is responsible for much of our present knowledge of how cyclic GMP functions in the brain. In addition, some of the first studies with regard to drug effects on *in vivo* steady-state levels of cyclic nucleotides in brain were conducted in his laboratory. One of the major thrusts of his current research has been to understand the molecular-biochemical actions of anticonvulsive drugs. His laboratory has provided evidence for a significant role for cyclic nucleotides during various stages of convulsive seizures, induced either by chemical or electrical means. Various experimental techniques utilizing *in vivo* or *in vitro* conditions have revealed a protective action of anticonvulsant drugs in preventing the rise in cyclic nucleotides induced by these convulsive conditions. A distinct contribution has thus been made toward our understanding of the basic underlying mechanisms of epilepsy.

Another prominent area in neurology, the study of the effects of anoxia and ischemic conditions, also has been linked experimentally to cyclic nucleotide actions. Drs. Lust and Passonneau additionally have investigated several other effects of the ischemic condition, notably those on high-energy compounds and intermediary metabolites. Taken together, their findings eventually may lead to a better understanding of related conditions in humans, notably stroke and its sequelae. Dr. Lust also worked on the initial experiments to determine a role for cyclic GMP in the brain.

Perhaps one of the most useful tools for neuropharmacological investigations requiring exact and meaningful measurements of metabolically labile compounds in the brain has resulted from the pioneering development and engineering modification of the high intensity-focused microwave apparatus by Drs. Jones and Stavinoha. This instrument is a useful tool in that precise measurements of cyclic nucleotide ratios can be carried out in discrete brain regions subsequent to drug injection. The resulting rapid fixation of the brain allows for a direct assessment of the effects of neuropharmacological agents under *in vivo* conditions. In their chapter the authors describe the extensive studies on cyclic nucleotides that were undertaken during the development of this instrument.

Given the rapid pace at which advances are being made, no book or symposium could adequately describe or encompass all the diverse aspects of even a highly specialized field such as this. However, an attempt was made to draw together a series of fresh ideas and conclusions about how certain drugs and conditions may affect brain function, with emphasis on the possible role of the cyclic nucleotides. Most readers can expect to gain some new insights into some old problems, as well as a clearer idea of where the results of future research are likely to lead.

G. Alan Robison,
and
Gene C. Palmer

Neuropharmacology of Cyclic Nucleotides

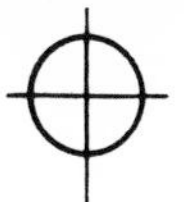

Perspectives on Dopamine-sensitive Adenylate Cyclase in the Brain

Michael J. Schmidt
Lilly Research Laboratories
Eli Lilly and Company
Indianapolis, Indiana 46206

INTRODUCTION

The dopamine theories of schizophrenia and parkinsonism have been useful in focusing research efforts in these areas. The hypothesis that schizophrenia is due to an overactive dopaminergic system in the brain (*Snyder et al.*, 1974) has fostered programs to find dopamine receptor blocking agents. Conversely, the lack of dopamine in the striatum of Parkinson patients (*Hornykiewicz*, 1966) has led to increased efforts to find direct-acting dopamine receptor stimulants. If these efforts to affect the dopamine systems in the brain are to be successful and the search for new compounds expedient, the experimental systems used to test compounds for dopamine agonist or antagonist activity must be valid models of the disease state and/or yield results that predict the clinical efficacy of new compounds.

Dopamine stimulation of adenylate cyclase in brain homogenates has been used widely in recent years to assess the dopamine agonist and antagonist activity of compounds. However, there is a growing list of correlative inconsistencies in the assay with regard to drug effects and physiological manipulations which alter dopamine receptor function *in vivo*. This review is an attempt to discuss and interpret these inconsistencies in order that a conclusion might be reached as to the validity and clinical predictability of results generated with this assay of dopamine receptor function.

HISTORICAL PERSPECTIVE

Shortly after the discovery of cyclic AMP (adenosine 3′,5′-monophosphate), it was found that the activity of adenylate cyclase [ATP pyro-

The author greatly appreciates the critical reading of this manuscript by Drs. Ray Fuller and Patrick Roffey of the Lilly Research Laboratories. He also thanks Ms. Ruth Leonard for her diligence and excellent help in preparing the text.

phosphatelyase (cyclizing):EC 4.6.1.1.] in the brain was greater than in other tissues (*Sutherland, Rall*, and *Menon*, 1962). This suggested that cyclic AMP might play an important role in brain function. Although the regional distribution of adenylate cyclase in the brain gave little insight as to the functional role of cyclic AMP in the central nervous system (*Weiss* and *Costa*, 1968), subcellular distribution studies were most intriguing. *DeRobertis et al.* (1967) observed that the major portion and highest specific activity of adenylate cyclase were found in the subcellular fraction which contained synaptosomes and membrane fragments. This supported the conjecture that cyclic AMP functions as a "second messenger" for neurotransmitters at the synapse, in an analogous fashion to the role postulated for cyclic AMP in peripheral organs such as the heart and liver.

The first search for the hormones which stimulated adenylate cyclase in the nervous system was undertaken by Sutherland and co-workers. These investigations revealed that epinephrine, norepinephrine and isoproterenol produced slight increases in the formation of cyclic AMP in cell-free preparations from brain regions of several species (*Klainer et al.*, 1962). A survey of the catecholamines showed isoproterenol to be more potent than epinephrine, which in turn was more potent than norepinephrine. *Greengard* and *Schmidt* (unpublished observations, 1968) observed a similar order of catechol potencies in the steer and rabbit cerebellum, and demonstrated that propranolol, but not phenoxybenzamine, blocked the rise in cyclic AMP in these tissues.

Other investigators, however, did not find significant hormonal effects in cell-free preparations from the brain. *Brown* and *Riley* (1965) were unable to observe any stimulation of adenylate cyclase in response to epinephrine, dopamine, histamine or norepinephrine. *Voigt* and *Krishna* (1967), *Schmidt et al.* (1970), *Weiss* and *Costa* (1968) and *Williams, Little,* and *Ensinck* (1969) all reported similar negative results in various areas of the brain in several animal species, including humans. The inability to replicate the findings of *Klainer et al.* (1962) was thought to reflect the extreme lability of adenylate cyclase in nervous tissue. This labile property of the cell-free enzyme stimulated workers to seek other tissue preparations in which to study the effects of hormones on brain cyclic AMP.

Kakiuchi and *Rall* (1968) were the first to report that in intact brain tissue, in the form of sliced tissue sections, reproducible and significant elevations of cyclic AMP occurred in the presence of several putative neurotransmitters. However, although norepinephrine produced a 5-fold rise in cyclic AMP concentrations in slices of the striatum, dopamine was without effect (*Palmer, Sulser,* and *Robison,* 1973). The inability of dopamine to increase cyclic AMP accumulation in striatal slices was an enigma. Also, the fact that neurohormones produced dramatic elevations of cyclic AMP in slices of brain but only slight percentage increases in homogenates discouraged further exploration of neurotransmitter-cyclic AMP interactions in homogenates. Most work on cyclic nucleotides in the CNS continued using the slice preparation.

Renewed interest in cell-free adenylate cyclase began when several investigators reported elevations of cyclic AMP synthesis in homogenates of nervous tissue in the presence of dopamine. *Kebabian* and *Greengard* (1971) observed that dopamine could stimulate cyclic AMP accumulation in homogenates of the superior cervical ganglia, and *Brown* and *Makman* (1972) detected dopamine stimulation of adenylate cyclase in the retina. Shortly thereafter, *Von Hungen* and *Roberts* (1973) demonstrated dopamine-stimulated adenylate cyclase in the crude mitochondrial fraction from rat cerebral cortex, and *Kebabian, Petzold,* and *Greengard* (1972) and *Von Hungen, Roberts,* and *Hill* (1974) detected dopamine stimulation in homogenates of the corpus striatum. In the latter preparation norepinephrine was also active, but to a lesser degree than dopamine. Apomorphine, a classical dopamine agonist, was able to partially activate cyclic AMP synthesis. Pharmacological characterization of the receptor indicated that the dopamine response could be inhibited by clinically effective antipsychotics (*Clement-Cormier et al.*, 1974; *Karobath* and *Leitich,* 1974; *Miller, Horn,* and *Iversen,* 1974a).

The dopamine-sensitive adenylate cyclase assay quickly gained acceptance as a useful model of dopamine receptor function, and a considerable amount of data has been accumulated in many laboratories on classical as well as new dopamine receptor antagonists and agonists. In most cases results agree and clinical efficacy might have been predicted from results of the cyclase assay. However, some discrepancies have appeared which have fostered a growing concern as to the relevance and predictive value of this assay. These discrepancies deserve consideration for two reasons: 1) to prevent "blind alley" research, and 2) to reveal subtle and potentially useful intricacies of the dopamine receptor. In the remainder of this discussion I hope to show that the dopamine-sensitive adenylate cyclase assay is useful for the latter reason.

CYCLIC AMP AND STRIATAL FUNCTION

At the outset it might be questioned whether cyclic AMP mediates the effects of dopamine in the striatum and other dopamine-rich areas of the brain. There is considerable evidence, both *in vitro* and *in vivo,* indicating a "second messenger" role for cyclic AMP in the striatum (Table 1). The enzymes responsible for cyclic AMP synthesis and metabolism, and cyclic AMP-dependent protein kinase and its endogenous substrates, are present in the striatum. Furthermore, cyclic AMP mimicks

Table 1. Evidence That Cyclic AMP Mediates the Effects of Dopamine in the Striatum

1. Dopamine-sensitive adenylate cyclase is present in brain areas which are rich in dopamine and have specific dopamine reuptake sites, e.g., the corpus striatum, olfactory tubercle, nucleus accumbens and frontal cortex (*Horn, Cuello,* and *Miller,* 1974; *Trabucchi et al.*, 1976a; *Clement-Cormier* and *Robison,* 1977).
2. The highest specific activity of dopamine-sensitive adenylate cyclase activity is found in subcellular fractions of the striatum which contain nerve endings (*Krueger, Forn,* and *Greengard,* 1975).
3. Dopamine is the most potent catechol stimulant of cyclic AMP synthesis in the striatum *in vitro* (*Kebabian, Petzold,* and *Greengard,* 1972; *Makman et al.*, 1975).
4. The dopamine agonists apomorphine, epinine and S 584 stimulate adenylate cyclase *in vitro* (*Makman et al.*, 1975).
5. The dopamine-induced elevation of cyclic AMP is specifically antagonized by dopamine receptor blocking agents; beta blockers are inactive and alpha blockers only partially effective (*Kebabian et al.*, 1972; *Clement-Cormier et al.*, 1974; *Karobath* and *Leitich,* 1974; *Iversen, Rogawski,* and *Miller,* 1976; *Schmidt* and *Hill,* 1977).
6. Apomorphine, amphetamine and L-DOPA elevate cyclic AMP in the striatum *in vivo* (*Carenzi et al.*, 1975a, 1976; *Garelis* and *Neff,* 1973; *Von Voigtlander, Boukma,* and *Johnson,* 1973).
7. Cyclic AMP phosphodiesterase is present in the striatum (*Weiss* and *Costa,* 1968; *Minneman,* 1976).
8. Phosphodiesterase inhibitors increase cyclic AMP concentrations in the striatum (*Krueger et al.*, 1975).
9. Apomorphine-, pyribedil- and L-DOPA-induced contralateral rotation in lesioned rats is potentiated by theophylline, IBMX, dipyridamol and caffeine, which are cyclic AMP phosphodiesterase inhibitors (*Fuxe* and *Ungerstedt,* 1974; *Satoh et al.*, 1976; *Fredholm, Fuxe,* and *Agnati,* 1976).
10. Injection of dibutyryl cyclic AMP into lesioned rats produces contralateral rotation in a manner similar to that of apomorphine (*Satoh et al.*, 1976).
11. The electrical effects of dopamine in the striatum are potentiated by co-administration of cyclic AMP phosphodiesterase inhibitors (*Siggins, Hoffer,* and *Ungerstedt,* 1974).
12. Dopamine and apomorphine-induced depression of neuronal activity in the striatum is mimicked by the iontophoretic application of dibutyryl cyclic AMP (*Siggins et al.*, 1974; *Woodruff, McCarthy,* and *Walker*, 1976).
13. Cyclic AMP-dependent protein kinase is present in the striatum (*Truex* and *Schmidt,* unpublished observations).
14. Specific proteins in synaptic membrane fragments from the striatum are phosphorylated in the presence of cyclic AMP or 3-isobutyl-1-methylxanthine, a phosphodiesterase inhibitor (*Krueger et al.*, 1975; *Ehrlich, Rabjohns,* and *Routtenberg,* 1977).
15. Phosphoprotein phosphatase is present in the caudate nucleus (*Maeno* and *Greengard,* 1972; *Truex* and *Schmidt,* unpublished observations).

the effects of dopamine and dopamine agonists both electrophysiologically and behaviorally. The pharmacology of the striatum also indicates that cyclic AMP mediates the effects of dopamine. Therefore, a large number of observations from independent laboratories have confirmed a role for cyclic AMP in the functioning of the striatum.

ENIGMAS OF THE DOPAMINE-SENSITIVE ADENYLATE CYCLASE ASSAY

Although there is overwhelming evidence that cyclic AMP and dopamine are acting in concert in the striatum *in vivo,* not all of the results from *in vitro* studies of dopamine-sensitive adenylate cyclase in the striatum are reconcilable with our current understanding and concept of the dopamine receptor (Table 2).

Homogenization

Dopamine-stimulated cyclic AMP synthesis is greatest and most reproducible in striatal tissue after homogenization. More than 30 laboratories have replicated the original findings of *Kebabian et al.* (1972) that dopamine is more active than norepinephrine or isoproterenol in stimulating cyclic AMP synthesis in striatal homogenates. However, only one laboratory (*Wilkening* and *Makman,* 1975) has replicated and published the experiments by *Forn, Krueger,* and *Greengard* (1974) showing that dopamine could elevate cyclic AMP levels in intact tissue slices of the striatum. Other workers have found dopamine not to be a potent stimulant in slices (*Harris,* 1976), and some have been unable to detect a dopamine-induced rise in cyclic AMP in slices of the caudate *in vitro* (*Palmer et al.,* 1973; *Schmidt,* unpublished observations; *J. H. Heltzer* and *H. M. Sarau,* personal communication; *Friedhoff, Bonnet,* and *Rosengarten,* 1977). An increase in cyclic AMP in striatal slices was observed by *Heltzer* and *Sarau* (personal communication), but only when the concentration of dopamine was in the millimolar range. This response was attenuated by theophylline, an adenosine receptor blocking agent, indicating that the high levels of dopamine probably induced the release of adenosine, which was the actual mediator of the increments observed. Dopamine also does not stimulate cyclic AMP accumulation

Table 2. Enigmatic Findings

1. Dopamine-stimulated cyclic AMP synthesis is greatest, most specific and most reproducible in caudates which have been homogenized.
2. Not all dopamine antagonists inhibit dopamine-sensitive adenylate cyclase in order of their clinical potency.
3. Some compounds which are direct-acting dopamine agonists *in vivo* do not activate dopamine-sensitive adenylate cyclase *in vitro.*
4. Denervation or pharmacologically induced supersensitivity is not always reflected in elevated dopamine-sensitive adenylate cyclase activity *in vitro.*
5. The pituitary gland has adenylate cyclase and dopamine, apomorphine and haloperidol binding sites, but dopamine does not stimulate cyclic AMP synthesis in the pituitary.

in slices of the limbic forebrain, which is also a dopamine-rich area of the brain (*Blumberg, Taylor,* and *Sulser,* 1975; *Christina et al.,* 1977).

The dilemma is that under the conditions in which dopamine is inactive, norepinephrine and isoproterenol elevate cyclic AMP 3- to 5-fold. Most organs lose hormonal responsiveness upon homogenization (*Robison, Schmidt,* and *Sutherland,* 1970), which is understandable in terms of mechanical disruption of the receptor sites coupled to adenylate cyclase. What is not easily understood is why a hormonal response (dopamine stimulation of striatal adenylate cyclase) is enhanced by homogenization. This occurs in the hypothalamus (*Palmer et al.*, 1973; *Gunaga* and *Menon*, 1973; *Ahn* and *Makman*, 1977) and limbic forebrain (*Christina et al.,* 1977; *Horn, Cuello*, and *Miller*, 1974) as well.

Activity of Dopamine Receptor Antagonists

In general, there has been good correlation between inhibition of dopamine-sensitive adenylate cyclase activity *in vitro* and the clinical effectiveness of antipsychotic agents, e.g., in the phenothiazine (fluphenazine vs promazine; *Clement-Cormier et al.,* 1974), thioxanthene (α-flupenthixol vs β-flupenthixol; *Miller et al.,* 1974a) and dibenzazepine (loxapine vs clozapine; *Iversen, Horn,* and *Miller,* 1975) classes, and with the (+) and (−) enantiomers of butaclamol (*Miller, Horn,* and *Iversen,* 1975). However, several investigators have found that butyrophenones such as haloperidol, pimozide and spiroperidol are very active clinically but are poor antagonists in the adenylate cyclase assay (*Clement-Cormier et al.,* 1974; *Karobath* and *Leitich,* 1974; *Iversen, Rogawski,* and *Miller*, 1976). This discrepancy has been ascribed to drug metabolism or distribution (*Clement-Cormier et al.*, 1974; *Soudjin* and *van Wijngaarden,* 1972); the poor solubility of this class of compound *in vitro* (*Laduron,* 1976); other therapeutically important effects of compounds *in vivo*, e.g., the antimuscarinic action of clozapine (*Anden* and *Stock,* 1973); the presynaptic blockade of dopamine release by butyrophenones (*Seeman* and *Lee,* 1975); or the inhibition of GABA uptake in synaptosomes (*Enna et al.,* 1976).

The butyrophenones also seem to exert greater activity on presynaptic receptor sites, as evidenced by greater antagonism of apomorphine-induced alterations in catechol synthesis compared to adenylate cyclase inhibition (*Iversen et al.,* 1976). Therefore, subtle differences in the topography of pre- and postsynaptic receptors might contribute to the lack of correlation between clinical effectiveness of antipsychotics and cyclase antagonism. In the latter case only postsynaptic receptor activity is reflected, while in the whole organism both pre- and postsynaptic effects contribute to the overall effectiveness of a given compound.

Strong support for the contention that subtle differences exist in receptor topographies has come from studies on the binding of radioactive ligands which represent dopamine agonists and antagonists. *Burt et al.* (1975) found that dopamine agonists had higher affinities and more easily displaced dopamine from striatal membranes than dopamine antagonists. The butyrophenones were especially weak in their ability to displace dopamine, just as they are weak inhibitors in the adenylate cyclase assay. However, dopamine antagonists proved to be especially active in displacing labeled haloperidol from its binding sites (*Burt, Creese*, and *Snyder*, 1976). In fact, the order of activity in the haloperidol binding assay closely resembles that of the clinical potencies of this class of compound. The latter studies indicate that it is not the solubilities of butyrophenones which limit their effectiveness in the adenylate cyclase assay *in vitro,* as has been previously suggested (*Laduron,* 1976), since clinical potencies still correlate among soluble (phenothiazines) and insoluble (butyrophenones) compounds in binding assays *in vitro.* It would seem, then, that dopamine antagonists in the butyrophenone class interact with the dopamine receptor sites in binding assays differently than in the dopamine-sensitive adenylate cyclase assay.

There are other examples of compounds displaying dopamine antagonist properties *in vivo* and showing little or no antagonism of dopamine-sensitive adenylate cyclase *in vitro. Pugsley, Merker,* and *Lippman* (1976) found that analogues of butaclamol differed in their ability to antagonize dopamine in the cyclase assay *in vitro* and to affect amine metabolism *in vivo.* We have found that fenfluramine and norfenfluramine, compounds which elevate serum prolactin concentrations and brain DOPAC levels (*Fuller, Perry,* and *Clemens,* 1976) as well as antagonize amphetamine-induced sterotypy and hyperactivity (*Berger, Brown,* and *Krantz,* 1973), do not block dopamine-stimulated cyclic AMP synthesis *in vitro* (*Schmidt* and *Hill,* unpublished observations; *Setler,* personal communication). Substituted benzamides (sulpiride, metoclopramide) also do not block cyclase *in vitro* yet exert a wide range of dopamine blocking activities *in vivo* (*Elliott et al.*, 1977).

The above examples illustrate that the dopamine antagonist capacity of several classes of compounds *in vivo* is not always reflected in the ability of these compounds to antagonize dopamine-induced stimulation of cyclic AMP accumulation in striatal homogenates *in vitro.* Although drug distribution and metabolism might account for some of the disparities, the *in vitro* experiments with ligand binding suggest that dopamine receptor sites *in vivo* and in the binding preparations might be different than the dopamine receptor sites responsible for activity in the adenylate cyclase assay.

Activity of Dopamine Receptor Agonists

Just as not all dopamine receptor antagonists inhibit adenylate cyclase (as might have been expected from clinical and other *in vivo* data), so also dopamine receptor agonist activity has not always been mirrored in the adenylate cyclase assay. The earliest account of dopamine-sensitive adenylate cyclase also gave the first indication that the dopamine receptor coupled to cyclase might not be the classical dopamine-apomorphine receptor seen in other test systems.

Kebabian et al. (1972) observed that apomorphine, which is considered to be a direct-acting dopamine agonist (*Anden et al.*, 1967), failed to increase adenylate cyclase activity to the degree seen in the presence of dopamine. Dopamine stimulated cyclic AMP synthesis 111% over control, while apomorphine increased activity only 41%. Others have also found apomorphine to be only a partial agonist in the cyclase assay in the frontal cortex (*Palmer* and *Palmer*, 1978), nucleus accumbens and olfactory tubercle (*Horn, Cuello,* and *Miller,* 1974), median eminence (*Clement-Cormier* and *Robison,* 1977), substantia nigra (*DiChiara et al.*, 1977a), and striatum of rats (*Baldessarini* and *Walton,* 1975; *Spano et al.*, 1976b; *Schmidt* and *Hill,* 1977) and monkeys (*Makman, Brown,* and *Mishra,* 1975). Activations produced by apomorphine have ranged from 16% to 40%, compared with 80% to 150% elevations seen with dopamine. Only *Makman et al.* (1975) and *Mishra et al.* (1975) have reported apomorphine to be equal in activity to dopamine in the adenylate cyclase assay. These workers suggested that impurities or oxidation products of apomorphine preparations account for the low activity reported by other workers. However, we have found that apomorphine, prepared in 0.1% ascorbic acid immediately before use still did not elevate cyclic AMP synthesis to the level seen with dopamine.

Furthermore, there are other reasons to believe that dopamine and apomorphine are not behaving in a similar manner in the adenylate cyclase assay. The dose-response curve for apomorphine, but not dopamine, in the cyclase assay is biphasic: stimulation at low concentrations and inhibition at higher concentrations (*Kebabian et al.*, 1972; *Clement-Cormier* and *Robison,* 1977; *Schmidt* and *Hill,* unpublished observations; *Miller et al.*, 1974b). In fact, concentrations of apomorphine greater than 10^{-5} M actually antagonize dopamine-stimulated synthesis of cyclic AMP (*Kebabian et al.*, 1972; *Miller et al.*, 1974b; *Schmidt,* unpublished observations). Thus, apomorphine behaves like a mixed agonist/antagonist in the adenylate cyclase assay, which is not usually observed in *in vivo* tests of apomorphine's dopaminergic activity.

Another dissimilarity between dopamine and apomorphine in the cyclase assay has been observed in studies on the development of the dopaminergic receptor system in the striatum. *Spano et al.* (1976b) reported that the maximum response to apomorphine, but not dopamine, declined during the first weeks after birth. The K_m for apomorphine also increased during development, indicating a decreased affinity of apomorphine for adenylate cyclase in more mature animals. Ontogenic studies have also suggested that the development of the dopamine binding capacity in the brain and dopamine-sensitive adenylate cyclase do not occur simultaneously. Development of dopamine-sensitive adenylate cyclase precedes the development of haloperidol binding activity, although both reach adult potential at the same time after birth (*Coyle* and *Campochiaro,* 1976; *Pardo et al.,* 1977).

An even greater disparity between the dopamine agonist profile of a compound *in vivo* and activity in the adenylate cyclase assay occurs among ergots and ergolines. Many of the clinical effects of ergots are compatible with the dopaminergic activity of these compounds, and several ergots mimic the actions of dopamine in animals (see *Schmidt* and *Hill,* 1977, for a brief review). However, none of the ergots tested thus far in the adenylate cyclase assay are equal in activity to dopamine (*Von Hungen, Roberts,* and *Hill,* 1975; *DaPrada et al.,* 1975; *Trabucchi et al.,* 1976a; *Schmidt* and *Hill,* 1977; *Govoni et al.,* 1977a; *Fuxe et al.,* 1978a, b & c). In many cases ergots were totally incapable of significantly elevating cyclic AMP synthesis in striatal homogenates (*Schmidt* and *Hill,* 1977; *Govoni et al.,* 1977a; *Trabucchi et al.,* 1976a), and most displayed the agonist/antagonist profile seen with apomorphine in the cyclase assay (*Von Hungen, Roberts,* and *Hill,* 1975; *Schmidt* and *Hill,* 1977; *Govoni et al.,* 1977a).

In summary, the incongruous results in the adenylate cyclase assay with a number of established dopamine agonists, and the indications that the development of dopamine-sensitive adenylate cyclase and haloperidol binding do not occur in concert, suggest that the dopamine receptor site in the adenylate cyclase assay might not be equivalent to the classical dopamine receptor or the dopamine receptor that mediates the behavior normally classed as dopaminergic.

Supersensitivity Phenomenon

Another discrepancy in the adenylate cyclase literature involves the changes in postsynaptic receptors which accompany denervation. Following interruption of dopamine pathways by electrolytic, surgical or chemical means, or by chronic administration of postsynaptic receptor blocking agents, a phenomenon referred to as supersensitivity develops. This is reflected in an enhanced stereotype response to apomorphine (*Tarsy* and *Baldessarini,* 1974), elevated apomorphine-induced locomotor activity (*Gianutsos* and *Moore,* 1977), reduced dopamine turnover

(*Gianutsos* and *Moore*, 1977), contralateral rotation in response to direct-acting dopamine agonists (*Ungerstedt*, 1971), increased sensitivity to iontophoretically applied dopamine and apomorphine in the striatum (*Siggins, Hoffer,* and *Ungerstedt,* 1974), and enhanced binding of ^{3}H-haloperidol (*Burt, Creese,* and *Snyder,* 1977; *Creese, Burt,* and *Snyder,* 1977a) and ^{3}H-dopamine (*Friedhoff, Bonnet,* and *Rosengarten,* 1977) to striatal membrane preparations. In the binding studies it was concluded that an increase in the number of specific binding sites, rather than an alteration in the affinity of the compounds for the receptors, accounted for the changes seen. If adenylate cyclase and the dopamine receptor are coupled or one and the same (*Kebabian et al.*, 1972), it would be expected that the denervation-induced increase in the number of receptor sites would be accompanied by an elevation of adenylate cyclase in the striatum. However, the majority of investigators who have studied adenylate cyclase in the supersensitivity paradigms have found that dopamine-adenylate cyclase is not changed.

Intraventricular injection of 6-hydroxydopamine or radiofrequency lesions of the substantia nigra produce a marked fall in dopamine concentrations in the striatum. Under these conditions *Von Voigtlander, Boukma,* and *Johnson* (1973) were unable to detect significant differences in dopamine-induced elevations of cyclic AMP in homogenates of the striatum. Dopamine depletion by α-methylparatyrosine also failed to enhance the dopamine response. *Krueger et al.* (1976) also were unable to detect an elevation in dopamine-stimulated cyclic AMP accumulation after radiofrequency or 6-hydroxydopamine lesioning. Others have found this to be the case as well, but have not formally reported their results (*Kebabian et al.; M. Goldstein* and *U. Ungerstedt*; *Iversen*, as noted by *Krueger et al.*, 1976). In all of the above studies, clear evidence of altered sensitivity was seen in behavioral testing.

In conflict with the above studies is the report of *Mishra et al.* (1974). These workers found that adenylate cyclase sensitivity was enhanced following either type of lesion. However, a close inspection of the data reveals that the major effect of the radiofrequency lesion was to depress "basal" adenylate cyclase activity without changing the amount of cyclic AMP produced in the presence of dopamine. Therefore, the calculated effect of dopamine (stimulated minus basal) was in fact greater following the lesion. It is not clear why the basal activity fell following the radiofrequency lesion or why *Krueger et al.* (1976) did not observe this effect. *Mishra et al.* (1974) saw no change in basal activity following 6-hydroxydopamine injection, but the dopamine response was clearly elevated on the lesioned side. Recently, *Fuxe et al.* (1978a, c) reported that denervation increased by 27% the dopamine-elicited synthesis of cyclic AMP in striatal homogenates. The activity of another dopamine agonist, lergotrile, was also increased. Yet three other dopamine agonists did not display enhanced activity in the denervated striatum. Thus far, these are the only reports of enhanced dopamine-stimulated adenylate cyclase activity in homogenates of the striatum following chemical or electrothermic lesions. It appears that the method of lesions is not a factor, since Krueger was unable to detect a consistent dopamine supersensitivity in rats prepared by the Makman group (*Krueger,* personal communication, 1977).

Striatal homogenates have also been employed to study supersensitivity developing after chronic administration of neuroleptics or agents which deplete the brain of dopamine. When mice were removed from a diet containing haloperidol, they displayed exaggerated locomotor responses when challenged with apomorphine or amphetamine (*Von Voigtlander, Losey,* and *Triezenberg,* 1975), yet adenylate cyclase in these mice did not become hyperresponsive to dopamine. Others also found that haloperidol or clozapine (*Lloyd et al.,* 1977) or chlorpromazine (*Rotrosen, Friedman,* and *Gershon,* 1975) did not alter basal or dopamine-stimulated adenylate cyclase activity in striatal homogenates. *Iwatsubo* and *Clouet* (1975) were able to detect a small increase in dopamine sensitivity of adenylate cyclase in striatal synaptosomal-mitochondrial preparations after acute administration of haloperidol (20–60 min) or after 3 days of daily injections. However, the differences with treatment were small, and a confounding fact was that morphine produced much the same effect.

Palmer and *Wagner* (1976) found that depleting brain amines by administration of reserpine for 4 days led to an enhanced (+60%) dopamine- or norepinephrine-induced elevation of cyclic AMP in homogenates of the cerebral cortex and striatum. However, these same authors (*Palmer, Wagner,* and *Putnam,* 1976) could detect only minor increases (10%–15%) in dopamine sensitivity in subsequent experiments aimed at determining if amine supersensitivity resided in the neuronal fraction. In fact, there was no increase in dopamine sensitivity in the neuronal fraction from chronically reserpinized rats. *Rotrosen, Friedman,* and *Gershon* (1975) also detected no change in dopamine-sensitive adenylate cyclase in the striatum of reserpinized rats.

Interesting findings in this area have recently been reported by *Gnegy, Uzunov,* and *Costa* (1977). Clear evidence of increased dopamine-sensitive adenylate cyclase activity in striatal homogenates was seen after chronic administration of haloperidol or d-butaclamol, but not after treatment with the inactive enantiomer of butaclamol. The changes in receptor sensitivity correlated with an increase in the amount of the calcium-dependent protein activator of adenylate cyclase present in the preparation. This work supports the contention that the amount of protein activator in the striatum might regulate the sensitivity of dopamine receptors in the striatum (*Gnegy, Uzunov,* and *Costa,* 1976).

Confirmation of these studies is important, for the findings might bear directly on possible mechanisms by which receptors regulate their sensitivity.

In summary, attempts to demonstrate increased dopamine-sensitive adenylate cyclase activity in cell-free preparations of the striatum following chronic amine depletion or receptor blockade have not always been successful. Only three groups have reported enhanced sensitivity, and in two instances the changes were small. On the other hand, at least seven groups have failed to observe alterations in adenylate cyclase under conditions in which behavioral or biochemical evidence of dopamine receptor supersensitivity was evident. There is obviously a need for further experimentation in this area.

Attempts to demonstrate dopamine supersensitivity have also been conducted using intact cyclic AMP generating systems. As previously mentioned, it is difficult to detect dopamine-induced elevations of cyclic AMP levels in slices of the striatum. However, *Krueger et al.* (1976) were able to measure this effect. They extended these studies into the supersensitivity area employing electrothermic lesions of the substantia nigra to induce striatal depletion of dopamine. Both dopamine and norepinephrine stimulated cyclic AMP accumulation in striatal slices, with norepinephrine being approximately twice as active as dopamine. Lesions produced a reduction in the half-maximally effective concentration of norepinephrine and dopamine, but the response characteristics to isoproterenol were unchanged by the lesion (*Krueger et al.,* 1976). It should be remembered that in similar animals these workers could not detect a supersensitivity to dopamine in homogenates of the striatum.

Tissue slices of the striatum were also used by *Friedhoff et al.* (1977) to study effects of prolonged neuroleptic treatment (haloperidol for 28 days) on dopamine receptor function. Dopamine binding was increased after treatment, and both basal and dopamine-stimulated cyclic AMP accumulation in tissue slices were elevated. Basal activity was increased 2-fold, and dopamine-stimulated concentrations of cyclic AMP were raised nearly 3-fold. L-DOPA was able to attenuate the supersensitivity-induced changes in dopamine binding and cyclic nucleotide accumulation. One disconcerting observation was that dopamine produced no detectable increase in cyclic AMP accumulation in normal control animals.

The inability to detect dopamine supersensitivity in striatal homogenates from denervated animals and the capacity of two groups to observe the phenomenon in striatal slices are difficult to reconcile. The fact that isoproterenol sensitivity is not changed indicates some specificity in the response. However, in other brain areas the ability to alter isoprotenerol- or norepinephrine-induced accumulations of cyclic AMP in tissue slices by chronic administration of psychoactive compounds is well documented (*Palmer,* 1972; *Schmidt* and *Thornberry,* 1977; *Dismukes* and *Daly,* 1976).

In summary, many measures of dopamine receptor function are increased by chronically depriving postsynaptic receptors of input or by reducing dopamine stores. Changes in dopamine agonist supersensitivity have been detected using behavioral as well as biochemical methods of assessing dopamine receptor function. However, in most all studies reported, parallel increases in dopamine-stimulated cyclic AMP accumulation have not been detected in the adenylate cyclase assay. Why this facet of dopamine receptor function would not be coupled to adenylate cyclase is unexplainable.

Dopamine Receptors in the Pituitary Gland

The secretion of prolactin from the pituitary gland is controlled by the hypothalamus (see *Meites et al.,* 1977, for a brief review). There is considerable evidence that one inhibitory substance is dopamine, which is able to act directly on the pituitary gland to reduce prolactin release (*Shaar* and *Clemens,* 1974). Other dopamine agonists such as apomorphine (*Smalstig, Sawyer,* and *Clemens,* 1974) and ergot alkaloids (*Clemens et al.,* 1974) also block the secretion of prolactin *in vitro* and *in vivo.* Dopamine and neuroleptic receptors are found in the pituitary gland, as evidenced by specific binding of radioactive dopamine or haloperidol to pituitary gland membranes (*Brown, Seeman,* and *Lee,* 1976; *Creese, Schneider,* and *Snyder,* 1977). The "specific binding" of ligands to dopamine receptors in the pituitary gland displays the same characteristics as previously seen in the striatum.

Based on the above evidence documenting the presence of dopamine receptors in the pituitary gland and reports that dopamine exerts a physiological role in pituitary function, it would be expected that if dopamine exerts its effects in tissues through the elevation of cyclic AMP synthesis, then dopamine-dependent adenylate cyclase should exist in the pituitary gland. However, we found that neither dopamine nor other dopamine agonists could stimulate the synthesis of cyclic AMP in the intact gland or homogenates of the pituitary *in vitro* (*Schmidt* and *Hill,* 1977). Adenylate cyclase is present in the tissue and responsive since cyclic AMP synthesis increased in the presence of sodium fluoride, a nonspecific stimulant of adenylate cyclase. Therefore, it appears that in the pituitary gland dopamine receptors are not coupled to adenylate cyclase and dopamine does not exert its control over prolactin secretion by elevating cyclic AMP in the gland. Physiological evidence for this exists. *Hill, Macleod,* and *Orcott* (1976) found that dibutyryl cyclic AMP actually antagonized or reversed the actions of dopamine in isolated pituitary glands.

These studies with the pituitary gland dopaminergic system indicate that the hypothesis that dopamine exerts its effects in all target cells by elevating cyclic AMP has limited generality. Alternatively, it is quite probable that prolactin secretion emanates from only a select population of cells in the anterior pituitary gland and that dopamine exerts its effects only on these cells. Possibly, measurement of adenylate cyclase activity in the entire pituitary gland reflects the global events in the system rather than specific changes in the appropriate cell population. If this is the case, the subtleties of the dopamine-dependent adenylate cyclase which controls prolactin secretion might be obscured by the larger mass of nonresponsive tissue. The studies which use specific binding ligands offer advantages in this situation since they specifically monitor only discrete stereospecific components in pituitary membrane fragments. However, the fact that cyclic AMP and dopamine oppose each other in the intact gland (*Hill et al.*, 1976) still indicates that those dopamine receptors in the pituitary gland which control prolactin secretion are not functioning through the synthesis of cyclic AMP in the tissue.

There are, then, instances where measurements of cyclic AMP synthesis either do not correlate with other tests of dopamine function or do not fit with current theories of dopamine receptor function. These enigmatic findings from a large number of independent investigators using a variety of test procedures raise the following questions: 1) Is the dopamine receptor in the adenylate cyclase assay a classical dopamine receptor? a homogenization-distorted adrenergic receptor? or a unique "$dopamine_2$" receptor? 2) Is the dopamine-sensitive adenylate cyclase assay predictive for clinically active dopamine receptor agonists and antagonists? and 3) What are the reasons for the disparate results generated with the assay among investigators studying the same phenomenon?

THE ADENYLATE CYCLASE ASSAY: SOURCES OF VARIABILITY

Source of Tissue

Some of the disparity among investigators is due to the tissue used as a source of adenylate cyclase (Table 3).

Animal Age

An important factor is the age of the animal. Dopamine-sensitive adenylate cyclase activity increases during development in both the

Table 3. Potential Sources of Variability in the Adenylate Cyclase Assay by Source of Tissue

Variable	Basal Activity	Dopamine Stimulation	% Dopamine Stimulation	References
Species and Strain	Variable	Variable	130–223%	*Makman, Brown,* and *Mishra,* 1975; *Tang* and *Cotzias,* 1977
Increased Animal Age	Affect not significant	↓	↓	*Von Hungen, Roberts,* and *Hill,* 1974; *Coyle* and *Campochiaro,* 1976; *Spano et al.,* 1976b; *Walker* and *Walker,* 1973; *Puri* and *Volicer,* 1977; *Schmidt* and *Thornberry,* 1978; *Govoni et al.,* 1977b
Region of Brain	Caudate, olfactory tubercle, nucleus accumbens, frontal cortex			*Clement-Cormier* and *Robison,* 1977; *Mishra et al.,* 1975
Region within Striatum	Caudal > rostral; medial > lateral			*Bockaert et al.,* 1976
Retina vs Caudate	Retina more responsive to apomorphine and ergots			*Schmidt* and *Hill,* 1977; *Schorderet,* 1976, 1978

retina (*Brown,* 1976; *Makman et al.,* 1975) and striatum (*Coyle* and *Campochiaro,* 1976; *Spano et al.,* 1976a). Maximal or adult activity is present at 10-15 days, after which a decline in activity has been reported. *Coyle* and *Campochiaro* (1976) noted that dopamine stimulation was greatest only during the first month of life, and *Spano et al.* (1976a) found that apomorphine activation of cyclic AMP synthesis also declined during this time. *Brown* (1976) and *Makman et al.* (1975) also saw a decrease in dopamine stimulation after the first weeks of life. Cortical adenylate cyclase from immature rats was found to be more heat stable than the enzyme in mature rats (*Von Hungen et al.,* 1974), which again suggests a difference in the system with age. Several investigators have reported significant, and sometimes a marked decline in striatal dopamine-stimulated adenylate cyclase activity in older rats (*Walker* and *Walker,* 1973; *Puri* and *Volicer,* 1977; *Govoni et al.,* 1977b; *Schmidt* and *Thornberry,* 1978). The above discussion emphasizes that adenylate cyclase does change during development and aging. Experiments with immature rats or aged subjects might produce misleading conclusions due to subtle differences in receptor characteristics at these ages.

Species

Species differences also exist. *Makman et al.* (1975) found that in the retina of rats dopamine increased adenylate cyclase activity approximately

47%, while in the retina of the Cebus monkey cyclic AMP synthesis increased over 10-fold. Less startling species differences occurred in the caudate; however, differences did exist (*Makman et al.*, 1975). Dopamine stimulation of adenylate cyclase also varies among strains of animals (*Tang* and *Cotzias,* 1977) and perhaps between sexes (*Tang, Cotzias,* and *Tumbleson,* 1978).

Brain Region

Dopamine-sensitive adenylate cyclase has been located in many areas of the central nervous system, including the striatum (*Kebabian et al.,* 1972), nucleus accumbens and olfactory tubercle (*Horn et al.,* 1974), areas of the cerebral and entorhinal cortex (*Von Hungen* and *Roberts,* 1973; *Trabucchi et al.,* 1976b; *Bockaert et al.,* 1977), amygdala and median eminence (*Clement-Cormier* and *Robison,* 1977), various areas of the primate cortex (*Mishra et al.,* 1975), the hypothalamus (*Ahn* and *Makman,* 1977), and the substantia nigra (*Kebabian* and *Saavedra,* 1976; *DiChiara et al.,* 1977a). Dopamine-sensitive adenylate cyclase is also present in the retina (*Makman et al.,* 1975) and superior cervical ganglion (*Kebabian* and *Greengard,* 1971). In these tissues there are considerable differences in the sensitivity of adenylate cyclase to stimulation by dopamine (*Clement-Cormier* and *Robison,* 1977). Dopamine stimulation ranges from 166% in the caudate to 25% in the entorhinal cortex (*Trabucchi et al.,* 1976b). However, the ability of dopamine antagonists to attenuate the dopamine effect appears to be relatively similar across brain regions (*Clement-Cormier et al.,* 1974; *Clement-Cormier* and *Robison,* 1977; *Miller et al.,* 1974a; *Bockaert et al.,* 1977). One marked difference between striatal and retinal adenylate cyclase is that the former is only slightly stimulated by most ergot-like compounds (*Schmidt* and *Hill,* 1977; *Trabucchi et al.,* 1976a; *Fuxe et al.,* 1978b, c), while adenylate cyclase in the retina responds to several compounds in the series (*Schorderet,* 1976, 1978).

Differences in adenylate cyclase activity also have been detected within discrete brain regions. *Bockaert et al.* (1976) found that both basal and dopamine- or LSD-stimulated adenylate cyclase activities were higher in the rostral-dorsal areas of the striatum compared to more caudal-ventral regions. Isoproterenol-sensitive adenylate cyclase was equally distributed throughout the striatum. Subtle differences also exist in subsections of the cerebral cortex (*Bockaert et al.,* 1977). A distinct distribution of adenylate cyclase also exists in the retina, where the majority of activity is found in the inner retina (*Makman et al.,* 1975). Fine dissection of the substantia nigra also revealed that although basal adenylate cyclase was equally distributed between the zona compacta, zona reticulata and pars lateralis, only the zona reticulata contains significant dopamine-sensitive adenylate cyclase activity (*Kebabian* and *Saavedra,* 1976).

There also have been attempts to define the cell type which is responsible for dopamine-induced cyclic AMP synthesis. *Palmer* and *Manian* (1976) reported that dopamine-sensitive adenylate cyclase activity was associated with the neuronal fraction from the caudate nucleus and not with the glial cell fraction. However, dopamine and haloperidol binding, as well as dopamine-stimulated adenylate cyclase activity, were found recently in the glial cell fraction of bovine caudate (*Henn, Anderson,* and *Sellstrom,* 1977). It is important to resolve the disparities of these two studies. Confirmation that adenylate cyclase activity and dopamine and antipsychotic binding activity are localized in glial cells would necessitate a major revision in our theoretical concept of dopamine function in the striatum.

Therefore, choice of species, strain, sex, brain region and age of animals used can markedly influence the final results generated in the dopamine-sensitive adenylate cyclase assay. Even within the same species and brain region, however, dopamine-sensitive adenylate cyclase activity is influenced by the way in which the tissue is handled and processed prior to addition to the reaction tube (Table 4).

Tissue Preparation

Postmortem Effects

Williams et al. (1969) reported that preparations of the cerebral cortex and cerebellum from human autopsy material contained adenylate cyclase activity which was activated by NaF. Activity in this study was within the range seen when surgical specimens were used for assay. However, we found that the ability of norepinephrine to stimulate cyclic AMP synthesis in tissue slices from brain regions is a function of the time interval between death and incubation of the tissues (*Schmidt, Truex,* and *Thornberry,* 1978). If the brain remains at 36°C for 3 hr after death, as in the rule with the majority of human cases, no detectable increase in cyclic AMP synthesis occurs in response to norepinephrine. However, if the animals are cooled within 1 hr after death, the system is stable for up to 18 hr. The dopaminergic stimulation in the caudate also appears to be labile. *Carenzi et al.* (1975b) noted that both basal and dopamine-stimulated adenylate cyclase activity declined 40% within 1 hr after death and fell 70% during the first 5 hr after decapitation. In the latter studies the degree of stimulation produced by dopamine remained constant throughout the postmortem period; therefore, these investigators felt valid estimates of dopamine-adenylate cyclase activity could be made on human autopsy material.

Table 4. Potential Sources of Variability in the Adenylate Cyclase Assay by Tissue Preparation

Tissue Condition	Basal Activity	Dopamine Stimulation	% Stimulation	References
Time post mortem	NS	NS	NS	*Williams, Little,* and *Ersinck,* 1969
(4 hrs at 4°C)	↓	↓	NS	*Carenzi et al.,* 1975a; *Clement-Cormier et al.,* 1975
Freezing and thawing tissue	↑	↓	↓	*Williams et al.,* 1969; *Perkins* and *Moore,* 1971
Freezing and thawing homogenate	NS	↓	↓	*Williams et al.,* 1969; *Bockaert et al.,* 1976
Homogenization*	NS	↑	↑	*Forn, Krueger,* and *Greengard,* 1974; *Bockaert et al.,* 1976; *Schmidt et al.,* 1970
Homogenization in H_2O	↑	↓↓	↓↓	*McCune et al.,* 1971
Storage of homogenate at 0°C (3 hr)	NS	NS	NS	*Clement-Cormier et al.,* 1975
Incubation of homogenate and reaction mixture (0°C)	↓↓	↓	↑	*Clement-Cormier et al.,* 1975
Subcellular fractionation (M_1 fraction)	↑↑	↑	↓	*Clement-Cormier et al.,* 1975
Heating homogenate	↓↓	↓↓	↓↓	*Clement-Cormier et al.,* 1975

*Norepinephrine and isoproterenol stimulations decrease
NS = activity not significantly affected

Clement-Cormier et al. (1975) also detected dopamine-sensitive adenylate cyclase activity in the human caudate and noted a significant loss in enzyme activity post mortem. It is obvious that rapid removal of the brain and preparation and incubation of the tissue are desirable for maximal cyclase activity. Postmortem decay factors appear not to affect the binding of various ligands, including dopamine, to receptor sites (*Enna et al.,* 1976). Freezing brain tissue also does not alter ligand-receptor binding (*Enna et al.,* 1976), but this variable influences adenylate cyclase activity in brain.

Temperature Effects

The freezing and thawing of brain tissue has for some time been known to alter the properties of adenylate cyclase. *Perkins* and *Moore* (1971) reported that freezing tissue prior to homogenization markedly elevated basal enzyme activity and eliminated any stimulatory effects of NaF. Hormonal responses also are lost after this treatment (*Williams et al.,* 1969; *Schmidt,* unpublished observations). Freezing tissue homogenates, however, does not reduce the affinity of dopamine for adenylate cyclase, but the level of maximal stimulation is reduced from 250% to 150% above basal activity (*Bockaert et al.,* 1976). Storage of the homogenate at 0-4°C for up to 3 hr did not alter dopamine stimulatory effects (*Clement-Cormier et al.,* 1975), but the majority of enzyme activity was lost from cortical homogenates after 4-7 days at 4°C (*Williams et al.,* 1969). Elevating the temperature of the homogenate to 45°C drastically reduced activity (*Clement-Cormier et al.,* 1975), which illustrates the need to maintain the homogenizer in an ice bath during grinding of the tissue.

Homogenization

The choice of homogenizing buffer appears not to be especially critical. Sizable dopamine stimulations have been detected in caudates homogenized in TRIS-acetate-EGTA, TRIS-maleate-EGTA, di-GLYCINE, HEPES, TES-EGTA or sucrose solutions. Homogenization in water, however, increased the basal activity of adenylate cyclase in the cerebral cortex and eliminated hormonal effects (*McCune et al.,* 1971).

Homogenization *per se* is an important variable in adenylate cyclase studies. As was previously pointed out, homogenization of the striatum increases basal activity, reduces norepinephrine- and isoproterenol-stimulated cyclic AMP synthesis, but enhances dopamine's ability to stimulate cyclic AMP accumulation (*Forn et al.,* 1974; *Bockaert et al.,* 1976; *Sheppard* and *Burghardt,* 1976; *Schmidt et al.,* unpublished observations). The deleterious effects of tissue homogenization on hormonal responses in the brain have been discussed (*Robison et al.*, 1970) and low hormonal sensitivity attributed to mechanical disruption of receptor-adenylate cyclase coupling. Other forms of membrane disruption, such as sonication and detergents, also elevate adenylate cyclase activity (*Perkins,* 1973). It is tempting to speculate that the binding of hormones to membrane receptors coupled to adenylate cyclase brings about changes which are similar to (although obviously of orders of magnitudes more subtle than) membrane changes produced by mechanical perturbation.

Choice of Cell Fraction

Once the homogenate is prepared, strikingly different results can be obtained depending on whether the homogenate or a subcellular fraction is used in the cyclase assay. *DeRobertis et al.* (1967) found that adenylate cyclase in the cerebral cortex was almost wholly particulate, with most activity located in the mitochondrial fraction and subfractions containing nerve endings. Similar studies were conducted by *Harris* (1976), *Krueger, Forn,* and *Greengard* (1975) and *Clement-Cormier et al.* (1975) on catecholamine-sensitive adenylate cyclase in the corpus striatum. Again, the majority of activity was found in the mitochondrial fraction when assays were conducted in the presence or absence of dopamine. The highest specific activity was present in the fraction containing nerve endings, but the percentage stimulation produced by dopamine was greatest in the crude homogenate (*Clement-Cormier et al.,* 1975). Dopamine stimulated cyclic AMP synthesis 168% in the crude homogenate but only 60% in the M_1 subfraction which had the highest specific activity. Therefore, it appears that little is gained by fractionation of homogenates; equally valid pharmacological studies can be carried out with the more simple tissue preparation.

The majority of work in the area has been carried out using crude homogenates. However, variable effects of dopamine—ranging from 30% activations to 250% stimulations—have been reported by investigators who have used striatal homogenates from young rats (see *Bockaert et al.,* 1976, for a listing). An analysis of the components of the reaction mixtures employed offers some insight into this dilemma (Table 5).

Components of The Reaction Mixture

Another source of variability in the dopamine-sensitive adenylate cyclase assay is the constituency of the reaction mixture used. Many investigators have used systems in which the activity of dopamine-sensitive adenylate cyclase is suboptimal. The most extensive attempts

Table 5. Potential Sources of Variability in the Adenylate Cyclase Assay by Components of Reaction Mixture

Condition	Basal Activity	Dopamine Stimulation	% Dopamine Stimulation	References
[MG^{++}] 1-10 mM	↑	↓	↓	*Bockaert et al.*, 1976; *Burkard*, 1977
EGTA	↓	NS	↑	*Von Hungen* and *Roberts*, 1973; *Makman, Brown*, and *Mishra*, 1975; *Clement-Cormier et al.*, 1975
pH 7-9	↑	↑↓	NS	*Clement-Cormier et al.*, 1975; *Bockaert et al.*, 1976
Guanyl nucleotides	NS or ↓↓	NS ↑	variable	*Clement-Cormier et al.*, 1975; *Burkard*, 1977; *Bockaert et al.*, 1976
Methylxanthines	↓	NS	↑	*Premont, Perez*, and *Bockaert*, 1977
Temperature 30°-37°C	↓↓	↓	↑	*Clement-Cormier et al.*, 1975
Adenosine	↑	NS	↓	*Premont et al.*, 1977

NS = activity not significantly affected

to optimize the striatal cyclase assay have been carried out by *Clement-Cormier et al.* (1975) and *Bockaert et al.* (1976). The absolute amounts of Mg^{++} and ATP, as well as the ratio of these chemicals, are important. *Bockaert et al.* (1976) found that dopamine stimulated adenylate cyclase 220% in the presence of 1 mM Mg^{++}, while elevating Mg^{++} to 5-10 mM reduced the dopamine response to 30% over baseline. This was attributed to an elevation in the basal level of cyclase activity without a concomitant rise in the dopamine effect. Increasing the concentration of ATP elevated both basal and dopamine-stimulated activity until a maximum was attained at a Mg/ATP between 4 and 10 (*Clement-Cormier et al.,* 1975).

The choice of ATP substrate is also important. Although the same data can be generated using radioactive ATP and measuring the end product via scintillation counting as by assaying in a binding assay the amount of endogenous "cold" cyclic AMP formed during the reaction (*Makman et al.,* 1975; *Schmidt,* unpublished observations), the latter is more time-consuming and perhaps more costly. However, use of "generally" labeled ATP (i.e., ^{3}H- or ^{14}C-adenosine triphosphate) instead of ^{32}P-α-ATP can lead to erroneously high blank values unless proper column purification procedures are employed to purify the end product, cyclic AMP. When the label is distributed throughout the compound, adenosine formed from ATP can co-purify with cyclic AMP if only an alumina column is used to purify the product. More reliable and precise results can be generated by including an ion exchange column either before or after the alumina column. The alumina removes over 99% of the ATP substrate and the Dowex 1 allows separation of adenosine contaminants from the synthesized cyclic AMP. Using this alumina-Dowex 1 technique, we have reduced enzyme "blank" activity to approximately 20 counts over instrument background.

Maintaining the levels of ATP high enough to saturate the system also can be a problem when using brain tissue which has a high concentration of ATPase (*Williams et al.,* 1969). This problem can be overcome by adding an ATP regenerating system to the reaction mixture (*Schmidt* and *Hill,* 1977). However, many workers find this an unnecessary precaution (*Kebabian* and *Greengard,* 1971; *Clement-Cormier et al.,* 1975; *Iversen et al.,* 1976), since less than 10% of the ATP is hydrolyzed during short-term incubations (*Clement-Cormier et al.,* 1975).

The optimal temperature for the cyclase reaction appears to be about 30°C, but the maximal dopamine stimulatory effect occurs at 37°C (*Clement-Cormier et al.,* 1975). *Bockaert et al.* (1976) reported that the maximal dopamine response decreased between pH 7-8, but *Clement-Cormier et al.* (1975) published a pH dependency curve showing that both basal and dopamine stimulation increased with increasing pH between 7 and 8. However, although a maximal absolute amount of cyclic AMP was generated at pH 8.2, the percentage stimulation by dopamine at this pH was 62% compared to a 117% stimulation at pH 7.4.

The stability of newly synthesized cyclic AMP has been examined under assay conditions. When 10 mM theophylline was included in the reaction, less than 10% of the cyclic AMP formed was hydrolyzed (*Kebabian et al.*, 1972). *Bockaert et al.* (1976) found that about 25% of the cyclic AMP formed in response to dopamine was degraded in their assay, even in the presence of 10 mM theophylline. *Williams et al.* (1969) reported that 20 mM aminophylline inhibited cyclic AMP catabolism 95% for 5 min. Use of more potent phosphodiesterase inhibitors such as isobutylmethylxanthine (IBMX) reduced cyclic AMP hydrolysis even further (*Schmidt*, unpublished observations). Adding an excess of unlabeled cyclic AMP also can be used to retard degradation of the newly formed radioactive cyclic AMP. The best assurance that one is measuring the total amount of synthesized cyclic AMP and recovering the end product quantitatively is to add to the reaction mixture a small amount of cyclic AMP tagged with a different label than the ATP precursor, i.e., ^{3}H-cyclic AMP added along with ^{32}P-ATP. In this manner the hydrolysis of the end product can be followed, and counting the ^{3}H in the final column eluate verifies the end product as cyclic AMP and allows for corrections in degradation or loss which might have occurred during incubation or purification.

It is equally important to stabilize the stimulants added to the reaction since both dopamine and apomorphine are readily oxidized. In fact, *Makman et al.* (1975) suggested that contamination of preparations of apomorphine with oxidation products may explain why most investigators have found apomorphine to be only a partial agonist of dopamine-sensitive adenylate cyclase in the striatum. Therefore, preparation of compounds immediately before use and addition of an antioxidant such as 0.1% ascorbic acid are recommended. We have found that even with the above precautions and preparing the solution of apomorphine from a previously unopened bottle, the stimulatory effects of apomorphine are not equal to those of dopamine (*Schmidt* and *Hill*, 1977). In the reaction itself, it appears that metabolic destruction of dopamine by monoamine oxidase is unimportant since the stimulations induced by dopamine were not enhanced in the presence of pargyline, an MAO inhibitor (*Bockaert et al.*, 1976), or when the enzyme was saturated with alternative substrates such as tyramine or serotonin (*Sheppard* and *Burghardt*, 1974). Chemical oxidation of dopamine also appears not to be a concern since incubations conducted in an atmosphere of N_2 did not increase the dopamine-induced synthesis of cyclic AMP (*Sheppard* and *Burghardt*, 1974).

An important variable, both technically and theoretically, is calcium.

Von Hungen and *Roberts* (1973) were the first to report that addition of EGTA markedly reduced basal activity of brain adenylate cyclase while not affecting the stimulatory effects of norepinephrine. Therefore, the percentage stimulation by the hormone was considerably increased. They also noted that readdition of small amounts of Ca^{++} enhanced activity of cortical adenylate cyclase, while higher concentrations reduced activity. *Makman et al.* (1975) also observed this phenomenon in retinal adenylate cyclase, although both basal and dopamine-stimulated activity were elevated in the presence of EGTA. They found EDTA worked equally well. *Premont, Perez,* and *Bockaert* (1977) also observed that calcium could slightly stimulate striatal adenylate cyclase. According to their calculations, there is enough calcium in homogenates of the striatum to totally activate the calcium-sensitive portion of the enzyme. The studies by *Clement-Cormier et al.* (1975) clearly show that in the striatum a stimulation of cyclic AMP synthesis by dopamine can only be demonstrated in the presence of EGTA. In these experiments dopamine stimulated cyclic AMP synthesis only 12% in the absence of EGTA, but an apparent stimulation of 105% took place in the presence of the compound. This calcium chelating agent actually does not affect the amount of cyclic AMP synthesized in the presence of dopamine but rather reduces basal activity 50%. Therefore, the net effect is the "appearance" of a dopamine receptor; only the EGTA-inhibited enzyme displays the characteristics of the dopamine receptor.

From a technical point of view, it is clear that if one wants to monitor dopamine or other hormonal responses, inhibition of the calcium component of enzyme activity is required. Under these conditions, however, NaF (which is routinely used as a nonspecific activator of adenylate cyclase from all tissues) does not stimulate cyclic AMP synthesis in the striatum. We found this is probably due to the fact that EGTA not only removes Ca^{++} from the system but also chelates Mg^{++} significantly. If the concentration of Mg^{++} is elevated to 10 mM, a significant fluoride effect can be detected in striatal homogenates (*Schmidt,* unpublished observations). As mentioned, however, the dopamine response under this high Mg^{++} condition is significantly reduced.

Adenosine is another endogenous substance that has recently been shown to affect striatal adenylate cyclase. *Makman et al.* (1975) reported that adenosine did not stimulate the formation of cyclic AMP in homogenates of the retina, but an adenosine-sensitive cyclase has been detected in homogenates of the striatum (*Premont et al.,* 1977), guinea pig hypothalamus (*Ahn* and *Makman,* 1977) and in caudate tissue slices (*Wilkening* and *Makman,* 1975). The concentration of adenosine in homogenates of the striatum was found by *Premont et al.* (1977) to be high enough only to activate the enzyme submaximally. However, during the course of the incubation enough ATP is broken down to contribute sufficient adenosine for maximal activation of cyclic AMP synthesis.

Adenosine enhances the cyclic AMP-stimulatory effects of neurotransmitters in tissue slices from brain regions (*Huang, Ho,* and *Daly*, 1973), including the caudate (*Wilkening* and *Makman,* 1975), but adenosine appears not to synergize with dopamine in the adenylate cyclase assay (*Roufogalis, Thornton,* and *Wade,* 1976). *Premont et al.* (1977) also observed that adenosine did not affect the affinity of dopamine for adenylate cyclase, but the stimulatory effects of adenosine and dopamine were additive. Both theophylline and isobutylmethylxanthine have the capacity to block adenosine receptors, and both agents reduced the activity of cyclase in striatal homogenates. Therefore, there are two reasons why methylxanthines are useful in the dopamine-sensitive adenylate cyclase assay: 1) to inhibit the destruction of newly formed cyclic AMP, and 2) to inhibit adenosine-sensitive adenylate cyclase in the striatum and thereby lower the basal activity in the assay.

Guanyl nucleotides interact with adenylate cyclase from a number of peripheral tissues and potentiate the effects of hormones (*Rodbell et al.*, 1975). Dopamine-sensitive adenylate cyclase in the frontal cortex was stimulated 3-fold if GMP-PNP (a stable analogue of GTP) was added 5 min prior to the addition of the ATP substrate (*Bockaert et al.*, 1977). Recently, *Palmer* and *Palmer* (1978) reported that adenylate cyclase activity in homogenates of the frontal cortex was increased 5-fold by GMP-PNP, and the effects of dopamine, histamine, isoproterenol, prostaglandins and norepinephrine were markedly potentiated. Guanyl nucleotides also increase cyclase activity in homogenates of the hypothalamus (*Ahn* and *Makman,* 1977).

In the corpus striatum a variety of effects have been observed in the presence of guanyl nucleotides. Addition of GTP, GDP, GMP or GMP-PNP to striatal homogenates under standard assay conditions either did not increase the synthesis of cyclic AMP (*Fuxe et al.*, 1978b) or reduced cyclic nucleotide accumulation (*Clement-Cormier et al.*, 1975; *Tell, Pasternak, Cuatrecasas,* 1975; *Burkard,* 1977). All of the guanyl nucleotides tested in the striatum by *Clement-Cormier et al.* (1975) lowered basal activity significantly but changed the dopamine-induced elevation of cyclic AMP in crude homogenates only when the ATP concentration was lowered. Therefore, the overall effect of guanyl nucleotides in the striatum appears to be to increase slightly dopamine-stimulated cyclic AMP synthesis (*Clement-Cormier et al.*, 1975; *Burkard*, 1977; *Bockaert et al.*, 1977). In fact, *Roufogalis et al.* (1976) could detect a dopamine-induced elevation of cyclic AMP in striatal synaptosomes only in the presence of a guanyl nucleotide. Similarly, the ergot-like dopamine agonists, lergotrile and bromocryptine, only markedly stimulate cyclic AMP synthesis in striatal homogenates in the presence of GMP-PNP (*Fuxe et al.,* 1978c).

Based on the experiments to date, it is difficult to determine if guanyl nucleotides play an important physiological role in regulating dopamine-sensitive adenylate cyclase in the striatum. The variable effects on basal adenylate cyclase activity also make interpretation of the data difficult. The majority of investigators have not included guanyl nucleotides in dopamine-sensitive adenylate cyclase assays. It is clear that the effects of these compounds in the striatum are not nearly as dramatic as reported in peripheral tissues (*Rodbell et al.,* 1975).

The adenylate cyclase assay which is used in our laboratory has taken into account the above variables. A TES-HCl buffer was chosen instead of TRIS since the pK_a of TES is 7.5, which is closer to our desired pH of 7.4. The complete reaction mixture and the experimental schema are shown in Table 6.

PHARMACOLOGICAL CHARACTERIZATION OF THE DOPAMINE RECEPTOR IN THE DOPAMINE-SENSITIVE ADENYLATE CYCLASE ASSAY

Even with the assay optimized, we have observed some unexplainable pharmacological effects. Compounds which are dopamine agonists *in vivo* and in other biochemical tests have not always stimulated the synthesis of cyclic AMP in striatal homogenates, and some agents which by several criteria are dopamine antagonists do not attenuate the dopamine-induced elevation of cyclic AMP in the striatum. We felt that by studying the structures of these "exceptions" we might gain insight as to why many of the results generated in the adenylate cyclase assay do not coincide with other dopaminergic tests or conform to theoretical expectations.

Dopamine Receptor Antagonists

It is not the purpose of this review to discuss in depth the structure-activity relationships of the system. There have been several thorough reviews of the structure-activity relationships of dopamine agonists and antagonists in the adenylate cyclase assay (*Iversen et al.,* 1975; *Makman et al.,* 1975; *Miller et al.,* 1974a & b; *Sheppard,* 1977). Almost all of the major classes of CNS active agents have been tested in the adenylate cyclase assay for antagonism of dopamine stimulation. At low concentrations, only clinically active neuroleptics block dopamine's stimulation of cyclic AMP synthesis (Table 7). Within some classes e.g., the pheno-

Table 6. The Assay of Dopamine-sensitive Adenylate Cyclase

Components of the Reaction		Experimental Protocol
TES·HCl Buffer	50 mM, pH 7.4	1. Homogenize tissue in 50 mM TES, 2 mM EGTA, pH 7.5
$MgCl_2$	2 mM	2. Initiate the reaction by the addition of homogenate
IBMX	0.5 mM	3. Incubate 200 μg protein in reaction mixture for 3 min at 30°C
Cyclic AMP	0.5 mM	4. Boil for 3 min
ATP	1 mM	5. Add 1 ml of H_2O containing ^{3}H-cyclic AMP as tracer for purification procedure
Creatine-PO_4	20 mM	6. Centrifuge 5 min at 1000 × g
Creatine kinase	10 units	7. Apply supernatant to alumina column; remaining ^{32}P-ATP substrate is adsorbed while cyclic AMP passes through
EGTA	0.2 mM	8. Effluent from alumina passes into Dowex 1 column; cyclic AMP is adsorbed
Isotope	^{32}P-ATP	9. Wash Dowex column with H_2O
Dopamine, etc.	10^{-7}–10^{-4} M	10. Elute cyclic AMP with 0.05N HCl; collect in counting vials; add scintillator
Striatal homogenate	200 μgm protein	11. Count simultaneously ^{32}P (newly synthesized cyclic AMP) and ^{3}H (tracer cyclic AMP to monitor loss over the columns)
		12. Results expressed as pmole of cyclic AMP synthesized/mg tissue protein/min

thiazines and thioxanthines, there is a good correlation between the ability of compounds to inhibit cyclase and their clinical potency. But, as mentioned previously, there are several agents with proven clinical antipsychotic activity which have little inhibitory activity in the adenylate cyclase assay. Compounds in the butyrophenone (haloperidol), benzamide (sulpiride) and dibenzazepine (clozapine) classes are the best known examples of such discrepancies. We and others (*P. E. Sutler,* personal communication) found that fenfluramine, and a number of similar derivatives, did not antagonize the dopamine-induced elevation of cyclic AMP in striatal homogenates. However, these compounds elevate prolactin *in vivo,* raise DOPAC levels in the brain (*Fuller et al.,* 1976) and antagonize amphetamine-induced hyperactivity and stereotypy (*Berger et al.,* 1973).

Table 7. Antagonists of Dopamine-sensitive Adenylate Cyclase

Class	Representative Compound(s)	Activity [Ki(nM)]
Phenothiazine	Chlorpromazine,[1]	48
	Fluphenazine[1]	4
Thioxanthenes	Flupenthixol[1]	1
Butyrophenones	Spiroperidol,[2] Pimozide,[1]	270, 140
	Haloperidol[3]	220
Dibenzazepines	Clozapine,[1] Loxapine[1]	170, 45
	Butaclamol[1]	9
Alpha blockers	Phentolamine[1]	>1000
Beta blockers	Propranolol[1]	>1000
Anticholinergic	Benztropine[1]	>1000
Antihistaminic	Promazine,[1] Diphenylhydramine[1]	>1000
Benzamides	Sulpiride[4]	>1000
Ergot alkaloids	LSD,[2] Lergotrile,[2] Bromocryptine[5]	330, 350, 214
Benzodiazepines	Chlordiazepoxide[6]	>1000
Tricyclic antidepressant	Desipramine,[1] Imipramine[6]	>1000
Tetrahydroxyisoquinolines[7]	—	>1000

References: [1]*Iversen, Horn,* and *Miller,* 1975; [2]*Schmidt* and *Hill,* 1977; [3]*Clement-Cormier et al.,* 1974; [4]*Elliott et al.,* 1977; [5]*Trabucchi et al.,* 1978; [6]*Karobath* and *Leitich,* 1974; [7]*Sheppard, Burghardt* and *Teitel,* 1976

As of yet, we have no explanation for these disparate results. The possibility that solubility problems account for the differences in activity seen *in vivo* and in the *in vitro* cyclase assay (*Laduron,* 1976) seems unlikely since the butyrophenones do show high activity in *in vitro* receptor binding assays (*Burt et al.,* 1976). It is also unlikely that the compounds require metabolic conversion to active agents since many of the antidopamine effects are evident in *in vitro* tests (*Burt et al.,* 1976; *Shaar* and *Clemens,* 1974) or when the compounds are administered directly into regions of the brain (*Elliott et al.,* 1977).

It has been suggested that some of the effects of dopamine antagonists may be due to effects on presynaptic dopamine receptors (*Seeman, Staiman,* and *Chau-Wong,* 1974). The changes in amine metabolism that are often used to quantitate the dopamine agonist or antagonist activity of compounds are thought to result from alterations occurring in presynaptic terminals. Adenylate cyclase in the corpus striatum is located postsynaptically. Dopamine-stimulated cyclase is not eliminated by destruction of presynaptic nerve endings into the area (*Mishra et al.,* 1974), and cyclase activity in fractions from density gradient studies did not partition with other enzymes which are markers for presynaptic terminals (*Laduron et al.,* 1976). Therefore, one might expect that there would be instances in which marked changes in DOPAC concentrations in the striatum might occur following administration of a dopamine antagonist, while the same compound would have little effect on adenylate cyclase activity in the striatum. There now is evidence for this concept. When kainic acid (a neurotoxic analogue of glutamic acid) is injected into the striatum, it causes destruction of cell bodies, but presynaptic input into the area is not disrupted (*Schwarcz* and *Coyle,* 1977). A recent report indicated that in kainate-lesioned rats dopamine-sensitive adenylate cyclase is destroyed, yet antipsychotics produced the same changes in DOPAC as seen in normal rats (*DiChiara et al.,* 1977b). Therefore, it is not surprising that correlations between data generated in the cyclase assay and that coming from *in vivo* determinations of amine metabolism in the striatum are not exact (*Scatton et al.,* 1977). However, the majority of evidence indicates that neuroleptics exert their therapeutic actions on postsynaptic dopamine receptors in the brain (*Iversen et al.,* 1976). Thus, it still is not possible to explain all of the disparities between results in the cyclase assay and data from other tests of dopamine receptor antagonism.

Dopamine Receptor Agonists

An intriguing possibility is that there are multiple types of dopamine receptors in the brain (*Cools et al.,* 1976), which serve multiple functions (*Goldberg, Volkman, Kohli,* and *Kotake,* 1977), and that adenylate cyclase is a unique subtype of receptor. Attempting to characterize the topography of the dopamine receptor through the use of receptor antagonists is difficult due to the structural diversity of the compounds which inhibit dopamine-sensitive adenylate cyclase (see *Iversen et al.,* 1975, for example). We felt a more fruitful approach might be to examine the agonist specificities of dopamine-sensitive adenylate cyclase, especially seeking answers as to why all dopamine agonists do not stimulate the synthesis of cyclic AMP in striatal homogenates.

The area of dopamine agonists and adenylate cyclase also has been recently and thoroughly reviewed (*Iversen et al.,* 1975; *Sheppard,* 1977) and will not be discussed in depth here. However, little attention has been given to the newest class of dopamine agonists, the ergolines. Compounds in this class include LSD, ergonovine, bromocriptine and lergotrile. The basic structural component of the family is an ergot-like nucleus. Many of the compounds in this group display dopamine agonist activities in a variety of tests, and compounds such as these may represent a significant advancement in the treatment of Parkinson's disease.

Lergotrile (2-chloro-6-methylergoline-8β-acetonitrile; Eli Lilly and Co. 83636) has undergone considerable testing for dopamine agonist activity in preparation for clinical trials in galactorrhea and parkinsonism (Fig. 1). There are clear indications that the compound is a direct

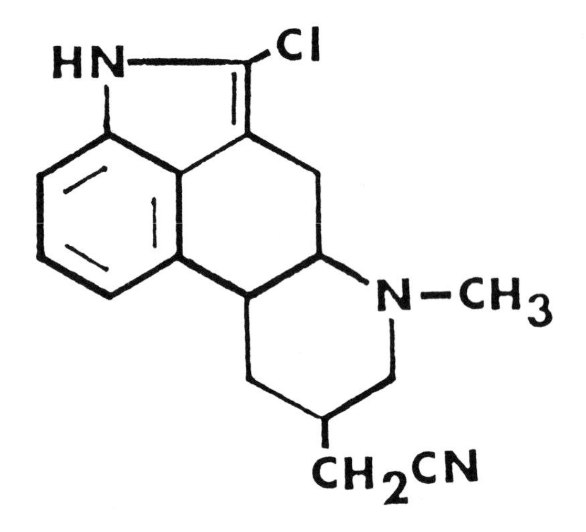

Fig. 1. Lergotrile (2-chloro-6-methylergoline-8β-acetonitrile).

dopamine receptor stimulating agent *in vivo* and *in vitro* (Table 8). Based on the above evidence, it seemed that, mechanistically, this agent should behave like other dopamine receptor agonists and elevate cyclic AMP in select target tissues. Lergotrile has two main sites of action: one in the pituitary gland where dopamine and dopamine agonists reduce the secretion of prolactin, and a second target in the

Table 8. Evidence that Lergotrile is a Direct-acting Dopamine Receptor Stimulant

1. Decreases prolactin secretion from isolated pituitary glands and *in vivo* (*Clemens, Smalstig,* and *Shaar,* 1975).
2. Causes rotation away from the lesioned side in rats with 6-hydroxydopamine-induced destruction of the substantia nigra (*Tye et al.*, 1977; *Fuxe et al.*, 1978c).
3. Displaces radioactive dopamine, apomorphine and haloperidol from binding sites on brain membrane fragments (*Lew et al.*, 1977; *Bymaster* and *Wong,* 1977).
4. Produces hyperactivity in mice which have had brain amines depleted by reserpine and catecholamine synthesis inhibited by α-methyl-paratyrosine (*Schmidt* and *Thornberry*, unpublished observations).
5. Decreases DOPAC levels and the turnover of dopamine in the brain (*Fuller* and *Perry,* 1978).
6. Does not stimulate the release of dopamine from synaptosomes (*Bymaster* and *Wong,* 1977).
7. Does not block the reuptake of dopamine in synaptosomes (*Bymaster* and *Wong,* 1977).
8. Reduces tremors in segmental lesioned monkeys (*Leiberman et al.*, 1975).
9. Is beneficial in the treatment of parkinsonism (*Leiberman et al.*, 1975; *D. B. Calne,* personal communication).

corpus striatum where there is evidence that dopamine might be an inhibitory neurotransmitter.

In our first series of experiments we used a high concentration of Mg^{++} ion, such that we could monitor the effects of NaF which non-specifically causes near-maximal activation of adenylate cyclase in tissues. As mentioned previously, dopamine stimulation is lower under these conditions. Three dopamine-rich brain areas were compared to activity in the pituitary gland (Table 9).

Adenylate cyclase was present in all tissues and significant increases in activity were produced by the addition of NaF. However, dopamine stimulated cyclic AMP synthesis in the brain but not in the pituitary gland homogenates. Apomorphine also did not activate pituitary adenylate cyclase (*Schmidt,* unpublished observations). We have not detected an increase in cyclic AMP in the pituitary incubated intact or minced (*Schmidt,* unpublished observations). Others too have reported that catecholamines do not elevate cyclic AMP significantly in the pituitary gland (*Sato et al.,* 1974; *Steiner et al.,* 1970). In fact, dibutyryl cyclic AMP and dopamine antagonize each other with respect to prolactin release (*Hill et al.,* 1976). Therefore, the pituitary gland is unique in that although it responds to dopamine (*Shaar* and *Clemens,* 1974), dopamine binding sites are present in the tissue (*Brown et al.,* 1976) and adenylate cyclase activity is detectable in the gland and is activated by NaF (Table 9), cyclic AMP concentrations do not rise in response to dopamine. It appears that those dopamine receptors in the pituitary gland that control prolactin secretion are not coupled to adenylate cyclase.

We also observed in these studies that lergotrile did not stimulate cyclic AMP synthesis in brain homogenates which did respond to dopamine (Table 9). When we lowered the concentration of magnesium to sensitize and maximize the assay, the dopamine response in the striatum

Table 9. Comparison of Effects of Dopamine and Lergotrile on Adenylate Cyclase Activities in Brain Regions and the Pituitary Gland

	Striatum	Frontal Cortex	Olfactory Tubercle	Pituitary
Ascorbate	128.2*	32.6	72.3	22.3
NaF, 10^{-2} M	206.2	347.0	185.1	180.0
Dopamine, 10^{-5} M	174.6	56.2	114.1	20.3
Dopamine, 10^{-4} M	183.6	71.2	157.7	23.7
Lergotrile, 10^{-5} M	76.3	54.2	74.2	25.9
Lergotrile, 10^{-4} M	96.0	55.8	69.4	20.1

*pmole cyclic AMP synthesized/mg protein/min
Values represent the means of duplicate homogenates

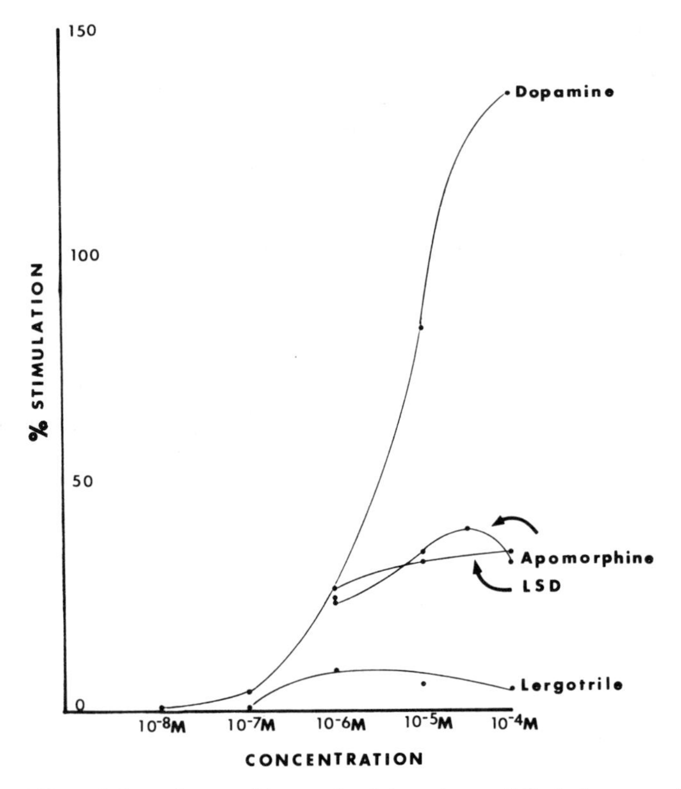

Fig. 2. Effect of dopamine agonists on adenylate cyclase activity in homogenates of the corpus striatum. Values represent the mean from three separate experiments, with duplicate or triplicate replications within each experiment. Variability was less than 10%.

was increased and apomorphine elevated cyclic AMP synthesis but lergotrile remained inactive (Fig. 2). However, another ergot, LSD, did increase adenylate cyclase activity as had been reported (*Von Hungen et al.,* 1975). Therefore, activity in our system could be demonstrated by ergots possessing particular structural characteristics.

Comparison of a series of ergots (Fig. 3) revealed that activities did differ and might be due to the size of the substituents at the 8 position (Fig. 4). One might have expected that large and bulky groups would hinder receptor interaction, yet LSD and ergocristine were more active than lergotrile.

Others also have found ergots not to be potent stimulants of dopamine-sensitive adenylate cyclase in the striatum (Table 10). *Kebabian, Calne,* and *Kebabian* (1977) and *Fuxe et al.* (1978a) have confirmed that lergotrile does not markedly stimulate cyclic AMP synthesis in the striatum, although, by running 20-40 replicates, they have been able to establish that the compound does produce a significant 10% elevation

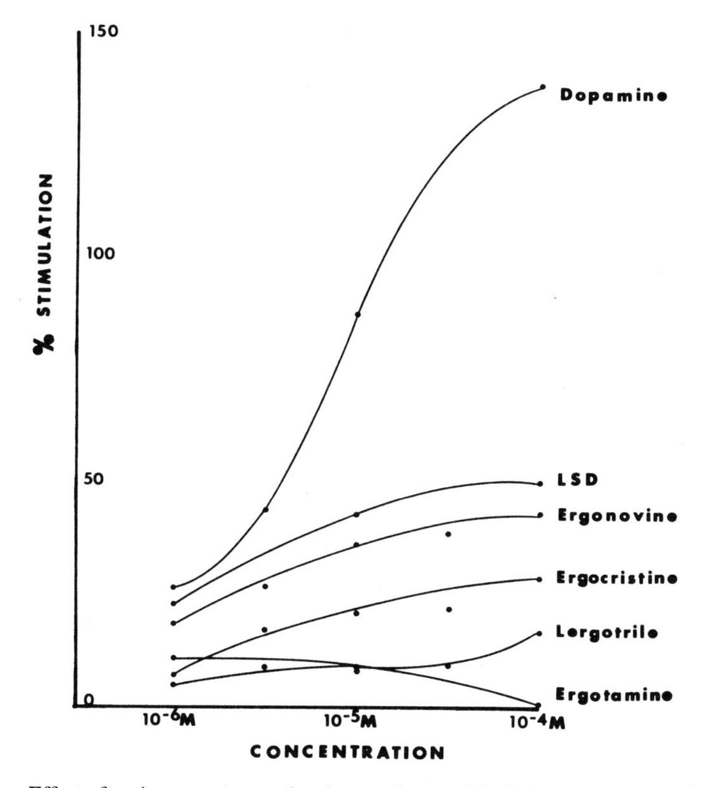

Fig. 3. Effect of various ergots on adenylate cyclase activity in homogenates of the corpus striatum. Values represent the mean from three experiments, with duplicate or triplicate replications within each experiment. Values across experiments differed by less than 5%. Activity in the absence of dopamine was 131.9 pmoles cyclic AMP synthesized/mg protein/min, and 308.3 pmoles cyclic AMP synthesized/mg protein/min in the presence of 10^{-4} M dopamine. From *Schmidt* and *Hill*, 1977.

in activity. This small elevation in activity is of questionable physiological relevance. Another ergot derivative which has shown promise in clinical trials is CB 154 (2-bromo-α-ergocryptine). There is agreement that this compound is not a potent stimulant of cyclase (*Trabucchi et al.,* 1976a; *Fuxe et al.,* 1978a).

Ergot compounds in the clavine class might well be the most active stimulants of adenylate cyclase in the striatum (*Fuxe et al.,* 1978a). Only agroclavine, CF 25-397, PRT-1742 and elmoclavine consistently stimulated cyclic AMP synthesis in striatal homogenates, yet setoclavin and festuclavine were inactive as cyclase stimulants. Therefore, the structure-activity relationship is not predictable even within families of ergots.

Even though some of the ergots consistently elevated cyclic AMP synthesis in the striatum, all were considerably less active than dopamine.

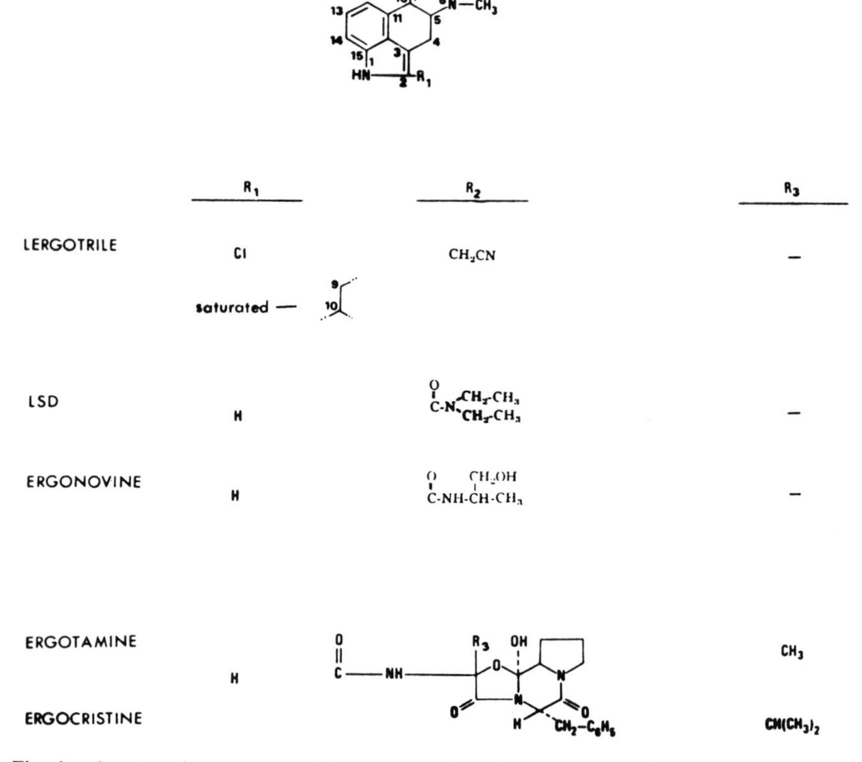

Fig. 4. A comparison of some of the ergots used in the present experiments.

One reason for this might have been the dual agonist-antagonist activity of these compounds. *Palmer* and *Burks* (1971) first reported that LSD could antagonize catecholamine-induced increases in cyclic AMP in the brain. *Von Hungen et al.* (1975) confirmed these findings and specifically showed that dopamine-sensitive adenylate cyclase could be inhibited by LSD. When we examined the series of ergots for dopamine blocking activity, we found that all were able to inhibit the dopamine-induced elevation of cyclic AMP in striatal homogenates (Fig. 5). Others have reported similar findings with other ergots, including those listed in Table 10. The dopamine antagonism appears to be specific since *Kebabian et al.* (1977) found that lergotrile did not antagonize the epinephrine-induced elevation of cyclic AMP accumulation in the rat cerebellum, which has been characterized as a β-receptor response. Lergotrile also does not antagonize the nonspecific activation of adenylate cyclase produced by NaF (*Schmidt* and *Hill*, 1977).

Table 10. Agonist and Antagonist Activities of Ergot-Related Compounds in the Adenylate Cyclase Assay

Compound	Maximum Percent Stimulation of Striatal Adenylate Cyclase	Antagonist Activity
LSD	60[1], 28[2], 40[5]	+++
Ergonovine	43[1]	++
Ergocristine	28[1], 0[10]	++
Lergotrile	17[1], 0[7], 10[8]	+++
Ergometrine	11[1]	+
Bromo-LSD	14[2]	+++
Methyl-LSD	8[2]	+++
Ergotamine	0[3], 0[5], 0[10]	+++
Dihydroergotamine	0[3], 0[10]	+++
Bromocryptine	0[6], 9[8], 0[10]	+++
Festuclavine	0[3]	not tested
Setoclavine	0[8]	not tested
Elmoclavine	34[8]	not tested
Agroclavine	30[8]	not tested
CF-25-397	23[9]	++
Ergocornine	10[9], 0[10]	+++++
PTR 17-402	30[8] (75% in limbic Cx)	not tested
Methergoline	0[10]	+++
Lisuride	0[10]	+++
Dihydroergocryptine	0[10]	++
Dihydroergotoxine	0[10]	++

References: [1]*Schmidt* and *Hill*, 1977; [2]*Von Hungen, Roberts*, and *Hill*, 1975; [3]*Govoni et al.*, 1977a; [4]*Makman et al.*, 1975; [5]*Bockaert et al.*, 1976; [6]*Trabucci et al.*, 1976b; [7]*Kebabian, Calne*, and *Kebabian*, 1977; [8]*Fuxe et al.*, 1978c; [9]*Fuxe et al.*, 1978a; [10]*Spano* and *Trabucchi*, 1978.

All the ergots tested thus far can inhibit dopamine-stimulated cyclic AMP synthesis (*Schmidt* and *Hill*, 1977; *Kebabian et al.*, 1977; *Spano* and *Trabucchi*, 1978). However, the antagonist properties of the compounds do not account for their marginal activity as dopamine receptor stimulants in the cyclase assay, for LSD was the most active stimulant as well as the most potent antagonist (*Schmidt* and *Hill*, 1977).

There are conditions under which some ergots become more active. The dopamine agonist activity of lergotrile and bromocryptine is increased in the presence of the GTP analogue GMP-PNP, but dopamine itself, agroclavine and ergocornine are not potentiated (*Fuxe et al.*, 1978b, c). On the other hand, agroclavine, dopamine and lergotrile are more active in striatum from denervated rats, whereas bromocryptine, ergocornine and CF 25-397 are not more active in the supersensitive preparations (*Fuxe et al.*, 1978b). Compound PRT 17402 increases

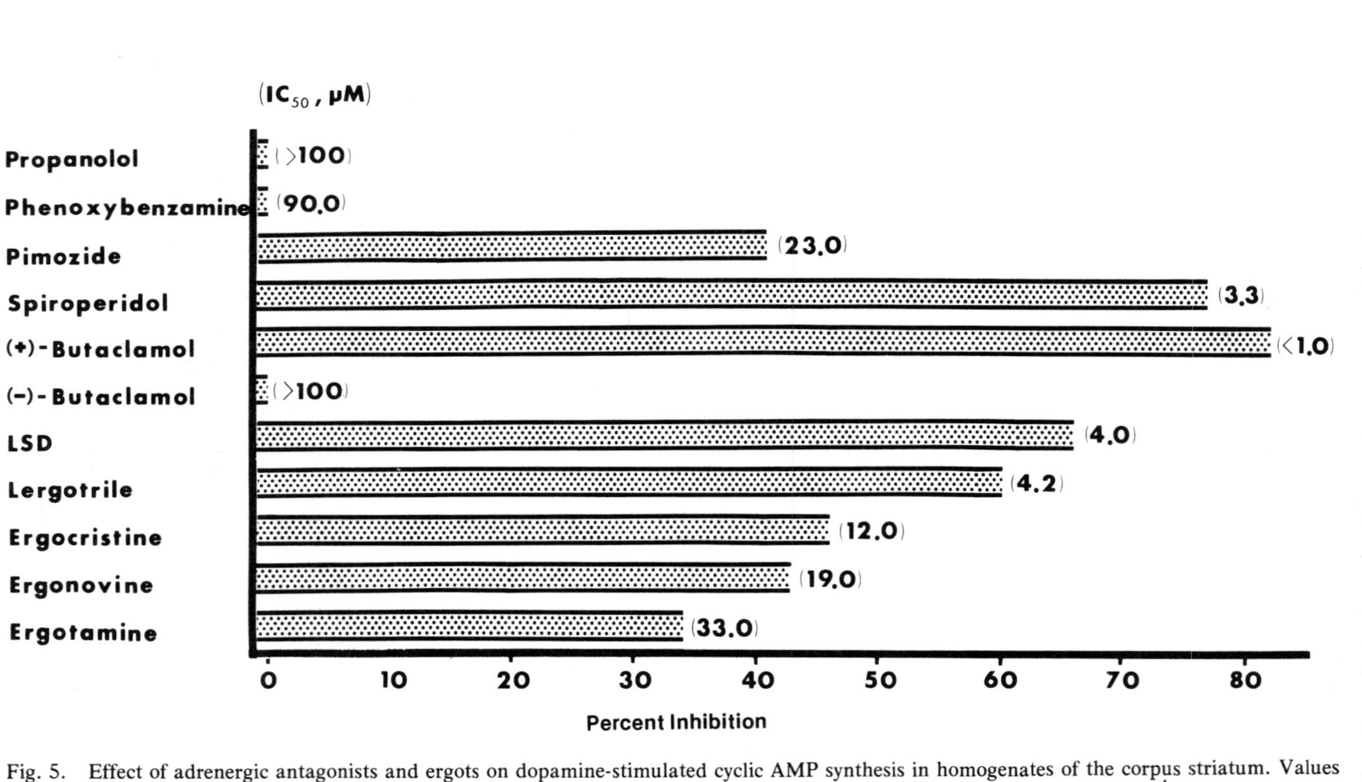

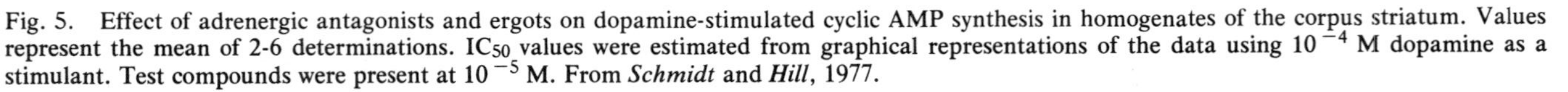
Fig. 5. Effect of adrenergic antagonists and ergots on dopamine-stimulated cyclic AMP synthesis in homogenates of the corpus striatum. Values represent the mean of 2-6 determinations. IC_{50} values were estimated from graphical representations of the data using 10^{-4} M dopamine as a stimulant. Test compounds were present at 10^{-5} M. From *Schmidt* and *Hill*, 1977.

cyclic AMP synthesis in the caudate to a maximum of 40% over control activity, but in homogenates of the limbic area activity is elevated 75% (*Fuxe et al.,* 1978a). Therefore, there may be special instances where ergots will display regional specificity or enhanced potency.

The inactivity of ergots in the adenylate cyclase assay is in clear contradistinction to the dopaminergic effects of these compounds in many other tests. One would have expected these agents to stimulate cyclic AMP synthesis in the caudate. It is all the more confusing that lergotrile-, bromocryptine- and ergocornine-induced rotation in denervated rats is potentiated by theophylline, a cyclic AMP-phosphodiesterase inhibitor (*Fuxe et al.,* 1978b). This suggests that the ergots are acting *in vivo* through cyclic AMP. However, none of the ergots tested (lergotrile, ergocornine, CF 25-397, agroclavine, bromocryptine) inhibit phosphodiesterase (*Fuxe et al.,* 1978b), and the presence or absence of phosphodiesterase inhibitors does not affect the activity of ergots in the adenylate cyclase assay (*Schmidt,* unpublished observations).

The observation that bromocryptine- and ergocornine-induced rotation in denervated rats is attenuated if the stores of amines are depleted by the administration of reserpine and a tyrosine hydroxylase inhibitor (*Corrodi et al.*, 1973) indicated that ergots might be acting indirectly. The finding by *Silbergeld* and *Pfeiffer* (1977) that bromocryptine stereotypy was reduced by tyrosine hydroxylase inhibition supported this contention. These workers also reported that lergotrile inhibited the reuptake and stimulated the release of dopamine from tissue minces. The compound does not influence these parameters in synaptosome preparations (*Bymaster* and *Wong*, 1977), however, and neither the rotatory behavior (*Fuxe et al.*, 1978a) nor the increased locomotor activity (*Schmidt* and *Thornberry*, unpublished observation) produced by lergotrile is attenuated by the depletion of amines. Furthermore, these ergots can directly interact with dopamine receptors *in vitro* (*Bymaster* and *Wong*, 1977; *Lew et al.*, 1977; *Fuxe et al.*, 1978a) and inhibit prolactin secretion in isolated anterior pituitary glands where there are no external sources of amines (*Clemens, Smalstig,* and *Shaar*, 1975).

It also seems unlikely, for many of the above reasons, that the *in vivo* dopaminergic effects of the ergots depend on their conversion to active metabolites. Furthermore, the main metabolites of lergotrile did not stimulate striatal adenylate cyclase (*Schmidt* and *Hill*, 1977), but some of the metabolites more actively displaced haloperidol and dopamine bound to striatal membranes (*D. Wong* and *F. Bymaster*, personal communication).

Topography of the Adenylate Cyclase Dopamine Receptor

It is clear that the dopamine-sensitive adenylate cyclase assay in the striatum does not reflect the dopamine agonist activity of all compounds, especially those in the ergot family. This indicates that the dopamine receptor in the cyclase assay might require different structural configurations than 1) the dopamine, apomorphine, haloperidol or spiroperidol binding sites *in vitro*, 2) the dopamine receptors in the pituitary gland which control prolactin secretion, or 3) the dopamine receptors present *in vivo* which mediate locomotor activity and stereotype behavior.

What then is the usefulness of the dopamine-sensitive adenylate cyclase assay? Is it an artifact or does it reflect a unique dopamine receptor site? It is not untenable that multiple dopamine receptors could exist throughout the body or in different brain regions (*Cools et al.,* 1976). It also is reasonable that dopamine might be able to fit multiple receptor sites with slightly different topographies.

Inspection of the dopamine molecule reveals the potential for considerable flexibility in the interactions of this amine with receptor sites. Rotation about the alpha carbon can place the nitrogen group in either an "extended" or "flexed" configuration with respect to the catechol moiety. Rotation about the beta carbon of the alkyl chain can position the catechol hydroxyls in what has been labeled an α- or β-rotomeric configuration with respect to the phenyl ring (Fig. 6). If one looks at the structure of lergotrile for the essence of dopamine, there is an intra-ergoline moiety, the tetrahydroaminonaphthalene (aminotetralin) group (Fig. 6), which might convey dopamine agonist properties on the molecule. This moiety is analogous to dopamine in the extended configuration. Interestingly, the aminotetralin moiety is also found in the classic dopamine agonist apomorphine. However, in apomorphine there is another conformer of dopamine, the isoquinoline moiety (Fig. 7). In the former configuration the nitrogen of dopamine would be in the extended position, while in the isoquinoline form dopamine would assume a flexed configuration.

Rigid analogues of these compounds have been synthesized and tested for dopamine agonist activity in the adenylate cyclase assay. *Miller, Horn,* and *Iversen* (1947b) reported an aminotetralin derivative (ADTN) to be as active as dopamine in stimulating cyclic AMP synthesis. However, *Sheppard* and *Burkhardt* (1974) found that isoquinoline equivalents were active stimulants of a β-receptor cyclase in erythrocytes, but not striatal dopamine-sensitive adenylate cyclase.

These findings indicate that the distance between the nitrogen group and the catechol hydroxyls is critical for cyclase activation (Fig. 7). Models of dopamine in the extended configuration reveal a spacing of 7.7 Å between the nitrogens and hydroxyls, and examination of crystalline dopamine indicates that the extended form is most preferred. There are obvious problems extrapolating the structure of crystalline dopamine to that in solution or at receptor sites in biomembranes, but the cyclase

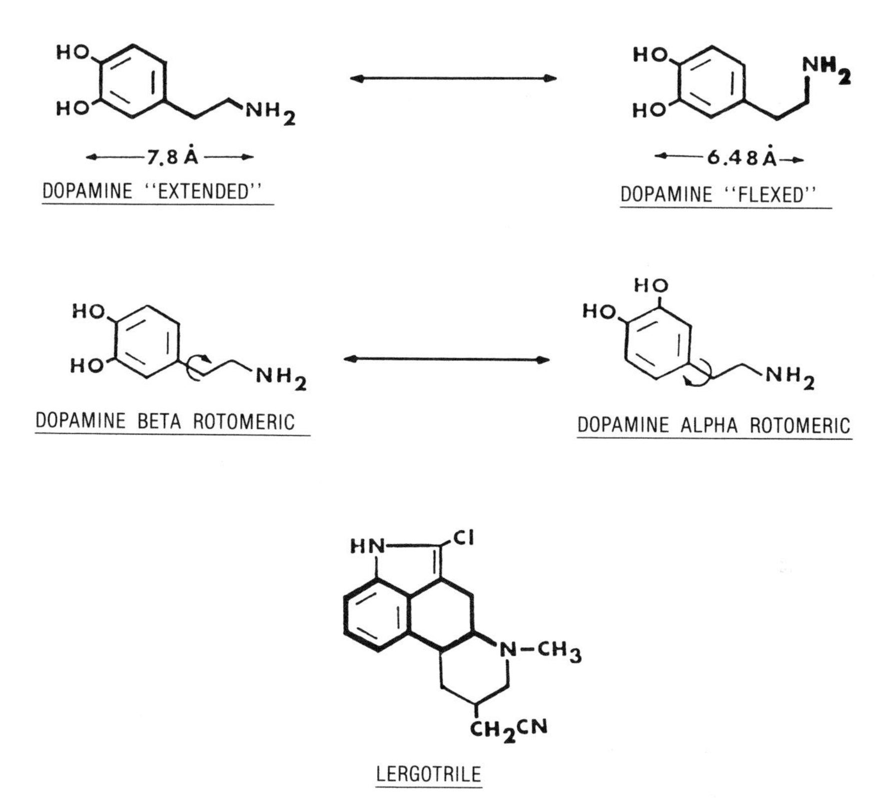

Fig. 6. Representations of potential configurations of dopamine compared to the structure of lergotrile.

data indicate that the extended form of dopamine is probably the active state of dopamine at the adenylate cyclase dopamine receptor.

If lergotrile, apomorphine and bromocryptine do have an aminotetralin within their structures which approximates dopamine in the preferred extended configuration, why don't they stimulate adenylate cyclase to the same extent as dopamine? Structural studies have given us some clue as to why this may be.

Dopamine, along with being able to assume an extended or flexed configuration, is also able to rotate about the beta carbon, thereby positioning the hydroxyl groups in different configurations with respect to the amine function (Fig. 8). A comparison of ADTN and apomorphine reveals that these two rigid compounds represent, respectively, the two hydroxyl configurations of freely rotating dopamine. Since ADTN was active as a cyclase stimulant (*Miller et al.*, 1974b) and apomorphine is only partially active (*Miller et al.*, 1974b; Fig. 2), it seemed that hydroxyl positioning might be critical for the adenylate cyclase receptor.

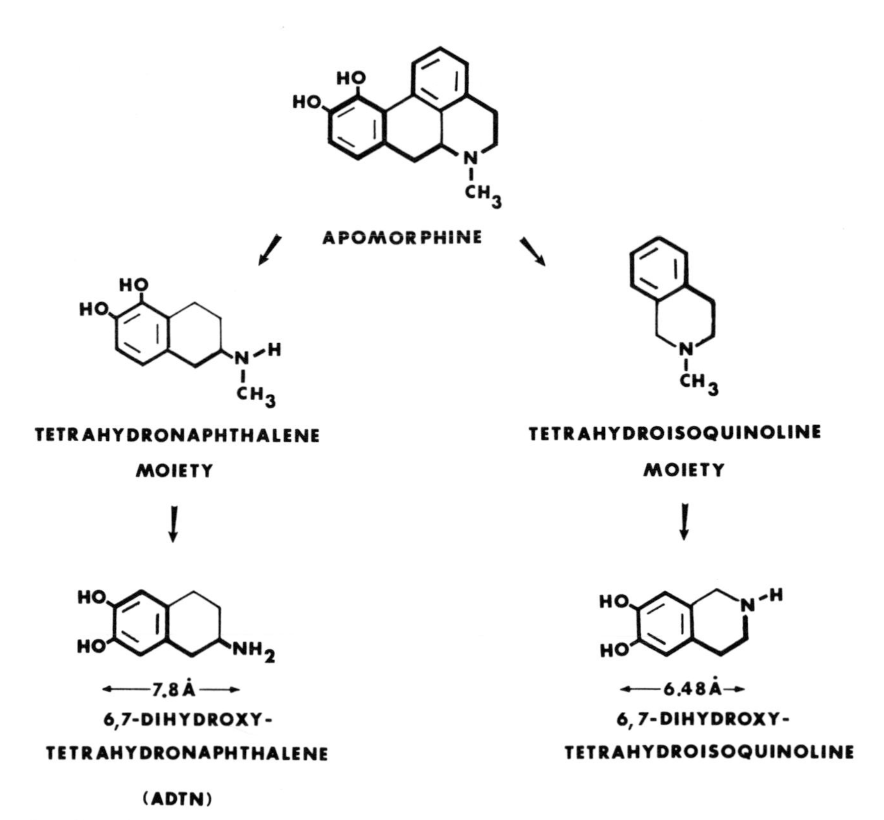

Fig. 7. The structure of apomorphine, showing potential intramolecular configurations of compounds similar to dopamine.

Two key compounds supplied by Dr. J. Cannon allowed us to test this hypothesis. These are shown as A and B in Figure 8. A comparison of the activities of these two compounds as stimulants of dopamine-sensitive adenylate cyclase in the striatum revealed marked differences. While the 6,7-dihydroxyaminotetralin was as active as dopamine, the 5,6-dihydroxy compound was lower in potency and comparable to apomorphine (Fig. 9). Similar observations have recently been made in the nucleus accumbens and striatum by *Woodruff et al.* (1977). Further evidence of the importance of the hydroxyls is shown by the fact that removal of the catechol hyroxyls, either singly or completely, from the aminotetralins or dopamine itself completely abolishes the ability of these compounds to activate adenylate cyclase (*Miller et al.*, 1974b; *Woodruff et al.*, 1977). This might explain why lergotrile and bromocryptine are inactive as cyclase stimulants; neither has hydroxyls, although both possess the necessary aminotetralin within their structures. The main metabolite of lergotrile (13-hydroxy lergotrile; *C. J. Parli,*

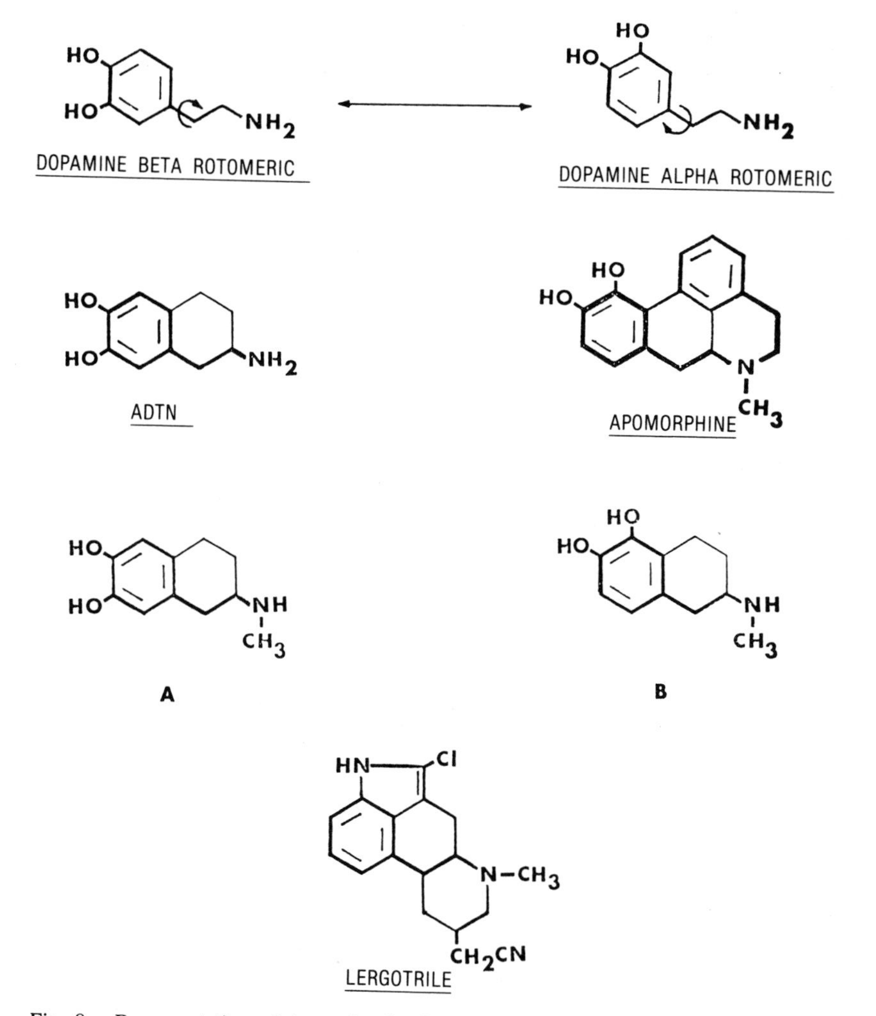

Fig. 8. Representation of dopamine in the alpha and beta rotomeric configurations in comparison to the analogous rigid analogues ADTN, apomorphine, 6,7-dihydroxyaminotetralin (A) or 5,6-dihydroxyaminotetralin (B).

personal communication) has only one hydroxyl, thereby ruling out conversion of lergotrile to an active cyclase stimulant *in vivo*.

Apomorphine, on the other hand, has catechol hydroxyls yet is only a partial agonist of adenylate cyclase. This is explained by the fact that the hydroxyls are in the improper configuration with respect to the amino

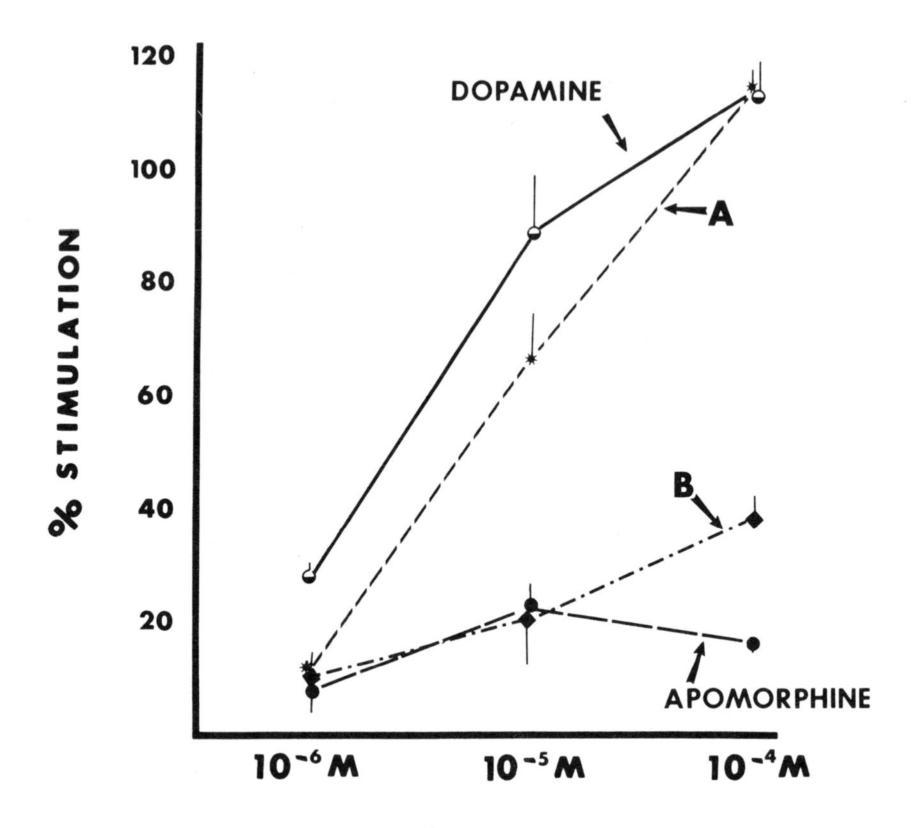

Fig. 9. Comparison of actions of compounds A (6,7-dihydroxyaminotetralin) or B (5,6-dihydroxyaminotetralin) with respect to their ability to activate adenylate cyclase in homogenates of rat striatum. Values represent the mean ± SEM of 4–5 replicates.

group. Extending this reasoning, one would assume that movement of the hydroxyls of apomorphine into a position equivalent to 6,7-dihydroxyaminotetralin would confer adenylate cyclase-stimulating activity on the compound. This compound, isoapomorphine, has not been tested as a stimulant of adenylate cyclase. However, there is reason to believe it would be inactive. Isoapomorphine was reported to be inactive on dopamine and ADTN-sensitive neurons in *Helix aspersa* (*Pinder, Buxton,* and *Woodruff,* 1972). Others (*Neumeyer, McCarthy,* and *Battista,* 1973; *Saari* and *King,* 1974) noted that isoapomorphine had little dopamine agonist activity *in vivo* in several animal tests.

The lack of activity of isoapomorphine as a dopamine agonist perhaps can be understood in terms of the stereochemical conformation of the molecule. *Grol* and *Rollema* (1977) suggested that the three-dimensional shape of isoapomorphine may hinder its association with the dopamine receptor due to the presence of bulky groups opposite the hydroxyl groups of the compound. There is evidence that substitution of bulky groups on dopamine itself attenuates its dopamine agonist activity (*Cannon, Khonje,* and *Long,* 1975).

We can say the following, then, with respect to the dopamine receptor which predominates in the adenylate cyclase assay: 1) catechol hydroxyls are required, 2) hydroxyl groups must be present in the 6,7-equivalent position, and 3) the hydroxyl-nitrogen positioning should be such that 7.7 Å is approximated, analogous to dopamine in the fully extended configuration.

SUMMARY AND CONCLUSIONS

The critical point is that *in vivo,* apomorphine, lergotrile and the 5,6-dihydroxyaminotetralins are active dopamine agonists (*Schmidt* and *Thornberry*, unpublished observations; *Fuxe et al.*, 1978a, b, c; *Cannon, Lee,* and *Goldman*, 1977; *McDermed, McKenzie,* and *Phillips*, 1975). In fact, *Cannon et al.* (1977) and *Costall et al.* (1977) have stressed that the 5,6-derivations are more active than the analogous 6,7-compounds. Therefore, if one were to use the dopamine-sensitive adenylate cyclase assay as a test for dopamine agonists for the treatment of parkinsonism, one would be pursuing the wrong structural track; apomorphine, lergotrile and bromocryptine would have been passed over. Also, dopamine antagonist antipsychotics in the butyrophenone class might well be discarded due to their low potency in blocking dopamine's stimulation of adenylate cyclase.

The easiest explanation for the discrepancies between results in the cyclase assay and the *in vivo* dopamine agonist properties of compounds is that homogenization of the striatum distorts dopamine receptors in the tissue such that they become abnormal receptor sites with different topographies and requirements for agonist interaction (*Harris*, 1976). An examination of the dopamine literature suggests that this might not be the case. Dopamine causes vasodilation of the renal artery *in vivo*, but apomorphine is inactive in this unperturbed preparation and so are bromocryptine and lergotrile (*Volkman* and *Goldberg,* 1976). Furthermore, 6,7-dihydroxyaminotetralins were equally as active as dopamine, but the 5,6-compounds were markedly less potent (*Volkman et al.*, 1977). Also, the vasodilation produced by dopamine or ADTN was not (*Crumly, Pinder,* and *Goldberg,* 1975) or was only partially (*Volkman et al.*, 1977) antagonized by haloperidol. Haloperidol is not a potent antagonist of dopamine-sensitive adenylate cyclase. Thus, the topography of the dopamine receptor in the renal artery *in vivo* closely resembles the dopamine receptor coupled to adenylate cyclase in the striatum (*Goldberg et al.,* 1977). Another intact dopamine receptor preparation, the *Helix aspersa* neuron, has pharmacological characteristics similar to the dopamine-sensitive adenylate cyclase (*Munday, Poat,* and *Woodruff,* 1976). It seems unlikely that homogenization of the striatum would by chance distort the dopamine or another adrenergic receptor into a configuration that exists unperturbed in nature, e.g., in the renal artery or in the CNS of *Helix aspersa.*

An alternative explanation is that adenylate cyclase in the striatum is revealing a unique dopamine receptor which is not the same as the classical dopamine receptor. Therapeutically, this might not be the important receptor as far as Parkinson's disease is concerned. However, what we do not know at this time are the functions of the dopamine receptors in other dopamine-rich areas of the brain (see *Costa* and *Gessa,* 1977, for review). Future experiments with rigid analogues of dopamine with hydroxyl variations might lead us to new dopamine agonists with selective functions outside the realm of parkinsonism.

REFERENCES

Ahn, H.S., and Makman, M.H. 1977. Neurotransmitter-sensitive adenylate cyclase in the hypothalami of guinea-pig, rat, and monkey. Brain Res. 138:125-138

Anden, N.E., Rubenson, A., Fuxe, K., and Hokfelt, T. 1967. Evidence for dopamine receptor stimulation by apomorphine. J Pharm Pharmacol 19:627-629

Anden, N.E., and Stock, G. 1973. Effect of clozapine on the turnover of dopamine in the corpus striatum and in the limbic system. J Pharm Pharmacol 25:346-348

Baldessarini, R.J., and Walton, K.G. 1975. Esters of apomorphine and N,N-dimethyldopamine as agonists of dopamine receptors in the rat brain *in vivo*. Neuropharmacology 14:725-731

Berger, H.J., Brown, C.C., and Krantz, J.C., Jr. 1973. Fenfluramine blockade of CNS stimulant effects of amphetamines. J Pharm Sci 62:788-791

Blumberg, J.B., Taylor, R.E., and Sulser, F. 1975. Blockade by pimozide of a noradrenaline sensitive adenylate cyclase in the limbic forebrain: Possible role of limbic noradrenergic mechanisms in the mode of action of antipsychotics. J Pharm Pharmacol 27:125-128

Bockaert, J., Premont, J., Glowinski, J., Thierry, A.M., and Tassin, J.P. 1976. Topographical distribution of dopaminergic innervation and of dopaminergic receptors in the rat striatum. II. Distribution and characteristics of dopamine adenylate cyclase-interaction of D-LSD with dopaminergic receptors. Brain Res 107:303-315

Bockaert, J., Tassin, J.P., Thierry, A.M., Glowinski, J., and Premont, J. 1977. Characteristics of dopamine and β-adrenergic sensitive adenylate cyclases in the frontal cerebral cortex of the rat. Comparative effects of neuroleptics on frontal cortex and striatal dopamine sensitive adenylate cyclases. Brain Res 122:71-86

Brown, J.H. 1976. Dopamine-sensitive adenylate cyclase of mammalian retina. Dissertation Intern B 37:(abs.)1638

Brown, J.H., and Makman, M.H. 1972. Stimulation by dopamine of adenylate cyclase in retinal homogenates and of adenosine 3′:5′-cyclic monophosphate formation in intact retina. Proc Nat Acad Sci USA 69:539-543

Brown, H.B., and Riley, G.A. 1965. Adenyl cyclase in the reticular formation of beef brain. Thesis submitted to the Western Reserve School of Medicine in partial fulfillment of the requirement for the Degree of Doctor of Medicine.

Brown, G.M., Seeman, P., and Lee, T. 1976. Dopamine/neuroleptic receptors in basal hypothalamus and pituitary. Endocrinology 99:1407-1410

Burkard, W.P. 1977. Different activation of striatal adenylate cyclase by dopamine and GTP-analogues. Experentia 33:788-789

Burt, D.R., Creese, I., and Snyder, S.H. 1976. Properties of [^{3}H] dopamine binding associated with dopamine receptors in calf brain membranes. Mol Pharmacol 12:800-812

Burt, D.R., Creese, I., and Synder, S.H. 1977. Antischizophrenic drugs: Chronic treatment elevates dopamine receptor binding in brain. Science 196:326-328

Burt, D.R., Enna, S.J., Creese, I., and Snyder, S.H. 1975. Dopamine receptor binding in the corpus striatum of mammalian brain. Proc Nat Acad Sci USA 72:4655-4659

Bymaster, F.P., and Wong, D.T. 1977. Effect of lergotrile (L) on dopamine (DA) receptor binding and on the uptake and release of catecholamines. Fed Proc 363:1006

Cannon, J.G., Khonje, P.R., and Long, J.P. 1975. Centrally acting emetics.9.Hofmann and Emde degradation products of nuciferine. J Med Chem 18:110-112

Cannon, J.G., Lee, T., and Goldman, H.D. 1977. Cerebral dopamine agonist properties of some 2-aminotetralin derivatives after peripheral and intracerebral administration. J Med Chem 20:1111-1116

Carenzi, A., Cheney, D.L., Costa, E., Guidotti, A., and Racagni, G. 1975a. Action of opiates, antipsychotics, amphetamine and apomorphine on dopamine receptors in rat striatum. *In vivo* changes of 3′,5′-cyclic AMP content and acetylcholine turnover rate. Neuropharmacology 14:927-939

Carenzi, A., Govoni, S., Spano, P.F., and Trabucchi, M. 1976. On the cyclic nucleotides involvement in rat striatal function. Pharmacol Res Comm 18:143-147

Carenzi, A., Gillin, J.C., Guidotti, A., Schwartz, M.A., Trabucchi, M., and Wyatt, R.J. 1975b. Dopamine-sensitive adenyl cyclase in human caudate nucleus. Arch Gen Psychiat 32:1056-1059

Christina, M., Sawaya, B., Dolphin, A., Jenner, P., Marsden, C.D., and Melbrum, B.S. 1977. Noradrenaline-sensitive adenylate cyclase in slices of mouse limbic forebrain. Characterization and effects of dopamine agonists. Biochem Pharmacol 26:1877-1884

Clemens, J.A., Shaar, C.J., Smalstig, F.B., Bach, N.J., and Kornfeld, E.C. 1974. Inhibition of prolactin secretion by ergolines. Endocrinology 94:1171-1176

Clemens, J.A., Smalstig, E.B., and Shaar, C.J. 1975. Inhibition of prolactin secretion by lergotrile mesylate: Mechanism of action. Acta Endocrinologia 7:230-237

Clement-Cormier, Y.C., Kebabian, J.W., Petzold, G.L., and Greengard, P. 1974. Dopamine-sensitive adenylate cyclase in mammalian brain. A possible site of action of antipsychotic drugs. Proc Nat Acad Sci USA 71:1113-1117

Clement-Cormier, Y.C., Parrish, R.G., Petzold, G.L., Kebabian, J.W., and Greengard, P. 1975. Characterization of a dopamine-sensitive adenylate cyclase in the rat caudate nucleus. J Neurochem 25:143-149

Clement-Cormier, Y.C., and Robison, G.A. 1977. Adenylate cyclase from various dopaminergic areas of the brain and the action of antipsychotic drugs. Biochem Pharmacol 26:1719-1722

Cools, A.R., Struyker Boudier, H.A., Jr., and Van Rossum, J.M. 1976. Dopamine receptors. Selective agonists and antagonists of functionally distinct types within the feline brain. Europ J Pharmacol 37:283-293

Corrodi, H., Fuxe, K., Hokfelt, T., Lidbrink, P., and Ungerstedt, U. 1973. Effect of ergot drugs on central catecholamine neurons. Evidence for a stimulation of central dopamine neurons. J Pharm Pharmacol 25:409-412

Costa, E., and Gessa, G.L. 1977. *Nonstriatal Dopaminergic Neurons*. New York:Raven Press

Costall, B., Naylor, R.J., Cannon, J.G., and Lee, T. 1977. Differential activation by some 2-aminotetraline derivatives of the receptor mechanisms in the nucleus accumbens of rat which mediate hyperactivity and stereotyped biting. Eur J Pharmacol 41:307-319

Coyle, J.T., and Compochiaro, P. 1976. Ontogenesis of dopaminergic-cholinergic interactions in the rat striatum. A neurochemical study. J Neurochem 27:673-678

Creese, I., Burt, D.R., and Snyder, S.H. 1977a. Dopamine receptor binding enhancement accompanies lesion-induced behavioral supersensitivity. Science 197:596–598
Creese, I., Schneider, R., and Snyder, S.H. 1977b. ^{3}H-spiroperidol labels dopamine receptors in pituitary and brain. Europ J Pharmacol 46:377–381
Crumly, H.J., Pinder, R., and Goldberg, L.I. 1975. Dopamine-like renal vasodilation caused by 2-amino-6, 7-dihydroxy-1,2,3,4-tetrahydronaphthalene (ADTN). Clin Res 23:505A
Da Prada, M., Saner, A., Burkhard, W.P., Bartholini, G., and Pletscher, A. 1975. Lysergic acid diethylamide. Evidence for stimulation of cerebral dopamine receptors. Brain Res 94:67–73
DeRobertis, E., Rodriquez, G., Arnaiz, D., Alberici, M., Butcher, R.W., and Sutherland, E.W. 1967. Subcellular distribution of adenyl cyclase and cyclic phosphodiesterase in rat brain cortex. J Biological Chem 242:3487–3493
DiChiara, G., Mereu, G.P., Vargiu, L., Porceddu, M., Mulas, A., Trabucchi, M., and Spano, P.F. 1977a. Evidence for the Existence of Regulatory DA Receptors in the Substania Nigra. In *Advances in Biochemical Psychopharmacology*, eds. E. Costa and G.L. Gessa, pp. 361–371. New York:Raven Press
DiChiara, J., Porceddu, M.L., Spano, P.F., and Gessa, G.L., 1977b. Haloperidol increases and apomorphine decreases striatal dopamine metabolism after destruction of striatal dopamine-sensitive adenylate cyclase by kainic acid. Brain Res 130:374–378
Dismukes, R.K., and Daly, J.W. 1976. Adaptive responses of brain cyclic AMP-generating systems to alterations in synaptic input. J Cyclic Nucleotide Res 2:321–336
Ehrlich, Y.H., Rabjohns, R.R., and Routtenberg, A. 1977. Experiental input alters the phosphorylation of specific proteins in brain membranes. Pharmacol Biochem and Behavior 169–174
Elliott, P.N.C., Jenner, P., Huizing, G., Marsden, C.D., and Miller, R. 1977. Substituted benzamides as cerebral dopamine antagonists in rodents. Neuropharmacology 16:333–342
Enna, S.J., Bird, E.D., Bennett, J.P., Bylund, D.B., Yamamura, H.I., Iversen, L.L., and Snyder, S.H. 1976. Huntington's chorea. Changes in neurotransmitter receptors in the brain. N Engl J Med 294:1305–1309
Forn, J., Krueger, B.K., and Greengard, P. 1974. Adenosine 3′,5′-monophosphate content in rat caudate nucleus. Demonstration of dopaminergic and adrenergic receptors. Science 174:1346–1349
Fredholm, B.B., Fuxe, K., and Agnati, L. 1976. Effects of some phosphodiesterase inhibitors on central dopamine mechanisms. Eur J Pharmacol 38:31–38
Friedhoff, A.J., Bonnet, K., and Rosengarten, H. 1977. Reversal of two manifestations of dopamine receptor supersensitivity by administration of L-DOPA. Res Comm Chem Path Pharmacol 16:411–423
Fuller, R.W., and Perry, K.W. 1978. Effect of lergotrile on 3,4-dihydroxyphenylacetic acid (DOPAC) concentration and dopamine turnover in rat brain. J Neurol Trans 42:23–35
Fuller, R.W., Perry, K.W., and Clemens, J.A. 1976. Elevation of 3,4-dihydroxyphenylacetic acid concentrations in rat brain and stimulation of prolactin secretion by fenfluramine. Evidence for antagonism at dopamine receptor sites. J Pharm Pharmacol 28:643–644
Fuxe, K., Fredholm, B.B., Agnati, L.F., Ogren S.-O., Everitt, B.J., Jonsson, G., and Gustafsson, J.-A. 1978a. Interaction of ergot drugs with central monoamine systems. Evidence for a high potential in the treatment of mental and neurological disorders in Pharmacology. 16:(Supp. 1)99–134
Fuxe, K., Fredholm, B.B., Ogren, S.-O, Agnati, L.F., Hokfelt, T., and Gustafsson, J.-A. 1978b. Pharmacological and biochemical evidence for the dopamine agonistic effect of ergot alkaloids particularly bromocriptine. Acta Endocrinologica, in press
Fuxe, K., Fredholm, B.B., Ogren, S.-O., Agnati, L.F., Hokfelt, T., and Gustafsson, J.-A. 1978c. Ergot drugs and central monoaminergic mechanisms. A histochemical, biochemical and behavioral analysis. Fed Proc 37:2181–2191
Fuxe, K., and Ungerstedt, U. 1974. Action of caffeine and theophylline on supersen-

sitive dopamine receptors. Considerable enhancement of receptor response to treatment with dopa and dopamine receptor agonists. Med Bull 52:48–54

Garelis, E., and Neff, N.H. 1973. Cyclic adenosine monophosphate: selective increase in caudate nucleus after administration of L-DOPA. Science 183:532–533

Gianutsos, G., and Moore, K.E. 1977. Dopaminergic supersensitivity in striatum and olfactory tubercle following chronic administration of haloperidol or clozapine. Life Sci 20:1585–1592

Gnegy, M., Uzunov, P., and Costa, E. 1977. Participation of an endogenous Ca^{++}-binding protein activator in the development of drug-induced supersensitivity of striatal dopamine receptors. J Pharmacol Exp Ther 202:558–564

Gnegy, M.E., Uzunov, P., and Costa, E. 1976. Regulation of dopamine stimulation of striatal adenylate cyclase by an endogenous Ca^{++}-binding protein. Proc Nat Acad Sci USA 73:3387–3890

Goldberg, L.I., Volkman, P.H., Kohli, J.D., and Kotake, A.N. 1977. Similarities and Difference of Dopamine Receptors in the Renal Vascular Bed and Elsewhere. In *Advances Biochemical Pharmacology*, Vol. 16, eds. E. Costa and G.L. Gessa, pp. 251–256. New York:Raven Press

Govoni, S., Iuliano, E., Spano, P.F., and Trabucchi, M. 1977a. Effect of ergotamine and dihydroergotamine on dopamine-stimulated adenylate cyclase in rat caudate nucleus. J Pharm Pharmacol 29:45–47

Govoni, S., Loddo, P., Spano, P.F., and Trabucchi, M. 1977b. Dopamine receptor sensitivity in brain and retina of rats during aging. Brain Res 138:565–570

Grol, C.J., and Rollema, H. 1977. Conformational analysis of dopamine by the INDO molecular orbital method. J Pharm Pharmacol 29:153–156

Gunaga, K.P., and Menon, K.M.J. 1973. Effect of catecholamines and ovarian hormones on cyclic AMP accumulation in rat hypothalamus. Biochem Biophys Res Commun 54:440–448

Harris, J.E. 1976. Beta adrenergic receptor-mediated adenosine cyclic 3′,5′-monophosphate accumulation in the rat corpus striatum. Mol Pharmacol 12:546–558

Henn, F.A., Anderson, D.J., and Sellstrom, A. 1977. Possible relationship between glial cells, dopamine and the effects of antipsychotic drugs. Nature 266:637–638

Hill, M.K., Macleod, R.M., and Orcott, P. 1976. Dibutyryl cyclic AMP, adenosine and guanosine blockade of the dopamine, ergocryptine and apomorphine inhibition of prolactin release *in vitro*. Endocrinology 99:1612–1617

Horn, A.S., Cuello, A.C., and Miller, R.J. 1974. Dopamine in the mesolimbic system of the rat brain. Endogenous levels and the effects of drugs on the uptake mechanism and stimulation of adenylate cyclase activity. J Neurochem 22:265–270

Hornykiewicz, O. 1966. Dopamine (3-hydroxytyramine) and brain function. Pharmacol Rev 18:925–964

Howell, S.L., and Montague, W. 1971. The mode of action of cyclic AMP on the rat anterior pituitary. FEBS Letters 18:293–296

Huang, M., Ho, A.K.S., and Daly, J.W. 1973. Accumulation of adenosine 3′,5′-monophosphate in rat cerebral cortical slices. Mol Pharmacol 9:711–717

Iversen, L.L., Horn, A.S., and Miller, K.J. 1975. Structure-activity Relationships for Agonists and Antagonist Drugs at Pre- and Post-synaptic Dopamine Receptor Sites in Brain. In *Pre- and Post-synaptic Receptors*, eds. E. Usdin and W. E. Bunney, Jr., pp. 207–243. New York:Marcel Dekker, Inc.

Iversen, L.L., Rogawski, M.A., and Miller, R.J. 1976. Comparison of the effects of neuroleptic drugs on pre- and post-synaptic dopaminergic mechanisms on the rat striatum. Mol Pharmacol 12:251–262

Iwatsubo, K., and Clouet, D.H. 1975. Dopamine-sensitive adenylate cyclase of the caudate nucleus of rats treated with morphine or haloperidol. Biochem Pharmacol 24:1499–1503

Kakiuchi, S., and Rall, T.W. 1968. The influence of chemical agents on the accumulation of adenosine 3′,5′-phosphate in slices of rabbit cerebellum. Mol Pharmacol 4:367–378

Karobath, M., and Leitich, H. 1974. Antipsychotic drugs and dopamine-stimulated adenylate cyclase prepared from corpus striatum of rat brain. Proc Natl Acad Sci USA 71:2915–2918

Kebabian, J.W., Calne, D.B., and Kebabian, P.R. 1977. Lergotrile mesylate. An *in vivo* dopamine agonist which blocks dopamine receptors *in vitro*. Comm Psychopharmacol 1:311–318

Kebabian, J.W., and Greengard, P. 1971. Dopamine-sensitive adenyl cyclase. Possible role in synaptic transmission. Science 174:1346–1349

Kebabian, J.W., Petzold, G.L., and Greengard, P. 1972. Dopamine-sensitive adenylate cyclase in caudate nucleus of rat brain, and its similarity to the "dopamine receptor". Proc Natl Acad Sci USA 69:2145–2149

Kebabian, J.W., and Saavedra, J.M. 1976. Dopamine-sensitive adenylate cyclase occurs in a region of substantia nigra containing dopaminergic dendrites. Science 193:683–685

Klainer, L.M., Chi, Y.-M., Freidberg, S.L., Rall, T.W., and Sutherland, E.W. 1962. Adenyl cyclase. IV. The effects of neurohormones on the formation of adenosine 3′,5′-phosphate by preparations from brain and other tissues. J Biol Chem 237:1239–1242

Krueger, B.K., Forn, J., and Greengard, P. 1975. Dopamine-sensitive Adenylate Cyclase and Protein Phosphorylation in the Rat Caudate Nucleus. In *Pre- and Postsynaptic Receptors,* eds. E. Usdin and W. E. Bunney, Jr., pp. 123–147. New York: Marcel Decker, Inc.

Krueger, B.K., Forn, J., Walters, J.R., Roth, P.H., and Greengard, P. 1976. Stimulation by dopamine of adenosine cyclic 3′,5′-monophosphate formation in rat caudate nucleus. Effect of lesions of the nigroneostriatal pathway. Mol Pharmacol 12:639–648

Laduron, P. 1976. Limiting factors in the antagonism of neuroleptics on dopamine-sensitive adenylate cyclase. J Pharm Pharmacol 28:250–251

Laduron, P., Verwimp, M., Janssen, P.F.M., and Leysen, J. 1976. Subcellular localization of dopamine-sensitive adenylate cyclase in rat brain striatum. Life Sci 18:433–440

Leiberman, A., Miyamoto, T., Battista, H., and Goldstein, M. 1975. Studies on the antiparkinsonian efficacy of lergotrile. Neurology 25:459–462

Lew, J.Y., Hata, F., Ohashi, T., and Goldstein, M. 1977. The interactions of bromocriptine and lergotrile and dopamine and α-adrenergic receptors. J Neural Trans 41: 109–121

Lloyd, K.G., Shibuya, M., Davidson, L., and Hornykiewicz, O. 1977. Chronic neuroleptic therapy. Tolerance and GABA systems. In *Advances in Biochemical Psychopharmacology*, Vol. 16, eds. E. Costa and G. L. Gessa, pp. 409–415. New York:Raven Press

Maeno, H., and Greengard, P. 1972. Phosphoprotein phosphatases from rat cerebral cortex. Subcellular distribution and characterization. J Biol Chem 247:3269–3277

Makman, M.H., Brown, J.H., and Mishra, R.K. 1975. Cyclic AMP in Retina and Caudate Nucleus: Influence of Dopamine and Other Agents. In *Advances in Cyclic Nucleotide Research,* eds. G.I. Drummond, P. Greengard, and G.A. Robison, pp. 661–679. New York: Raven Press

McCune, R.W., Gill, J.H., VonHungen, K., and Roberts, S. 1971. Catecholamine-sensitive adenyl cyclase in cell-free preparations from rat cerebral cortex. Life Sci 10:443–450

McDermed, J.D., McKenzie, G.M., and Phillips, A.P. 1975. Synthesis and pharmacology of some 2-aminotetralins. Dopamine receptor agonists. J Med Chem 18:362–367

Mcites, J., Simpkins, J., Bruai, J., and Advis, J. 1977. Role of biogenic amines in control of anterior pituitary hormones. IRCS Med Sci 5:1–7

Miller, R.J., Horn, A.S., and Iversen, L.L. 1975. Effect of butaclamol on dopamine-sensitive adenylate cyclase in the rat striatum. J Pharm Pharmacol 27:212–213

Miller, R.J., Horn, A.S., and Iversen, L.L. 1974a. The Action of neuroleptic drugs on dopamine-stimulated adenosine cyclic 3′,5′-monophosphate production in rat neostriatum and limbic forebrain. Mol Pharmacol 10:759–766

Miller, R., Horn, A., and Iversen, L. 1974b. Effects of dopamine-like drugs on rat striatal adenyl cyclase have implications for CNS dopamine receptor topography. Nature 250:238–241

Minneman, K.P. 1976. Cyclic nucleotide phosphodiesterase in rat neostriatum. Multiple isoelectric forms with similar kinetic properties. J Neurochem 27:1181–1189

Mishra, R.K., Demirjian, C., Katzman, R., and Makman, M.H. 1975. A dopamine-sensitive adenylate cyclase in anterior limbic cortex and mesolimbic region of primate brain. Brain Res 96:395–399

Mishra, R.K., Gardner, E.L., Katzman, R., and Makman, M.H. 1974. Enhancement of dopamine-stimulated adenylate cyclase activity in rat caudate after lesions in substantia nigra. Evidence for denervation supersensitivity. Proc Natl Acad Sci USA 71: 3883–3887

Munday, K.A., Poat, J.A., and Woodruff, G.N. 1976. Structure-activity studies on dopamine receptors; a comparison between rat striatal adenylate cyclase and *Helix aspersa* neurons. Brit J Pharm Pharmacol 57:452p

Neumeyer, J.L., McCarthy, M.M., and Battista, S.P. 1973. Aporphines.9. Synthesis and pharmacological evaluation of (±)-9,10-dihydroxyaporphine [(±)-isoapomorphine], (±), (−), and (±)-1,2-dihydroxyaporphine, and (±)-1,2,9,10-tetrahydroxyapormorphine. J Med Chem 16:1228–1233

Palmer, G.C. 1972. Increased cyclic AMP response to norepinephrine in the rat brain following 6-hydroxydopamine. Neuropharmacology 11:145–149

Palmer, G.C., and Burks, T.F. 1971. Central and peripheral adrenergic blocking actions of LSD and BOL. Eur J Pharmacol 16:113–116

Palmer, G.C., and Manian, A.A. 1976. Actions of phenothiazine analogues on dopamine-sensitive adenylate cyclase in neuronal and glial enriched fractions from rat brain. Biochem Pharmacol 25:63–71

Palmer, G.C., and Palmer, S.J. 1978. 5-guanylyl-imidodiphosphate potentiation of adenylate cyclase in homogenates plus neuronal and capillary fractions of rat cerebral cortex. Life Sci 23:207–216

Palmer, G.C., Sulser, F. and Robison, G.A. 1973. Effects of neurohumoral and adrenergic agents on cyclic AMP levels in various areas of the rat brain *in vitro*. Neuropharmacology 12:327–337

Palmer, G.C., and Wagner, H.R. 1976. Supersensitivity of striatal and cortical adenylate cyclase following reserpine. Lack of effect of chronic haloperidol. Res Comm Psychol Psychiat Behav 1:567–570

Palmer, G.C., Wagner, H.R., and Putnam, R.W. 1976. Neuronal localization of the enhanced adenylate cyclase responsiveness to catecholamines in the rat cerebral cortex following reserpine injections. Neuropharmacology 15:695–702

Pardo, J.V., Creese, I., Burt, D.R., and Snyder, S.H. 1977. Ontogenesis of dopamine receptor binding in the corpus striatum of the rat. Brain Res 125:376–382

Perkins, J.P. 1973. Adenyl cyclase. Adv Cyclic Nucleotide Res 3:1–64

Perkins, J.P., and Moore, M.M. 1971. Adenyl cyclase of rat cerebral cortex activation by sodium fluoride and detergents. J Biol Chem 246:62–68

Pinder, R.M., Buxton, D.A., and Woodruff, G.N. 1972. On apomorphine and dopamine receptors. J Pharm Pharmacol 24:903–904

Premont, J., Perez, M., and Bockaert, J. 1977. Adenosine-sensitive adenylate cyclase in rat striatal homogenates and its relationship to dopamine- and Ca^{2+}-sensitive adenylate cyclases. Mol Pharmacol 13:662–670

Pugsley, T.A., Merker, J., and Lippman, W. 1976. Effect of structural analogs of butaclamol (a new antipsychotic drug) on striatal homovanillic acid and adenyl cyclase of olfactory tubercle in rats. Can J Physiol Pharmacol 54:510–515

Puri, S.K., and Volicer, L. 1977. Effect of aging on cAMP levels and adenyl cyclase and phosphodiesterase activities in the rat corpus striatum. Mech Ageing 6:53–58

Robison, G.A., Schmidt, M.J., and Sutherland, E.W. 1970. On the development and properties of the brain adenyl cyclase system. In *Advances in Biochemical Psychopharmacology*, eds. P. Greengard and E. Costa, pp. 11–30. New York: Raven Press

Rodbell, M., Lin, M.C., Solomon, V., Londos, C., Harwood, J.P., Martin, B.R., Rendell, M., and Berman, M. 1975. Role of adenine and guanine nucleotides in the activity and response of adenylate cyclase systems to hormones. Evidence for Multisite transition states. In *Advances in Cyclic Nucleotide Research*, eds. G. I. Drummond, P. Greengard, G.A. Robison, pp. 3–29. New York: Raven Press

Rotrosen, J., Friedman, E., and Gershon, S. 1975. Striatal adenylate cyclase activity following reserpine and chronic chlorpromazine administration in rats. Life Sci 17: 563–568

Roufogalis, B.D., Thornton, M., and Wade, D.N. 1976. Nucleotide requirement of dopamine sensitive adenylate cyclase in synaptosomal membranes from the striatum of rat brain. J Neurochem 27:1533–1535
Saari, W.S., and King, S.W. 1974. Synthesis and biological activity of some aporphine derivatives related to apomorphine. J Med Chem 17:1086–1090
Sato, A., Onaya, T., Kotani, M., Haroda, A., and Yamada, T. 1974. Effects of biogenic amines on the formation of adenosine 3′,5′-monophosphate in the porcine cerebral cortex, hypothalamus and anterior pituitary slices. Endocrinology 94:1311–1317
Satoh, H., Satoh, Y., Notsu, Y., and Honda, F. 1976. Adenosine 3′,5′-cyclic monophosphate as a possible mediator of rotational behavior induced by dopaminergic receptor stimulation in rats lesioned unilaterally in the substantia nigra. Eur J Pharmacol 39:365–377
Scatton, B., Bischoff, S., Dedek, J., and Korf, J. 1977. Regional effects of neuroleptics on dopamine metabolism and dopamine-sensitive adenylate cyclase activity. Eur J Pharmacol 44:287–292
Schmidt, M.J., and Hill, L.E. 1977. Effects of ergots on adenylate cyclase activity in the corpus striatum and pituitary gland. Life Sci 20:789–798
Schmidt, M.J., Palmer, E.C., Dettbarn, W-D., and Robison, G.A. 1970. Cyclic AMP and adenyl cyclase in the developing rat brain. Develop Psychobiol 3:53–67
Schmidt, M.J., and Thornberry, J.F. 1978. Cyclic AMP and cyclic GMP accumulation *in vitro* in brain regions of young, old and aged rats. Brain Res 139:169–177
Schmidt, M.J., and Thornberry, J.F. 1977. Norepinephrine-stimulated cyclic AMP accumulation in brain slices *in vitro* after serotonin depletion or chronic administration of selective amine reuptake inhibitors. Arch Int Pharmacodyn Exper Therap, in press
Schmidt, M.J., Truex, L.L., and Thornberry, J.F. 1978. Cyclic nucleotides and protein kinase activity in the rat brain postmortem. J Neurochem, in press
Schorderet, M. 1978. Dopamine-mimetic activity of ergot derivatives as measured by the production of cyclic AMP in isolated retinae of the rabbit. Gerontology 24:86–93
Schorderet, M. 1976. Direct evidence for the stimulation of rabbit retina dopamine receptors by ergot alkaloids. Neurosci Letters 2:87–91
Schwarcz, R., and Coyle, J.T. 1977. Neurochemical sequelae of kainate injections in corpus striatum and substantia nigra of the rat. Life Sci 20:431–436
Seeman, P., and Lee, T. 1975. Antipsychotic drugs. Direct correlation between clinical potency and presynaptic action on dopamine neurons. Science 188:1217–1219
Seeman, P., Staiman, A., and Chau-Wong, M. 1974. The nerve impulse blocking actions of tranquilizers and the binding of neuroleptics to synaptosome membranes. J Pharmacol Exp Ther 190:123–130
Shaar, C.J., and Clemens, J.A. 1974. The role of catecholamines in the release of anterior pituitary prolactin *in vitro*. Endocrinology 95:1202–1212
Sheppard, H. 1977. Activities of dopamine and β-adrenergic adenylate cyclases. In *Annual Reports in Medicinal Chemistry*, Vol. 12, Chapter 18
Sheppard, H., and Burghardt, C.R. 1976. The demonstration of beta as well as dopamine type systems for the generation of cyclic AMP in homogenates of the rat caudate nucleus. Fed Proc 35:424
Sheppard, H., and Burghardt, C.R. 1974. The dopamine-sensitive adenylate cyclase of rat caudate nucleus. 1. Comparison with the isoproterenol-sensitive adenylate cyclase (beta receptor system) of rat erythrocytes in response to dopamine derivatives. Mol Pharmacol 10:721–726
Sheppard, H., Burghardt, C.R., and Teitel, S. 1976. The Dopamine-sensitivie adenylate cyclase of the rat caudate nucleus. II. A comparison with the isoproterenol-sensitive (beta) adenylate cyclase of the rat erythrocyte for inhibition or stimulation by tetrahydroisoquinolines. Mol Pharmacol 12:854–861
Siggins, G.R., Hoffer, B.J., and Ungerstedt, U. 1974. Electrophysiological evidence for involvement of cyclic adenosine monophosphate in dopamine responses of caudate neurons. Life Sci 15:779–792
Silbergeld, E.K., and Pfeiffer, R.F. 1977. Differential effects of three dopamine agonists. Apomorphine, bromocriptine and lergotrile. J Neurochem 28:1323–1326

Smalstig, E.B., Sawyer, B.D., and Clemens, J.A. 1974. Inhibition of rat prolactin release by apomorphine *in vivo* and *in vitro*. Endocrinology 5:123-129

Snyder, S.H., Banerjee, S.P., Yamamura, H.I., and Greenberg, D. 1974. Drugs, neurotransmitters and schizophrenia. Science 184:1243-1253

Soudjin, W., and van Wijngaarden, I. 1972. Localization of (^{3}H) pimozide in the rat brain in relation to its anti-amphetamine potency. J Pharm Pharmacol 24:773-780

Spano, P.F., DiChiara, G., Tonon, G.C., and Trabucchi, M. 1976a. A dopamine-stimulated adenylate cyclase in rat substantia nigra. J Neurochem 27:1565-1568

Spano, P.F., Kumakura, K., Govoni, S., and Trabucchi, M. 1976b. Ontogenetic development of neostriatal dopamine receptors in the rat. J Neurochem 27:621-624

Spano, P.F., and Trabucchi, M. 1978. Interaction of ergot alkaloids with dopaminergic receptors in the rat striatum and nucleus accumbens. Gerontology 24:106-114

Steiner, A.L., Peake, G.T., Utiger, R.D., Karl, I.E., and Kipnis, D.M. 1970. Hypothalamic stimulation of growth hormone and thyrotropin release *in vitro* and pituitary 3′,5′-adenosine cyclic monophosphate. Endocrinology 86:1345-1360

Sutherland, E.W., Rall, T.W., and Menon, T. 1962. Adenyl cyclase I. Distribution, preparation, and properties. J Biol Chem 237:1220-1227

Tang, L.C., and Cotzias, G. 1977. Quantitative correlation of dopamine-dependent adenylate cyclase with responses to levodopa in various mice. Proc Natl Acad Sci USA 74:1242-1244

Tang, L., Cotzias, G., and Tumbleson, M. 1978. Sex differences in dopamine activated adenylate cyclase in mouse caudate. Trans Amer Soc Neurochem 9:72

Tarsy, D., and Baldessarini, R.J. 1974. Behavioral supersensitivity to apomorphine following chronic treatment with drugs which interfere with the synaptic function of catecholamines. Neuropharmacology 13:927-940

Tell, G.P., Pasternak, G. W., and Cuatrecasas, P. 1975. Brain and caudate nucleus adenylate cyclase. Effects of dopamine, GTP, E prostaglandins and morphine. FEBS Letters 51:242-245

Trabucchi, M., Govoni, S., Tonon, G.C., and Spano, P.F. 1976a. Localization of dopamine receptors in the rat cerebral cortex. J Pharm Pharmacol 28:244-245

Trabucchi, M., Hofmann, M., Montefusco, O., and Spano, P.F. 1978. Ergot alkaloids and cyclic nucleotides in the CNS. Pharmacology 16 (Supp. 1):150-155

Trabucchi, M., Spano, P.F., Tonon, G.C., and Frattola, L. 1976b. Effect of bromocriptine on central dopaminergic receptors. Life Sci 19:225-232

Tye, N.C., Horsman, L., Wright, F.C., Lange, B.T., and Pullar, I.A. 1977. Two dopamine receptors. Supportive evidence with the rat rotational model. Eur J Pharmacol 45:87-90

Ungerstedt, U. 1971. Postsynaptic supersensitivity after 6-hydroxydopamine induced degeneration of the nigra-striatal dopamine system. Acta Physiol Scand 367:69-93

Voigt, K.M., and Krishna, G. 1967. Correlation between the distribution of adenyl cyclase (AC) and cyclic 3′,5′-AMP phosphodiesterase (PD) and various biological amines in various areas of the brain. Pharmacologist 9:239

Volkman, P.H., and Goldberg, L.I. 1976. Lack of correlation between inhibition of prolactin release and stimulation of dopaminergic renal vasodilation. Pharmacologist 18:130

Volkman, P.H., Kohli, J.D., Goldberg, L.F., Cannon, J.G., and Lee, T. 1977. Conformational requirements for dopamine-induced vasodilation. Proc Nat Acad Sci USA 74:3602-3606

von Hungen, K., and Roberts, S. 1973. Catecholamine and Ca^{2+} activation of adenylate cyclase systems in synaptosomal fractions from rat cerebral cortex. Nature New Biology 242:58-60

von Hungen, K., Roberts, S., and Hill, D.F. 1975. Interactions between lysergic acid diethylamide and dopamine-sensitive adenylate cyclase systems in rat brain. Brain Res 94:57-66

von Hungen, K., Roberts, S., and Hill, D.E. 1974. Developmental and regional variations in neurotransmitter-sensitive adenylate cyclase systems in cell-free preparations from rat brain. J Neurochem 22:811-819

Von Voigtlander, P.F., Boukma, S.J., and Johnson, G.A. 1973. Dopaminergic denervation supersensitivity and dopamine stimulated adenyl cyclase activity. Neuropharmacology 12:1081-1086

Von Voigtlander, P.F., Losey, E.G., and Triezenberg, H.J. 1975. Increased sensitivity to dopaminergic agents after chronic neuroleptic treatment. J Pharmacol Exp Ther 193:88-94

Walker, J.B., and Walker, J.P. 1973. Neurohumoral regulation of adenylate cyclase activity in rat striatum. Brain Res 54:386-390

Weiss, B., and Costa, E. 1968. Regional and subcellular distribution of adenyl cyclase and 3′,5′-cyclic nucleotide phosphodiesterase in brain and pineal gland. Biochem Pharmacol 17:2107-2116

Wilkening, D., and Makman, M.H. 1975. 2-chloroadenosine-dependent elevation of adenosine 3′,5′-cyclic monophosphate levels in rat caudate nucleus slices. Brain Res 92:522-528

Williams, R.H., Little, S.A., and Ensinck, J.W. 1969. Adenyl cyclase and phosphodiesterase activities in brain areas of man, monkey and rat. Amer J Med Sci 258: 190-202

Woodruff, G.N., McCarthy, P.S., and Walker, R.J. 1976. Studies on the pharmacology of neurones in the nucleus accumbens of the rat. Brain Res 115:233-242

Woodruff, G.N., Watling, K.J., Andrews, C.D., Poat, J.A., and McDermed, J.D. 1977. Dopamine receptors in rat striatum and nucleus accumbens. Conformational studies using rigid analogues of dopamine. J Pharm Pharmacol 29:422-427

Actions of Neuroleptic Agents on Central Cyclic Nucleotide Systems

Gene C. Palmer and Albert A. Manian***
*Dept. of Pharmacology
University of South Alabama
College of Medicine
Mobile, Alabama 36688
**Psychopharmacology Research Branch
National Institute of Mental Health
Rockville, Maryland 20857

INTRODUCTION

Ample evidence exists to indicate that norepinephrine (NE) and dopamine (DA) play important roles as neurotransmitters in the central nervous system. High concentrations of these catecholamines are found at discrete loci within particular brain regions. By using drugs as tools, as well as behavioral, electrophysiological and biochemical methods, the findings to date strongly suggest that central neural pathways mediating the action of biogenic amines are involved in mood and behavior. Clinical and laboratory findings have further demonstrated relationships between central monoamine levels and affective disorders. In this regard, fluctuations of central NE levels have been associated with endogenous depression (decreased amounts) and mania or schizophrenia (elevated functional levels). Increased physiological concentrations of DA at particular central synaptic regions have been related to psychotic behavior and schizophrenia, while decreased functional levels of DA in the striatal region are associated with parkinsonism. Furthermore, long-term treatment of humans with antipsychotic drugs leads to the syndrome of tardive dyskinesia, which has been postulated to result from a DA receptor supersensitivity so initiated to compensate for the chronic receptor blockade of striatal neurons. Similarly, drugs (amphetamines) effecting a release of central DA and NE produce hyperactivity and a psychosis indistinguishable from paranoid schizophrenia. Likewise, L-DOPA, a pre-

The authors wish to express appreciation to the Epilepsy Foundation of America for research support that was utilized to complete the most recent investigations. In addition, the authors would like to express a note of thanks to the following individuals for either technical or collaborative help in many of these investigations: Dr. F. Sulser, Dr. G. Alan Robison, Mr. Rufus Putnam, Mr. Halbert Scott, Mr. George Baca, Dr. H. Ryan Wagner, Ms. Maura Spiker, Mr. Gregory Cotter, Mrs. Shelby Jo Palmer, Dr. David Jones, Dr. Miguel Medina, Dr. William Stavinoha and Mr. Andrew Deam.

cursor of DA, is used to treat the functional deficiency associated with parkinsonism. Drugs which deplete central monoamines produce depression and sedation in animals and humans, while drugs inhibiting the destruction or reuptake of catecholamines alleviate depression and may in turn produce mania and psychosis. (For reviews, see *Schildkraut* and *Kety*, 1967; *Matthysse* and *Lipinski*, 1975; *Snyder et al.*, 1974a; *Iversen*, 1975; *Baldessarini*, 1977; *Mendels, Stern,* and *Frazer*, 1976).

It is generally agreed that neuroleptic agents including the phenothiazine and butyrophenone derivatives exert specific antipsychotic actions on the fundamental mechanisms of schizophrenia. The actions are not merely to produce sedation since some agents do not possess this property and most antipsychotic drugs will activate withdrawn patients and calm hyperactive ones as well. The most likely primary site of action to effect antipsychotic action was postulated by *Carlsson* and *Lindqvist* (1963) to be an antagonism of central DA receptors. Stereochemical and X-ray diffraction studies have confirmed that the conformation of the phenothiazine and butyrophenone molecules fits the theory of DA receptor blockade (*McDowell*, 1974; *Kaufman* and *Kerman*, 1974; *Feinberg* and *Snyder*, 1975; *Snyder*, 1976). However, the NE receptor would also be influenced by these molecules, albeit to a lesser extent. The postulated central sites for antipsychotic actions manifested by the neuroleptic agents are the mesolimbic region (nucleus accumbens and olfactory tubercle), the septal region including the amygdala, the hippocampus and the frontal cortex. These regions receive DA nerve endings from the substantia nigra and interpeduncular regions (A-10). The extrapyramidal site of action of phenothiazines is at the striatum, particularly the caudate nucleus which receives DA input principally via the substantia nigra.

A central receptor enzyme, adenylate cyclase, discovered by Earl Sutherland (see *Sutherland* and *Robison*, 1966), is located at postsynaptic regions within the central nervous system and is sensitive to stimulation via separate receptor components to both NE and DA. In some species and within particular brain regions, histamine, prostaglandin, substance P and serotonin activation of the enzyme occurs. The enzyme itself is a complex lipoprotein(s) whose structure is not entirely known. Basically, several different receptor components are thought to reside in the postsynaptic plasma membrane and face the exterior of the cell. Upon presynaptic release of biogenic amine neurotransmitters, a pattern of forces (most likely chemical interactions) involving an action of calcium is set in motion at the highly specific receptor site which recognizes the molecular configuration of the neurotransmitters. By unexplained mechanisms, the occupied receptor initiates a series of actions through an intermediate(s) transducer-coupling reaction that requires guanosine triphosphate (GTP). In turn, a catalytic moiety of the enzyme is activated which then catalyzes the formation of intracellular adenosine cyclic 3′,5′-monophosphate (cyclic AMP). The latter enzyme component is postulated to face the interior of the cell and, for presently unexplained reasons, is stimulated by fluoride ion. One or more regulator proteins, activated by calcium, also appear to be necessary for neurohumoral stimulation of adenylate cyclase. (For recent reviews, see *Rodbell et al.*, 1975; *Nathanson*, 1977; *Daly*, 1976, 1977; for original work, see *Robison*, *Butcher*, and *Sutherland*, 1968; see also Fig. 1.) Taken together, the data indicate a role for cyclic nucleotides in neurotransmission, but controversial findings do exist (*Phillis*, 1977).

The DA or NE receptor sites of adenylate cyclase, both of which are found in the striatum, mesolimbic area, septal region, hippocampus and frontal cortex, as well as other brain structures, are likely sites for the therapeutic actions manifested by neuroleptic agents. Whether the ultimate metabolic actions of phenothiazines on psychosis or schizophrenia are related to intracellular biochemical mechanisms affected by cyclic AMP is completely unknown. Therefore, additional mechanisms may be activated when central receptors are stimulated by cyclic AMP. Greengard and associates (*Greengard* and *Kebabian*, 1974) have suggested that phosphorylation of subsynaptic membranes, an event mediated by cyclic AMP-induced protein kinases, may be ultimately associated with postsynaptic neuronal functions.

Another naturally occurring cyclic nucleotide, guanosine cyclic 3′,5′-monophosphate (cyclic GMP), may in certain instances be related to central cholinergic (muscarinic) actions. Cyclic GMP has been shown in several tissues, including the brain, to modulate cyclic AMP levels. The enzyme for its activity has an absolute requirement for calcium. Relatively little information is available with respect to any alteration in levels of cyclic GMP to behavioral processes. However, as will be described later in this chapter, the neuroleptics do influence guanylate cyclase-cyclic GMP in the brain and additionally inhibit specific molecular forms of the enzymes (phosphodiesterases) responsible for metabolism of these cyclic nucleotides.

The data, presented and discussed in detail elsewhere (*Iversen*, 1975; *Snyder et al.*, 1974a; *Greengard* and *Kebabian*, 1974; *Daly*, 1976, 1977; *Nathanson*, 1977), indicate that DA- and, to a lesser extent, NE-sensitive adenylate cyclases represent the primary therapeutic sites of action of the neuroleptics. On the other hand, many neuroleptic agents possess a wide variety of side-effects which may influence their therapeutic outcome as well. In addition, many metabolites of these compounds display a wide array of behavioral, physiological and adverse reactions. Seeman and coworkers (*Seeman et al.*, 1974, 1975; *Seeman* and *Lee*, 1975; *Seeman*, 1977) have conducted extensive research as to the specific therapeutic or secondary actions of a wide variety of phenothiazine and butyrophenone analogues. Many of these actions are diagramed in Figure 2.

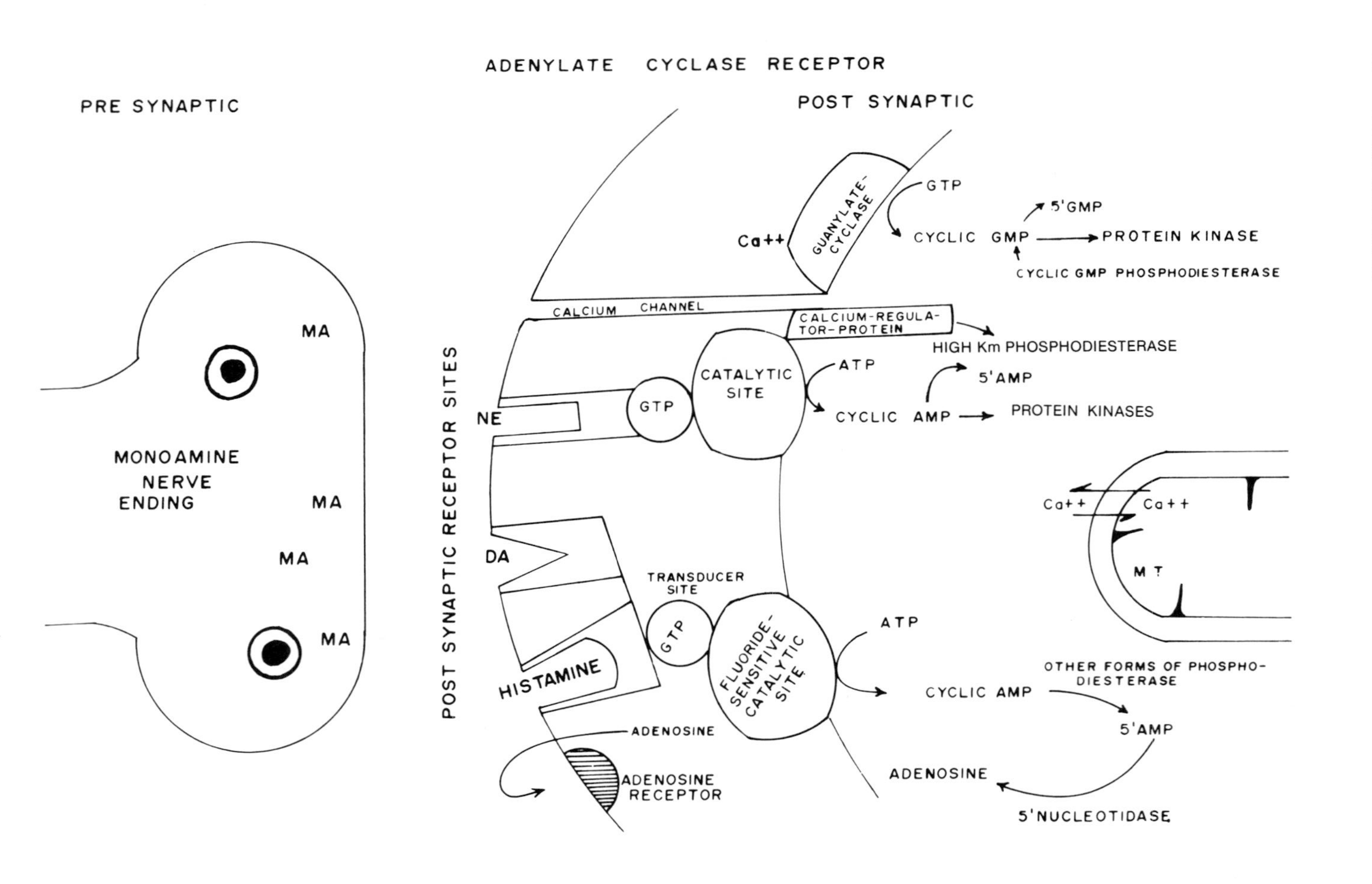
ADENYLATE CYCLASE RECEPTOR
PRE SYNAPTIC
POST SYNAPTIC
MONOAMINE NERVE ENDING
MA
MA
MA
MA
POST SYNAPTIC RECEPTOR SITES
NE
DA
HISTAMINE
ADENOSINE
ADENOSINE RECEPTOR
GTP
TRANSDUCER SITE
GTP
CATALYTIC SITE
FLUORIDE-SENSITIVE CATALYTIC SITE
CALCIUM CHANNEL
CALCIUM-REGULATOR-PROTEIN
GUANYLATE-CYCLASE
Ca++
GTP
CYCLIC GMP
5'GMP
PROTEIN KINASE
CYCLIC GMP PHOSPHODIESTERASE
HIGH Km PHOSPHODIESTERASE
ATP
5'AMP
CYCLIC AMP
PROTEIN KINASES
Ca++
Ca++
M T
ATP
OTHER FORMS OF PHOSPHO-DIESTERASE
CYCLIC AMP
5'AMP
ADENOSINE
5'NUCLEOTIDASE

ACTIONS OF NEUROLEPTICS ON DOPAMINE-SENSITIVE ADENYLATE CYCLASE SYSTEMS IN THE CENTRAL NERVOUS SYSTEM

Background

Within the past few years substantial evidence from a variety of sources has been accumulated to indicate an important role for DA as a central neurotransmitter. Abnormalities of functional DA actions in the central nervous system have been implicated in the affective disorders to include some psychoses, such as schizophrenia, and parkinsonism. Moreover, the drugs used to treat schizophrenia possess the structural configurations necessary to be effective antagonists of DA receptor actions in the brain. This chronic blockage of DA receptors within particular brain regions (striatum) is associated with neuroleptic-induced parkinsonism. Neuroleptics, however, have a wide variety of other primary and secondary actions. At the forefront of these investigations the action of one subtype of DA receptor in the brain appears to be mediated via an interaction of the adenylate cyclase-cyclic AMP system. Thus, the formation of cyclic AMP, the end product of an interaction of presynaptically released DA to a postsynaptic action on adenylate cyclase, may reflect a functional correlation within the brain in regard to behavioral changes. There is little evidence, however, to support this latter contention. Cyclic AMP does play a key metabolic role in central tissues by activating specific protein kinases which have been shown to phosphorylate postsynaptic membranes (possible control of ionic gradients), phosphorylate neuro-

---

Fig. 1. Hypothetical diagram of the postsynaptic monoamine-responsive nerve ending which depicts the interrelationships of the various components of adenylate cyclase, guanylate cyclase and the phosphodiesterases. The neurotransmitter is released from presynaptic terminals and combines in a sterospecific interaction at highly specified postsynaptic membrane sites. Following a conformation change at the receptor, a sequence of postsynaptic events may occur. Receptor action is coupled through intermediate steps, i.e., GTP-sensitive transducer sites, Ca^{++} translocation, the release of the Ca^{++} regulator protein and release of adenosine. The catalytic component of adenylate cyclase becomes activated via the transducer site and release of the Ca^{++} regulator protein. Thus, ATP is converted to cyclic AMP, which in turn binds to a subcomponent of protein kinase thereby releasing a catalytic moiety of this enzyme. The translocation of Ca^{++} activates the Ca^{++} regulator protein-dependent high Km form of phosphodiesterase and this enzyme, along with the other forms of phosphodiesterase, inactivates cyclic AMP to 5′-AMP. Moreover, the translocation of calcium in these processes may be influenced by phosphorylation of subsynaptic membranes by the activated form of protein kinases. In addition, Ca^{++} activates guanylate cyclase, and cyclic GMP may in turn modulate the level of cyclic AMP, as well as activate specific cyclic GMP-dependent protein kinases. The release of adenosine stimulates specific adenosine receptors which also elevate cyclic AMP through unknown mechanisms. The 5′-AMP may be recycled via the postsynaptic 5′-nucleotidase to adenosine, which is either available for release or reincorporated into ATP (*Suran*, personal communication).

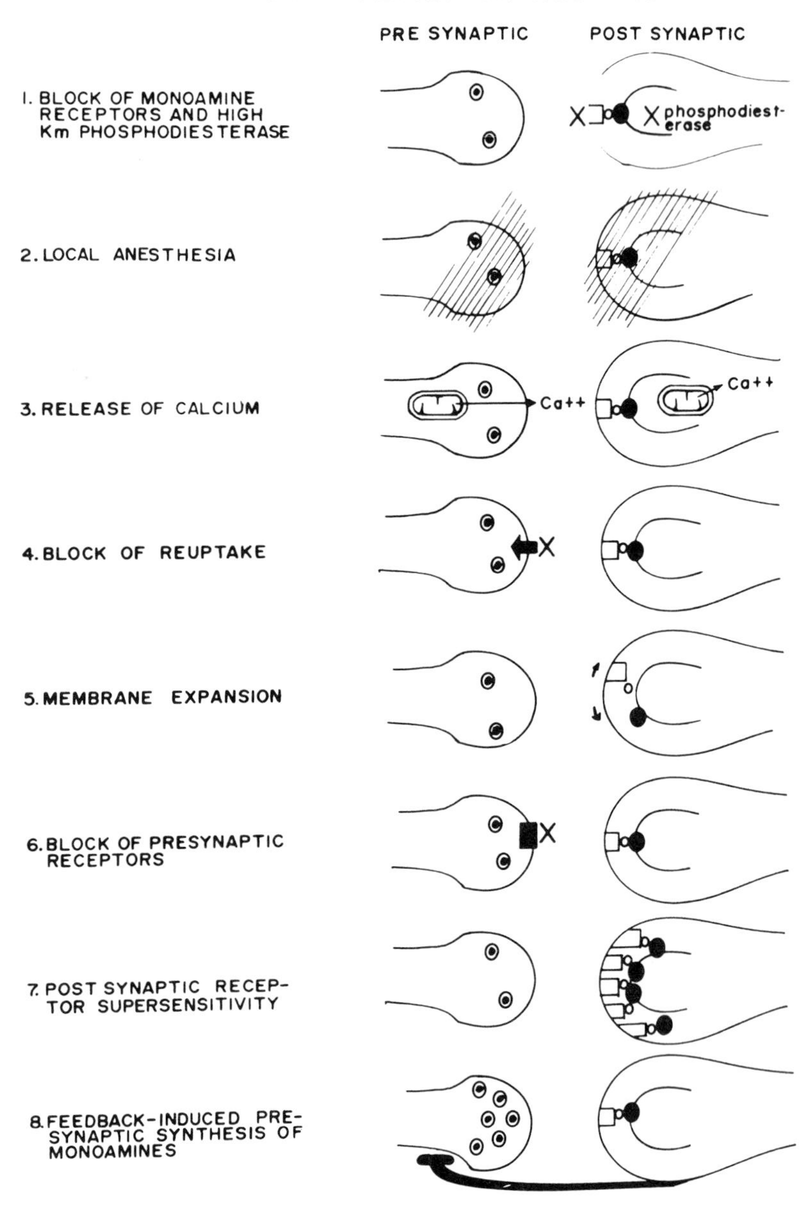

Fig. 2. Diagram of the postulated multiple actions of phenothiazines on pre- and post-synaptic nerve endings.

tubules (axonal transport may be involved), phosphorylate histones (influence transcription and translation) and activate phosphorylase (immediate energy source with glycogen breakdown). Moreover, the DA activation of adenylate cyclase is not solely confined to dendritic nerve endings because neuronal soma, glia and capillaries contain similar systems as well. Dopamine-sensitive adenylate cyclases do have pharmacological properties that distinguish them from the classical α- and β-adrenergic receptors. The most prominent aspect of DA-responsive adenylate cyclases is their extreme sensitivity to a competitive antagonism by neuroleptic drugs, as will be evident in the following paragraphs. (For reviews of DA actions and hypotheses, see *Snyder et al.*, 1974a; *Feinberg* and *Snyder*, 1975; *Iversen*, 1975; *Seeman*, 1977; *Nathanson*, 1977.)

Striatum

The initial observation by *Kebabian*, *Petzold*, and *Greengard* (1972) that broken cell preparations of rat caudate nucleus would respond to DA activation of adenylate cyclase led to an inundation of investigations in an attempt to identify neuroleptic actions at DA receptors. There is little argument that adenylate cyclase from tissue homogenates of rat striatum will respond to DA. A few groups of investigators have reported an elevation of cyclic AMP by DA in incubated tissue slices, i.e., intact cellular preparations of striatal tissue (*Walker* and *Walker*, 1973; *Forn*, *Krueger*, and *Greengard*, 1974). Among others, *Shimizu*, *Daly*, and *Creveling* (1969), *Palmer*, *Sulser*, and *Robison* (1973) and *Harris* (1976) were only able to show a positive action of NE and isoproterenol, but not DA, to elicit the accumulation of cyclic AMP in tissue slices of the caudate nucleus. Continuing arguments are put forth to establish with a degree of certainty whether or not the DA-sensitive adenylate cyclase represents a separate unique DA receptor. Some authors feel that DA responses are mediated instead through either a weak α- or β-agonist interaction at noradrenergic receptors (*Nathanson*, 1977; *Daly*, 1976, 1977; *Harris*, 1976).

Attempts to localize the site of DA-receptive adenylate cyclases in the striatum have again yielded conflicting results. In the rat, apparently only neuronal preparations will respond to DA (*Palmer* and *Manian*, 1976). However, the calf glial cells seem to possess this ability as well (*Henn*, *Anderson*, and *Sellstrom*, 1977). Subcellular fractionation of the caudate yields both DA and adrenergic activation of adenylate cyclase in the nuclear and mitochrondrial fractions. The latter contains the nerve endings, and further separation of the mitochrondrial fraction yields the highest DA responses in subfractions consisting mainly of nerve ending membranes (*von Hungen* and *Roberts*, 1974; *Clement-Cormier et al.*,

1975; *Harris*, 1976). On the other hand, *Laduron et al.* (1976) and *Leysen* and *Laduron* (1977) could not correlate the DA-sensitive adenylate cyclase to nerve endings. Following subfractionation of the heavy mitochrondrial fraction and sterospecific displacement of haloperidol binding by (+) and (−) butaclamol, neuroleptic receptors and DA-adenylate cyclases were not associated with the same subcellular particles.

The studies by the following investigators attest to the fact that DA-sensitive adenylate cyclases in homogenates of the rat caudate are influenced by a host of various neuroleptic agents and their corresponding derivatives and metabolites: *Kebabian et al.* (1972), *Miller* and *Iversen* (1974), *Miller* and *Hiley* (1974), *Miller, Horn*, and *Iversen* (1974), *von Hungen, Roberts*, and *Hill* (1974, 1975), *Clement-Cormier et al.* (1974, 1975), *Clement-Cormier* and *Robison* (1977), *Karobath* and *Leitich* (1974), *Karobath* (1975a & b), *Iwatsubo* and *Clouet* (1975), *Palmer* and *Manian* (1975, 1976), *Coupet, Szucs,* and *Greenblatt* (1976), *Roufogalis, Thornton*, and *Wade* (1976), *Laduron et al.* (1976), *Harris* (1976), *Seeber* and *Kuschinsky* (1976), *Weinryb* and *Michel* (1976), *Burkard* (1977), *Bockaert et al.* (1977), *Mishra* (1977), *Leysen* and *Laduron* (1977), *Scatton et al.* (1977) and *Palmer et al.* (1978a).

For the approximate IC_{50} and Ki values of this antagonism by neuroleptics of DA-sensitive adenylate cyclases, see Table 1. Several agents display a high degree of potency with regard to the IC_{50} value ($< 5\ \mu M$): CPZ; β-OH-CPZ; 7,8,β-triOH-CPZ; 7,8-dioxo-β-OH-CPZ; fluphenazine; trifluoperazine; promazine; triflupromazine; prochlorperazine; thiothixene; α-clopenthixol; α-chlorprothixene; α,β flupenthixol, α-flupenthixol; haloperidol; pipamperone; spiroperidol; clozapine; thioridazine; trimeprazine; mesoridazine and (±) butaclamol. Not all investigators reported Ki values in their publications. Neuroleptics displaying the lowest Ki values (< 10 nM) were fluphenazine; trifluoperazine; α,β-flupenthixol; α-flupenthixol; loxapine and 7-OH-loxapine. Agents with

Table 1. Concentrations of Neuroleptic Agents Required to Inhibit DA-sensitive Adenylate Cyclase in Rat Caudate Nucleus

Compound	References	$IC_{50}(\mu M)$	Ki(nM)
Phenothiazine Derivatives			
CPZ	1-4, 7, 13	1-6	33-115
Desmethyl-CPZ	11	6	NA
Didesmethyl-CPZ	11	11	NA
7-OH-CPZ	11	14	NA
7-OH-CPZ-MeI	10	6	NA
CPZ-NO	11	50	NA
CPZ-SO	11, 12	>100	NA
1-Cl-Promazine	9	50	NA

continued

Table 1. (*continued*)

Compound	References	IC_{50}(μM)	Ki(nM)
3-Cl-Promazine	9	16	NA
4-Cl-Promazine	9	8	NA
CPZ-MeCl	10	>100	NA
β-OH-CPZ	10	0.4	NA
7,8-DiOH-CPZ	10	20	NA
7,8-Dioxo-CPZ	10	7	NA
3,7,8-TriOH-CPZ	10	10	NA
7,8,βTriOH-CPZ	10	5	NA
7,8-Dioxo,β-OH-CPZ	10	2	NA
Promethazine	1, 7, 12	15->100	112-5000
Trimeprazine	4	0.5	NA
Promazine	1, 2	0.4-60	39-2800
Triflupromazine	2.4	0.3-0.4	40
Thioridazine	1, 2, 4, 7, 13	0.5-3	25-130
Mesoridazine	4	0.4	NA
Prochlorperazine	1, 2, 4	NA	NA
Trifluoperazine	1, 2, 4	0.1-0.6	10-19
Fluphenazine	1, 2, 4, 7	0.07-0.4	4.3-10
Methdilazine	4	8	
Thioxanthines			
Thiothixene	4	1.5	NA
α-Clopenthixol	1	0.3	16
β-Clopenthixol	1	60	2800
α-Chlorprothixene	1, 4	0.3-0.8	37
β-Chlorprothixene	1	20	950
Thioproperazine	7	NA	112
Thiethylperazine	4	6.5	NA
α-Flupenthixol	1	0.02	1
β-Flupenthixol	1	>100	>5000
α,β-Flupenthixol	1	0.08	3.5
Butyrophenones			
Haloperidol	1-7, 13	0.2-2	25-220
Droperidol	1, 2, 4	3-75	40
Spiroperidol	1	2	95
Penfluridol	5	8	NA
Pipamperone	5	2.8	NA
Azeperone	5	7.2	NA
Pimozide	1, 2, 5	3-15	140-1220
Clopimozide	5	6	NA
Sulpiride	13	700	14,000
Loxapines			
Loxapine	8	<30	10
7-OH-Loxapine	8	<30	8
8-OH-Loxapine	8	>30	NA
Desmethyl-loxapine	8	>30	NA
7-OH-Desmethyl-loxapine	8	<30	23
8-OH-Desmethyl-loxapine	8	>30	NA

Table 1. (*continued*)

Compound	References	$IC_{50}(\mu M)$	Ki(nM)
Dibenzodiazepines			
Clozapine	1, 2, 4, 7	0.6–3.5	61–170
(±)Butaclamol	10	6.0	NA

NA = not available
References: 1) *Miller, Horn,* and *Iversen,* 1974; 2) *Clement-Cormier et al.,* 1974; 3) *Kebabian, Petzold,* and *Greengard,* 1972; 4) *Karobath* and *Leitch,* 1974; 5) *Laduron,* 1976; 6) *Iwatsubo* and *Clouet,* 1975; 7) *Bockaert et al.,* 1977; 8) *Coupet, Szucs,* and *Greenblatt,* 1976; 9) *Roufogalis, Thornton,* and *Wade,* 1976; 10) *Palmer et al.,* 1978a; 11) *Miller* and *Iversen,* 1974; 12) *Palmer* and *Manian,* 1976; 13) *Scatton et al.,* 1977.

little or no action on DA adenylate cyclase were CPZ-SO, promethazine, sulpiride, β-flupenthixol, quaternary analogues of CPZ (except 7-OH-CPZ-MeI), perphenazine, prochlorperazine and haloperidol (*Langer et al.*, 1977).

It can readily be seen that the therapeutic potency of many antipsychotic agents, namely haloperidol, does not relate to their appropriate values for antagonism of DA-sensitive adenylate cyclase in the rat striatum, especially when compared to weaker antipsychotic agents such as clozapine and thioridazine. Another important point to draw from these investigations is that many metabolites of CPZ and loxapine possess potent effects with regard to central DA receptors. Similarly, stereospecificity is important for positive actions because only (+) and (cis) enantiomers were highly effective. Another agent, sulpiride, which supposedly has strong antidopaminergic actions in tests relative to other biochemical and behavioral parameters, is ineffective with respect to adenylate cyclase antagonism. It has been postulated that sulpiride may act on another subclass of DA receptors, especially at presynaptic sites (*Scatton et al.*, 1977). Interestingly, neuroleptic antagonism of DA-responsive adenylate cyclase appears to be rather specific for the receptor sites on the enzyme as *Burkard* (1977) found haloperidol to block only DA activation of the enzyme but not DA plus 5′-guanylyl imidophosphate-induced activation. In many cases, except with diOH, triOH and dioxo derivatives of pharmacologically active phenothiazines, the basal activity of adenylate cyclase was unaffected by highest concentrations of the drugs.

In the case of intact cellular preparations of rat caudate nucleus, *Forn et al.* (1974) were able to show an effective blockade by fluphenazine, CPZ and haloperidol of the DA-induced accumulation of cyclic AMP in incubated striatal tissue slices, an event not observed by *Harris* (1976). In a similar context *Siggins, Hoffer,* and *Ungerstedt* (1974) showed that when DA was applied to rat caudate neurons by microiontophoresis an inhibition of neuronal firing rates resulted. Fluphenazine and CPZ blocked the inhibition induced by DA but not by cyclic AMP or the inhibitory transmitter, γ-aminobutyric acid (GABA). Such data are similar to effects presented for noradrenergic actions in the cerebellum (*Freedman*, 1977). The findings do indicate the actions of the neuroleptics are associated solely through an interaction of a DA receptor. In support of this contention, *Creese et al.* (1978) evaluated the action of CPZ and its derivatives to displace the ^{3}H-haloperidol binding to DA receptors in the rat striatum. The sulfoxide, side chain-demethylated, mono- and dimethoxylated analogues which possess weak or no neuroleptic activity in animals were inactive in competing for ^{3}H-haloperidol binding. Hydroxylation at the 3 position and 3,7-dihydroxylation of CPZ double the affinity for receptor sites, while 7-OH-CPZ is similar to the parent compound. It should be noted, however, that during subcellular fraction studies *Laduron et al.* (1976) and *Leysen* and *Laduron* (1977) were unable to correlate ^{3}H-haloperidol binding to DA adenylate cyclase in the same tissue.

In one related experiment with a different species (mouse), *Tang* and *Cotzias* (1977) reported a competitive antagonism of DA-induced striatal adenylate cyclase by atropine and acetycholine. The enzyme preparation acted to bind both the cholinergic and the anticholinergic compounds, albeit with different constants. Their evidence reveals that other unexplained regulatory functions of adenylate cyclase might exist within the postsynaptic membrane-receptor complex. This interesting data should provide both a framework and a direction for future research.

In an interesting study *Laduron* (1976) presented evidence to explain some of the discrepancies observed which relate to a correlation between neuroleptic potency toward DA-stimulated adenylate cyclase and their pharmacological and clinical potency seen in other studies. For example, as observed in Table 1, strong neuroleptic agents like haloperidol and pimozide have inhibitory actions on striatal DA adenylate cyclase similar or less than compounds with considerably weaker therapeutic profiles. The salient findings of this investigation show that comparisons of a drug toward DA adenylate cyclase can only be made if the drugs have similar physicochemical properties. For example, butyrophenones (haloperidol) and diphenylbutylpiperidines (pimozide) differ markedly, as shown by a 1000- or 10,000-fold difference in the partition coefficients and a more than 1000-fold difference in solubility in water. The extremely low solubility of pimozide may explain its weak actions *in vitro*. However, these data do not explain all the discrepancies seen because other factors, especially in the *in vivo* state, need to be considered: degree of absorption into the blood if taken orally, degree of protein binding, capacity to cross the blood-brain barrier, ability to be taken up preferentially in specific brain regions, rate of metabolism both centrally and peripherally, rate of excretion and the presence of active vs inactive metabolites (for discussion, see *Iversen*, 1975; *Laduron*, 1976; *Soudijn* and *van Wijngaarden*, 1972; *Nathanson*, 1977).

Substantia Nigra

Within the past year evidence from independent laboratories has been presented describing an activation of adenylate cyclase by DA in the substantia nigra of rats. The source of the nerve endings releasing the DA is presumably from recurrent collaterals arising from the DA nerve cell bodies within the substantia nigra itself. The function is most likely to inhibit or modulate nigral firing rates. The exact localization of the nerve endings is within the nigral zone reticulata (*Kebabian* and *Saavedra*, 1976; *Phillipson* and *Horn*, 1976; *Traficante et al.*, 1976; *Spano et al.*, 1976; *Seeber* and *Kuschinsky*, 1976). *Kebabian* and *Saavedra* (1976) showed CPZ to inhibit this DA-sensitive activation of the zona reticulata enzyme with the same degree of affinity (66 nM) as adenylate cyclase in the caudate. However, in similar experiments it was shown that penfluridol was less active in the substantia nigra than in the caudate (*Seeber* and *Kuschinsky*, 1976). *Spano et al.* (1976) found haloperidol and fluphenazine (1 μM) to completely abolish both DA- and apomorphine-induced enzymatic stimulation without affecting the basal enzyme activity. Trifluperazine inhibited DA-sensitive adenylate cyclase with an IC_{50} of 0.1 μM (*Traficante et al.*, 1976), while *Phillipson* and *Horn* (1976) reported the following respective IC_{50} (μM) and Ki (nM) values for phenothiazine inhibition of the DA receptor mediating the elevation in nigral adenylate cyclase: α-flupenthixol, 0.01 and 0.5; CPZ, 0.032 and 1.4; clozapine, 0.4 and 17; pimozide, 0.41 and 18. With the exception of α-flupenthixol, these agents generally displayed a more potent antagonism of the nigral enzyme when compared to the caudate (see Table 1). In conclusion, phenothiazines not only antagonize DA receptors in the caudate but similarly reduce the action of the recurrent collaterals on the neurons of origin.

Retina

Depending upon the species under investigation, homogenates, cell membrane fractions and isolated intact retinas from several species, notably rat, chick, bovine and rabbit, possess adenylate cyclase systems which are readily activated by DA, apomorphine, NE, isoproterenol and epinephrine (*Brown* and *Makman*, 1972; *Schwarcz* and *Coyle*, 1976; *Bucher* and *Schorderet*, 1974a & b, 1975; *Schorderet*, 1976; *Schorderet* and *Frangaki*, 1976). Cell-free membrane fragments from chick retina are stimulated by apomorphine, DA, NE and isoproterenol, an event antagonized by 1 μM fluphenazine. On the other hand, tissue slices of chick retina only accumulate cyclic AMP in response to isoproterenol, an action blocked by propranolol but not by fluphenazine (*Schwarcz* and *Coyle,* 1976). In contrast, in a series of experiments both intact preparations and homogenates of rabbit retina display an adenylate cyclase that is highly responsive to DA, apomorphine, NE and epinephrine but not to isoproterenol. Likewise, ergot alkaloids, L-DOPA and N-methyl-DA elevate cyclic AMP in intact rabbit retina. These actions are all blocked by high concentrations of haloperidol, CPZ and fluphenazine. The authors concluded that the agonists acting on this system possess highly active DA-mimetic actions (*Bucher* and *Schorderet*, 1974a & b, 1975; *Schorderet*, 1976; *Schorderet* and *Frangaki*, 1976). Even though retinal adenylate cyclases appear to be rather biochemically dynamic processes, their function in the metabolic events related to retinal physiology is unknown.

Mesolimbic-Limbic Structures

Dopamine-sensitive adenylate cyclases have been described for the rat olfactory tubercle, nucleus accumbens, amygdala and hippocampus (*Scatton et al.,* 1977; *Clement-Cormier et al.,* 1974; *Horn, Cuello,* and *Miller,* 1974; *Pugsley, Merker,* and *Lippmann,* 1976; *Weinryb* and *Michel*, 1976; *Laduron et al.*, 1976; *von Hungen et al.*, 1975). Along with the frontal cortex, it is in these brain structures where the major tranquilizers are thought to exert their therapeutic actions with respect to alleviating the symptoms of schizophrenia and psychosis (*Iversen,* 1975; *Horn et al.*, 1974; *Nathanson*, 1977). The following table (Table 2) illustrates that many antipsychotic agents act to antagonize the mesolimbic activation of adenylate cyclase by DA in a manner corresponding to their action in rat striatal tissue. Dopamine also is capable of stimulating adenylate cyclase in the anterior limbic cortex, nucleus accumbens and olfactory tubercle in two species of primates. The latter DA action is readily antagonized by fluphenazine (*Mishra et al.*, 1975; *Weinryb* and *Michel*, 1976).

Cerebral Cortex

The activation of adenylate cyclase by DA in the cerebral cortex has been described for the rat (*von Hungen et al.*, 1975; *Palmer* and *Manian*, 1976; *Palmer* and *Wagner*, 1976; *Laduron*, 1976; *Bockaert et al.*, 1977), the primate (*Ahn et al.*, 1976; *Weinryb* and *Michel*, 1976) and the mouse (*Palmer et al.*, 1978b; *Cotter et al.*, 1978). Furthermore, when the cellular elements of the rat cerebral cortex are separated into neuronal, glial and capillary elements, these cell types all contain ad-

Table 2. Concentrations of Neuroleptic Agents Required to Inhibit DA-sensitive Adenylate Cyclase in Rat Mesolimbic Structures

Drug	References	Olfactory Tubercle		Nucleus Accumbens	
		$IC_{50}/\mu M$	$Ki/\mu M$	$IC_{50}/\mu M$	$Ki/\mu M$
CPZ	1-3, 5	0.9, 1.5	0.05-.06	1.8	0.075
Haloperidol	1, 5	1.1, 1.3	0.04, 0.14	1.0	0.03
Thioridazine	1, 5	2.5, 5	0.03, 0.05	0.7	0.02
Fluphenazine	2, 4-6	0.4, 0.1	0.007-0.01	NA	0.007
Clozapine	2, 5	1.0	0.060-0.11	NA	0.059
Trifluoperazine	5	0.2	0.02	NA	NA
Loxapine	2	NA	0.014	NA	0.01
Promethazine	3	>100	NA	40	NA
Sulpiride	1	>20	23	>60	23
(±)Butaclamol	6	0.53	0.106	NA	NA
(−)Butaclamol	6	>100	>20	NA	NA
(+)Butaclamol	6	0.12	0.024	NA	NA
*(+)3-Phenylbutaclamol	6	0.41	0.082	NA	NA
*(+)3-Cyclohexylbutaclamol	6	0.60	0.12	NA	NA
*(+)Isopropylbutaclamol	6	0.43	0.09	NA	NA

*The (±) derivatives are 2–4 times weaker than the derivatives shown, while the (−) derivatives act exactly like butaclamol.
NA = not available
References: 1) *Scatton et al.*, 1977; 2) *Clement-Cormier* and *Robison*, 1977; 3) *Horn, Cuello*, and *Miller*, 1974; 4) *Weinryb* and *Michel*, 1976; 5) *Clement-Cormier et al.*, 1974; 6) *Pugsley, Merker*, and *Lippmann*, 1976.

enylate cyclase systems which respond to a variety of neurohumoral agents including DA (*Palmer et al.*, 1976a; *Palmer* and *Manian*, 1976; *Baca* and *Palmer*, 1978). The enzymes from various species of primate frontal and sensorimotor cortex are inhibited in a potent manner by haloperidol, fluphenazine > clozapine, thioridazine > pimozide. Moreover, fluphenazine inhibited both DA- and DA plus 5′-guanylyl imidodiphosphate-induced activation of the cortical enzyme (*Ahn et al.*, 1976; *Weinryb* and *Michel*, 1976). Using a similar DA-responsive enzyme preparation from the rat frontal cortex, *Bockaert et al.* (1977) reported a pattern of phenothiazine antagonism which was essentially similar to that for the primate. Dopamine responses in the mouse frontal cortex were more susceptible to antagonism by 7-OH-CPZ derivatives than were NE responses. Chlorpromazine was the most potent (IC_{50} = 0.6 nM) > 7-OH-CPZ > 7-OH-CPZ-MeI > 7-OH-CPZ-glucuronide (*Cotter et al.*, 1978).

When adenylate cyclase inhibition was evaluated in isolated neuronal and glial fractions of rat cerebrum, rather large concentrations of phenothiazines were required to block the effects of DA. One consideration of these findings is the lack of synapses to the dendrites which are lost during the neuronal isolation procedure. To what extent axodendritic synapses contribute to the total amount of DA-sensitive receptor sites in the intact functional cerebrum is unknown. Furthermore, the degree of stimulation of neuronal adenylate cyclase elicited by DA is considerably less than with the total cerebral homogenate. The most potent (IC_{50} = 0.1–2 μM) derivatives toward these cellular enzymes are the 7,8-diOH analogues of CPZ, perphenazine and prochlorperazine. Likewise 7,8-dioxo-CPZ and 2,3-diOH-promazine were equally as potent. These diOH and dioxo compounds acted additionally to antagonize the basal activity of the enzyme. Other relatively potent agents (IC_{50} = 4–16 μM) were 7-OH-prochlorperazine > prochlorperazine > 8-OH-prochlorperazine > fluphenazine > 8-OH-fluphenazine > perphenazine > 8-OH-perphenazine > 8-OH-CPZ. The 8-OH analogues were usually more potent than corresponding 7-OH phenothiazine analogues; a contrast was demonstrated with regard to NE-sensitive adenylate cyclases. The remaining cortical element, the capillary enzyme, was strongly antagonized by haloperidol, while phentolamine (α antagonist) and propranolol (β antagonist) were active only at highest concentrations (*Palmer* and *Manian*, 1975, 1976; *Baca* and *Palmer*, 1978).

Other Neural Tissues

Neuronal and glial fractions from rat thalamic regions contain a DA-sensitive adenylate cyclase system which is blocked by CPZ, 8-OH-CPZ, 7,8-diOH-CPZ and 7,8-dioxo-CPZ (*Palmer* and *Manian*, 1976). *Clement-*

Cormier (1977), *Clement-Cormier* and *Robison* (1977) and *Clement-Cormier et al.* (1977) demonstrated in the rat median eminence an inhibition of DA actions on the cyclic AMP system by fluphenazine > loxapine > clozapine > CPZ. The pattern of inhibition was identical to similar DA-responsive brain regions, namely the caudate, olfactory tubercle, nucleus accumbens and amygdala. *Clement-Cormier* (1977) explained these findings with regard to the manner in which phenothiazines may precipitate hyperprolactinemia. It is postulated that a DA-sensitive adenylate cyclase in the median eminence mediated by cyclic AMP acts to inhibit pituitary prolactin release. Cultured glial cells from three strains (B19-C14, B92-C15 and B111-C11) respond to a dopaminergic stimulation by increasing their intracellular levels of cyclic AMP. In these investigations the action of DA was diminished by triflupromazine, thioridazine and promethazine, while haloperidol was barely active (*Schubert, Tarikas*, and *LaCorbiere*, 1976). The thoracic ganglion of insects contains a DA-responsive adenylate cyclase, an action strongly inhibited by haloperidol (*Nathanson*, 1976, 1977). In homogenates of snail central ganglia, DA adenylate cyclase was blocked by fluphenazine > haloperidol > CPZ (*Osborne*, 1977).

ACTIONS OF NEUROLEPTICS ON NOREPINEPHRINE-SENSITIVE ADENYLATE CYCLASE SYSTEMS IN THE CENTRAL NERVOUS SYSTEM

Background

Direct clinical and experimental evidence is lacking to define a role for NE as a biological substrate responsible for the mediation of affective disorders. The current thrust of investigative methods to date attempts to relegate such a function to DA (*Snyder et al.*, 1974a). Proponents of this latter hypothesis feel the major clinical and molecular actions of the antipsychotic drugs are manifested by a selective antagonism of central DA receptors. Arguments have been put forth to suggest the caudate nucleus and mesolimbic cortex as the structures receiving innervation from DA neurons, structures which subsequently mediate sensory, motor and cognitive disorders of schizophrenia (*Matthysse*, 1974). However, noradrenergic fibers arising from the locus ceruleus and lateral reticular nuclei innervate not only the cerebellum but many areas of the brain, namely the brain stem, hypothalamus, hippocampus, mesolimbic structures and the frontal cortex, all areas which in one fashion or another modulate behavior (*Ungerstedt*, 1971). Along this hypothetical line of thought, NE acts in a potent manner to activate central adenylate cyclase. Increased levels of cyclic AMP resulting from the action of this catecholamine occur in most brain regions and preparations which include intact cells (incubated tissue slices), broken cells (homogenates, subcellular fractions, neuronal, glial and capillary fractions) and following intracisternal injection of the amine (see *von Hungen* and *Roberts*, 1974; *Daly*, 1976, 1977; *Nathanson*, 1977; *Palmer* and *Manian*, 1974a; *Palmer, Wagner,* and *Putnam*, 1976a; *Baca* and *Palmer*, 1978). Moreover, ample evidence exists that the major brain regions in which antipsychotic drugs antagonize DA activation of adenylate cyclase are identical with respect to inhibition of the NE-sensitive enzyme. Thus, the number of neuronal structures which could be considered to be active in schizophrenia or other psychosis is greatly increased by the evidence for an interaction between NE and antipsychotic agents (*Freedman*, 1977).

The following discussion will focus on the individual species and particular brain regions in which neuroleptic drugs have been used with respect to adenylate cyclase systems responsive to NE. Sufficient evidence exists to draw conclusions with regard to structure-activity relationships of not only the parent neuroleptic compounds but many of their metabolites and derivatives.

Rabbit Brain

Only limited studies have been conducted concerning an action of phenothiazines on cyclic nucleotide systems responsive to NE in rabbit brain. The initial investigation was with the incubated tissue slice procedure in which a twofold increase in levels of cyclic AMP was seen in the cerebellum following addition of NE (50 μm). Preincubation of the tissue with 5 or 50 μM CPZ reduced the NE response by 21% and 47%, respectively (*Kakuichi* and *Rall,* 1968a). In another study *Sebens* and *Korf* (1975) showed that injected haloperidol did not influence the elevation in cerebral spinal fluid levels of cyclic AMP elicited by intracisternally administered NE and isoproterenol.

Rat Brain

The majority of experimental manipulations have evaluated the role of neuroleptic agents and their metabolic derivatives on NE-sensitive adenylate cyclases in various regions of the rat brain. In many instances there are conflicts in the experimental findings among different laboratories.

Hypothalamus

In the initial experiments tissue slices of rat hypothalamus which are potently activated by d'l-NE (3- to 5-fold at 10 or 50 μM) via α and β

Table 3. Approximate Concentration (μM) of Neuroleptic to Inhibit by 50% (EC_{50}) the d'l-norepinephrine (50 μM)-induced Accumulation of Cyclic AMP in Incubated Tissue Slices of Rat Brain

	Region		
Compound	Lateral Hypothalamus	Medial Hypothalamus	Cerebral Cortex
CPZ	90	80	30
7-OH-CPZ	5	2	14
7-OH-CPZ-MeI	NA	NA	50
8-OH-CPZ	>100	>100	>100
8-OH-7-MeO-CPZ	30	8	>100
3-OH-CPZ	140	30	14
β-OH-CPZ	NA	NA	>100
7,8-DiOH-CPZ	>100	>100	>100
7,8-Dioxo-CPZ	>100	90	>100
3,7,8-TriOH-CPZ	NA	NA	>100
7,8,β-TriOH-CPZ	NA	NA	>100
7,8-Dioxo-β-OH-CPZ	NA	NA	>100
7-MeO-CPZ	>100	>100	10
CPZ-CH_3Cl	NA	NA	52
3,7-DiMeO-CPZ	>100	>100	45
7,8-DiMeO-CPZ	>100	>100	>100
CPZ-SO	>100	>100	90
Promethazine	20	38	>100
Phenothiazine	>100	>100	>100
3-OH-Phenothiazine	>100	>100	>100
2-Cl,7,8-Dioxo-Phenothiazine	>100	>100	>100
Promazine	5	2	2
2-OH-Promazine	20	14	100
3-OH-Promazine	>100	>100	45
2,3-DiOH-Promazine	20	>100	>100
Prochlorperazine	3	0.7	50
7-OH-Prochlorperazine	30	4.5	8
8-OH-Prochlorperazine	>100	>100	>100
7,8-DiOH-Prochlorperazine	100	>100	>100
Perphenazine	4.5	0.6	1.3
7-OH-Perphenazine	70	28	>100
8-OH-Perphenazine	>100	>100	>100
7,8-DiOH-Perphenazine	15	100	>100
Fluphenazine	3.8	1.5	10
7-OH-Fluphenazine	0.7	1.8	100
8-OH-Fluphenazine	6	3	4.5
Thiothixene	20	25	100
(±)Butaclamol	NA	NA	6
Haloperidol	>100	>100	25

From *Palmer* and *Manian* (1974b) and *Palmer et al.*, 1978a.

receptors were antagonized by 10 μM CPZ, 7-OH-CPZ > prochlorperazine > 8-OH-CPZ while CPZ-SO was ineffective (*Palmer, Robison,* and *Sulser,* 1971; *Palmer et al.,* 1972; *Palmer, Sulser,* and *Robison,* 1973). The hypothalamus was further divided into medial and lateral components because there are functional, behavioral and pharmacological differences between the two brain regions (*Palmer* and *Manian,* 1974b). The medial hypothalamus displayed a greater sensitivity to NE than the lateral structure. Using incubated tissue slices, 32 neuroleptic agents were evaluated with respect to an inhibition of NE-induced cyclic AMP accumulation. As seen in Table 3, 7-OH-fluphenazine was clearly the most potent compound in the lateral hypothalamus, followed by, in descending order of potency, prochlorperazine > fluphenazine > perphenazine > 7-OH-CPZ > promazine > 8-OH-fluphenazine > 7,8-diOH-perphenazine > thiothixene > promethazine > 2-OH-promazine > 2,3-diOH-promazine > 7-OH-prochlorperazine > 7-OH-perphenazine. Dihydroxylated, 8-OH, methoxy and sulfoxide analogues of their parent compounds were usually ineffective. Likewise, phenothiazine analogues lacking the side chain were inactive. The parent compound, CPZ, was considerably weaker than the metabolites. In the medial hypothalamus, which appeared to possess a greater sensitivity to antagonism, perphenazine was the most active compound, followed closely by prochlorperazine > fluphenazine > 7-OH-fluphenazine > 7-OH-CPZ > promazine > 8-OH-fluphenazine > 7-OH-prochlorperazine > 8-OH-7-CH_3O-CPZ > 2-OH-promazine > thiothixene > 7-OH-perphenazine. In general, compounds inactive in the lateral hypothalamus acted likewise in the medial region. In conclusion, parent compounds, perphenazine, prochlorperazine, fluphenazine, CPZ and notably promazine, plus the respective 7-OH metabolites, possess the greatest ability to inhibit hypothalamic adenylate cyclase in response to an NE stimulus.

Cerebral Cortex

Uzunov and *Weiss* (1971) first demonstrated an inhibition by trifluperazine of the enhanced accumulation of cyclic AMP elicited by NE in incubated tissue slices of rat cerebral cortex. In follow-up investigations, we (*Palmer* and *Manian*, 1974b; *Palmer et al.*, 1978a) evaluated the structure-activity relationships with regard to the action of a series of parent neuroleptic agents and their derivatives on a similar NE-responsive system. As observed in Table 3, the following compounds acted in a potent manner to antagonize the NE actions on cyclic AMP in the cerebral cortex: perphenazine > promazine > 8-OH-fluphenazine > ± butaclamol > 7-OH-prochlorperazine > fluphenazine, 7-MeO-CPZ > 3-OH-CPZ > haloperidol > CPZ. These findings did not exactly coincide with those observed in the hypothalamus.

In a recent experiment *Bockaert et al.* (1977) used homogenates of rat frontal cortex (receives both adrenergic and dopaminergic innervation) and demonstrated an activation of either basal adenylate cyclase or 5′-guanylyl imidodiphosphate-induced enzyme activity with DA, NE and isoproterenol. The effects of DA and NE, but not that of isoproterenol, were inhibited by fluphenazine. The action of NE was only reduced 50% by the phenothiazine and a combination of fluphenazine and pindolol (β antagonist) was required to totally block this action of NE. The authors concluded that NE possesses the ability to activate both β-adrenergic and dopaminergic receptors.

Brain Stem

In incubated tissue slices of rat brain stem, which essentially contain the noradrenergic neuronal cell bodies and limited numbers of adrenergic nerve endings, CPZ, haloperidol, 7-OH-CPZ, prochlorperazine, chlorprothixene and trifluperazine at 10 and 50 μM inhibit the NE-induced accumulation of cyclic AMP. In these studies CPZ-SO, promethazine and trifluperazine sulfoxide were either inactive or acted to a considerably lesser extent (*Palmer et al.*, 1971, 1972; *Uzunov* and *Weiss*, 1971, 1972; *Weiss* and *Greenberg*, 1975).

Cerebellum

Introductory experiments with this brain structure were performed by *Uzunov* and *Weiss* (1971). High concentrations of trifluperazine (500 μM) antagonized the action of NE (100 μM) on cyclic AMP levels in tissue slices.

A series of investigations with both whole and broken cellular preparations revealed the only neurohumoral agents capable of adenylate cyclase activation were NE and the β agonist, isoproterenol. Iontophoretic application of these agents to Purkinje cells of the cerebellum resulted in a decreased firing rate, an event mimicked by cyclic AMP. Furthermore, stimulation of the locus ceruleus caused by enhanced localization of fluorescent antibody-labeled cyclic AMP only in the cell bodies of cerebellar Purkinje cells (*Daly,* 1976; *Nathanson,* 1977; *Siggins et al.*, 1973; *Skolnick et al.*, 1976; *Palmer et al.*, 1978b).

In experiments with phenothiazine agents *Freedman* and *Hoffer* (1974), *Freedman, Hoffer,* and *Siggins* (1974) and *Freedman* (1977) showed that when iontophoretically applied NE reduced the discharge rate of cerebellar Purkinje cells, ejection of fluphenazine or flupenthixol reversed the action of NE. The inhibition of Purkinje cell firing by either γ-aminobutyric acid or cyclic AMP was unaffected by the two phenothiazines. Injections of neuroleptics blocked the subsequent inhibition of

Purkinje cells induced by stimulation of the locus ceruleus but did not affect inhibition of firing rates of basket and stellate cells. These findings indicate the action of the phenothiazines was confined to the adrenergic receptor component of adenylate cyclase in the Purkinje cells alone. Studies with tissue slices suggested the possible existence of different β-type adrenergic receptors within the cerebellum. The unique type confined to the Purkinje cells was influenced by the phenothiazines. In support of this contention, fluphenazine and α-flupenthixol, but not promethazine or β-flupenthixol, blocked both the NE- and isoproterenol-induced elevation of cyclic AMP in tissue slices and reversed the NE and isoproterenol inhibition of Purkinje cell discharge. High concentrations (100 μM) completely antagonized the actions of isoproterenol and NE.

In further experiments, young rats were exposed to X-irradiation which destroyed the majority of the granule, basket and stellate cells of the cerebellum. In control tissue, NE elevated cerebellar cyclic AMP by 4-fold, while in irradiated animals the response was 2.5-fold. In this experiment fluphenazine decreased NE-induced cyclic AMP levels by 50% in control tissue but completely antagonized this response in cerebellar tissue slices from irradiated animals. Apparently X-irradiation destroyed β receptors in cells resistant to fluphenazine. With iontophoretic studies the neuroleptic acted similarly in control and X-irradiated rats to reverse NE depression of Purkinje cell firing rates. These innovative findings reveal that phenothiazines have a direct action on specific types of β-adrenergic adenylate cyclase receptors in the cerebellum. Moreover, these cerebellar actions do not appear to include an action on the phosphodiesterases because the specific form of the enzyme influenced by neuroleptics is not present in the rat cerebellum (*Hoffer et al.*, 1976; *Skolnick et al.*, 1976; *Freedman* and *Hoffer*, 1974; see also the subsequent discussion on phosphodiesterase).

Limbic Forebrain

The limbic forebrain of the rat, consisting of olfactory tubercles, nucleus accumbens, septal nuclei, rostral limbic nuclei and anterior amygdaloid nuclei, receives both dopaminergic and noradrenergic inputs. Lately the DA contribution has been the object of intensive investigations with respect to a molecular involvement into the pathogenesis of schizophrenia and affective disorders. It is principally here that the phenothiazines are postulated to exert their therapeutic actions, especially with regard to an inhibition of DA-sensitive adenylate cyclase (*Synder et al.*, 1974a; *Miller et al.*, 1974). *Blumberg, Taylor* and *Sulser* (1975) and *Blumberg et al.* (1976), using incubated tissue slices of this brain region, found NE to be an extremely potent (EC_{50} = 5.00 μM) activator of adenylate cyclase. This response was highly sensitive to inhibition by phenothiazines, namely the DA antagonist, pimozide (0.08 μM), clozapine (0.06 μM), thioridazine (1.2 μM), CPZ (9 μM) and to a lesser extent by haloperidol (10 μM) and promethazine (> 10 μM). Their findings indicate that NE may play an equally important role with DA with regard to phenothiazine modification of behavior in the mesolimbic system.

Striatum

Both adrenergic and dopaminergic agonists activate adenylate cyclase in either tissue slices or homogenates of the rat striatum (*Palmer et al.*, 1973; *Walker* and *Walker*, 1973; *Harris*, 1976; *Forn et al.*, 1974). There is a general disagreement as to whether DA acts with any degree of potency in incubated striatal slices (see *Daly,* 1976; *Nathanson,* 1977). Using tissue slices, however, *Forn et al.* (1974) showed that fluphenazine would reduce the action of NE on cyclic AMP by only 25%. Furthermore, fluphenazine did not influence the effects of isoproterenol. *Walker* and *Walker* (1973), using striatal homogenates, found no action of haloperidol to antagonize NE-sensitive adenylate cyclase. These findings were confirmed by *Harris* (1976) using trifluperazine. The results of these investigations suggest that a separate β-adrenergic receptor is present in striatal tissue that appears to be resistant to the actions of neuroleptic agents. This is in contrast to the DA receptor in the striatum which is highly susceptible to the actions of phenothiazines (see preceding section).

Pineal

Homogenates of rat pineal contain an adenylate cyclase system that is highly responsive to β-adrenergic agonists. Trifluperazine was more reactive in diminishing the NE activation of adenylate cyclase than CPZ, while the sulfoxide analogue of trifluperazine was without effect (*Uzunov* and *Weiss*, 1971).

Neurons and Glial-enriched Fractions

When NE was added to disrupted cellular elements (neurons and glia) from rat cerebral cortex there was a 30–40% elevation in adenylate cyclase activity. Rather high levels of CPZ, 7-OH-CPZ and 8-OH-CPZ (10^{-4} M) were necessary to prevent this action of NE. 7-OH-CPZ was the most potent in this regard. On the other hand, 7,8-diOH and 7,8-dioxo derivatives of CPZ, which were relatively inactive in intact cells (incubated tissue slices), now displayed an extremely powerful blockade of enzyme activation by NE. Moreover, these two derivatives blocked basal adenylate cyclase activity. The pattern of inhibition by 7,8-diOH-CPZ is not overcome by adding excess ATP, 5′-guanylyl imidodiphosphate, Mg^{++} or Ca^{++} *(Palmer* and *Manian*, 1974a, unpublished findings). Apparently the diOH and dioxo analogues of CPZ form free radicals and bind to a host of cellular proteins because neural ATPases are also inhibited by 7,8-diOH-CPZ and seem to involve sulfhydryl group interactions (*Akera et al.*, 1974). Most likely the charge on the molecule prevents it from entering the intact cell. However, rabbit liver tissue does have the capability to synthesize the compounds under *in vitro* conditions (*Daly* and *Manian*, 1967). Whether the brain has this capacity has not as yet been demonstrated.

Other Species

In the mouse cerebral cortex, using both incubated tissue slices and homogenates of frontal cortex, adenylate cyclase was stimulated by NE in the intact cell preparation and by both NE and DA in the broken cellular preparation. This observation is analogous to that observed for the rat brain. With the incubated tissue slices, the action of NE (50 μM) was significantly antagonized by 7-OH-CPZ (0.1 to 100 μM) > CPZ (1 to 100 μM) > 7-OH-CPZ-MeI and 7-OH-CPZ-glucuronide (100 μM). Basal levels of cyclic AMP were unaffected by the highest levels (100 μM) of any compound. In broken cell preparations CPZ was the most potent (IC_{50} = 5 μM) followed by 7-OH-CPZ (25 μM) > 7-OH-CPZ-MeI (250 μM), while the glucuronide metabolite was inactive at 1000 μM (*Cotter et al.*, 1978). In a similar experiment with both the cerebral cortex and cerebellum, haloperidol was barely active (100 μM) in blocking NE-(100 μM) sensitive adenylate cyclase in both whole and broken cell preparations. Furthermore, CPZ was at least 10-fold less potent in inhibiting tissue slice accumulation of cyclic AMP elicited by maximal concentrations of norepinephrine (100 μM) (*Palmer et al.*, 1978b). Tissue slices of mouse limbic forebrain readily synthesized cyclic AMP when NE was added to the medium. Pimozide, a DA antagonist, potently antagonized the action of NE (IC_{50} = 0.13 μM) (*Sawaya et al.*, 1977).

Homogenates of monkey frontal cortex contain an adenylate cyclase system sensitive to a variety of agents, including DA, apomorphine, isoproterenol, NE and the α agonist, clonidine. Fluphenazine did not influence the action of isoproterenol but acted at 5 μM to inhibit clonidine and NE. Dopamine was potently affected by the phenothiazine. As discussed previously, it was indicated that NE possibly has the capacity to activate a host of receptors, i.e., α, β and DA, in the frontal cortex. Fluphenazine did not reduce the elevation of the transducer site of adenylate cyclase by 5′-guanylyl imidodiphosphate (Ahn *et al.*, 1976).

The guinea pig cerebral cortex has been studied extensively with respect to agents that elevate cyclic AMP (see *Daly*, 1976). At concentrations of 100 μM and 50 μM, CPZ inhibited the stimulation of cyclic

AMP in tissue slices by the following agonist combinations: 43 mM KCl + 100 μM NE; adenosine-NE (100 μM each); and NE-histamine (100 μM each) (*Huang* and *Daly*, 1972, 1974).

Conclusion of Norepinephrine Effects

The preceding information does indicate a role for the NE receptor with respect to a central action of neuroleptic drugs. The studies bring to light some interesting findings in that in some brain regions (cerebellum) specific β-adrenergic receptors are present that are relatively resistant to inhibition by the phenothiazines. In other brain regions (striatum and frontal cortex) the antagonism of NE-induced adenylate cyclase by particular neuroleptic agents might be explained via an interaction of NE at DA receptor loci. On the other hand, in the limbic forebrain neuroleptic agents presumably with a high affinity for DA receptors readily inhibit the NE-induced accumulation of cyclic AMP in intact cell preparations in which DA is not active. Thus, it is at times difficult to make comparisons of phenothiazine actions between intact and broken cellular preparations, especially with diOH and dioxo analogues of pharmacologically active parent phenothiazines. When structure-activity relationships are made, parent compounds of clinically active antipsychotic phenothiazines and their corresponding monohydroxy metabolites, notably at the 7 position, are the most powerful to block the stimulation of adenylate cyclase by NE. The least potent agents in this regard appear to be the sulfoxide, methoxy, dimethoxy and side chain-demethylated analogues of the 2-substituted phenothiazine neuroleptics, as well as promethazine. Butyrophenones such as haloperidol are relatively weak inhibitors of NE-sensitive adenylate cyclase in central preparations which include either tissue slices or broken cells. On the other hand, pimozide, another butyrophenone with a strong affinity for central DA receptors, readily inhibited the accumulation of cyclic AMP in response to an NE stimulus in incubated tissue slices of rat limbic forebrain (*Blumberg et al.*, 1976).

ACTIONS OF NEUROLEPTICS ON HISTAMINE-SENSITIVE ADENYLATE CYCLASE IN CENTRAL TISSUE

Histamine readily elevates the accumulation of cyclic AMP in both incubated tissue slices and broken cellular preparations (including in some instances neuronal or glial fractions) of rabbit, rat and guinea pig brain (*Kakiuchi* and *Rall*, 1968a & b; *Spiker*, *Palmer*, and *Manian*, 1976; *Free, Paik*, and *Shada*, 1974; *Huang* and *Daly*, 1974; *Daly*, 1976;

Forn and *Krishna*, 1971; *Suran*, 1977). Furthermore, these data, those of *Garbarg et al.* (1974) which show a histaminergic pathway to the frontal cortex, and the remaining evidence discussed by *Schwartz* (1975), *Calcutt* (1976) and *Chand* and *Eyre* (1975) strongly suggest a role for histamine as a central neurotransmitter. Likewise, preliminary evidence has implicated histamine in neural mechanisms of behavior and affective disorders (*Matthysse* and *Lipinski*, 1975). In this context, the antipsychotic drugs modify the activation of adenylate cyclase by histamine. Whether these observations with phenothiazines can be extended to clinical situations is presently a rather tenuous hypothesis.

The initial studies which showed an antagonism by CPZ of the histamine-induced accumulation of cyclic AMP in tissue slices of rabbit cerebellum were conducted by *Kakiuchi* and *Rall* (1968a). Similar tissue slice preparations from guinea pig hippocampus (*Free et al.*, 1974) and rat lung (*Palmer*, 1971) (see following section) yielded analogous findings. Using the technique of labeled adenine incorporation into ATP, *Huang* and *Daly* (1972, 1974) reported that high concentrations of CPZ (5×10^{-4} M) increased the turnover of ATP into cyclic AMP in the guinea pig cerebrum. The latter effect was blocked by theophylline, which inhibits the stimulation of cyclic AMP by adenosine. At this CPZ level, however, theophylline was shown to release adenosine into the incubation medium and additionally blocked the uptake of labeled adenosine into the tissue. Histamine, histamine-KCl, histamine-adenosine and histamine-NE combinations which all elevated cyclic AMP were inhibited by lower concentrations of CPZ (5×10^{-5} M).

In a more detailed investigation *Spiker et al.* (1976) examined the role of several neuroleptic agents with respect to the histamine-responsive adenylate cyclase in incubated tissue slices and broken cellular preparations consisting of neurons or glia from the rabbit cerebral cortex. In the incubated tissue slices, histamine elevated the cyclic nucleotide 5-fold; in the broken cellular fractions the monoamine was capable only of eliciting a 30–70% increase in enzyme activity. With the tissue slices, clozapine, quaternary-CPZ, ($\pm$) butaclamol and thioridazine were the most potent antipsychotic agents with respect to antagonism of the histamine-induced accumulation of cyclic AMP. The antihistaminic phenothiazine, promethazine, despite possessing rather weak antipsychotic actions, was equally effective, followed by 7-OH-CPZ > 8-OH-CPZ > β-OH-CPZ > CPZ > haloperidol. Additional metabolites or derivatives of CPZ, namely CPZ-SO, 7,8-diMeO-CPZ, 7-OH-CPZ-MeI, 3,7,8-triOH-CPZ and 7,8-diOH-CPZ derivatives, were effective only at highest concentrations (10^{-4} M) (also see *Palmer et al.*, 1978a). These data were not extrapolated to the broken cell fractions. For example, in the cortical neurons thioridazine was again the most potent ($IC_{50} = 2 \times 10^{-5}$ M) > 8-OH-CPZ = CPZ = 7-OH-CPZ > 7,8-diOH-CPZ > 7,8-

dioxo-CPZ = clozapine = haloperidol, while promethazine and CPZ-SO were inactive. Basal adenylate cyclase activity was also inhibited to a limited extent by the drugs in the neuronal homogenates. The histamine-responsive enzyme in the glial-enriched fraction was more sensitive to antagonism by neuroleptic agents. Thioridazine was again the most potent ($IC_{50} = 5 \times 10^{-6}$ M) and the remainder of the drugs acted in somewhat the same fashion as with the neurons, except promethazine was now active and clozapine was inactive.

Taken together, the data show that high concentrations of CPZ act both to release adenosine from and to prevent uptake of adenosine into neural tissues, which then causes levels of cyclic AMP to increase. On the other hand, smaller concentrations of CPZ presumably act to antagonize the histamine-sensitive receptor sites of adenylate cyclase. This pattern of adenylate cyclase inhibition by neuroleptic drugs does not closely parallel findings with either DA or NE. With the histamine-sensitive enzyme, neuroleptics with either antihistaminic (promethazine) or antimuscarinic (clozapine and thioridazine) actions were more potent (for further discussion, see *Miller* and *Hiley*, 1974; *Snyder, Greenberg*, and *Yamamura*, 1974b). Haloperidol and CPZ were considerably less potent with regard to histamine but acted in a stronger manner toward DA- and NE-sensitive enzymes, as discussed in the previous sections. Moreover, findings with the phenothiazines on histamine-sensitive adenylate cyclases in broken cell preparations do not coincide with incubated tissue slices. For example, clozapine and promethazine exerted strong blocking actions in whole-cell preparations but were scarcely effective when the structural integrity of the cell was disrupted. Since our present understanding of a role for histamine in the brain is not clear, it would be difficult to speculate just how phenothiazines might act with regard to clinical situations. However, the well-known sedative actions of most phenothiazines could be related in part to their antihistaminic actions. Whether or not this sedation is reflected via a particular central adenylate cyclase system is a topic for further investigation.

INFLUENCES OF INJECTED NEUROLEPTICS ON SUBSEQUENT ALTERATIONS IN ADENYLATE CYCLASE SYSTEMS

Supersensitivity Studies

Investigations conducted by various laboratories have demonstrated a direct involvement of adenylate cyclase as a molecular component of the underlying mechanisms as related to the phenomenon of adrenergic denervation supersensitivity. Following depletion of central catecholamines, destruction of adrenergic nerve endings by 6-hydroxydopamine and/or surgical procedures, adenylate cyclase in several brain regions responds in a hypersensitive fashion to NE, isoproterenol and DA. The process appears to be mediated by β-adrenergic receptors and apparently occurs only in neurons (for reviews, see *Daly*, 1976, 1977; *Wolfe, Hardin*, and *Molinoff*, 1977; *Dismukes* and *Daly*, 1976; see also *Palmer*, 1972; *Palmer et al.*, 1973, 1976a; *Palmer* and *Wagner*, 1976; *Weiss* and *Costa*, 1967; *Blumberg et al.*, 1976; *Kalisker, Rutledge*, and *Perkins*, 1973). In addition, behavioral, electrophysiological and biochemical studies in laboratory animals show that a DA-induced supersensitivity becomes manifest subsequent to a sequence of injections with neuroleptic agents. These findings, especially after chronic neuroleptic injections, indicate that the extrapyramidal supersensitivity to DA may serve as a molecular model of the syndrome of tardive dyskinesia (*Yarbrough*, 1975; *Sayers et al.*, 1975; *Gianutsos et al.*, 1974; *Smith* and *Davis*, 1976; *Dunstan* and *Jackson*, 1976; *Christensen, Fjalland*, and *Moller-Nielsen*, 1976). Preliminary evidence both for and against will be presented in the following discussion to suggest an involvement of neurohumorally sensitive adenylate cyclases which relate to such changes produced by *in vivo* injections of neuroleptic agents.

Iwatsubo and *Clouet* (1975) first reported an acute injection of haloperidol (20–60 min) produced a twofold increase in DA-sensitive adenylate cyclase in a mitochrondrial-synaptosomal fraction from the rat caudate region. The heightened sensitivity returned to normal within 4 hr. By employing a technique of rapidly removing and freezing "core samples" of rat cerebral cortex, *Berndt* and *Schwabe* (1973) demonstrated a transient rise (1 min) of cyclic AMP levels in response to an i.v. injection of either CPZ (5 mg/kg) or haloperidol (0.25 mg/kg). A similar response was evident in the liver but not in heart, kidney or fat tissue. Within 5 min, cyclic AMP levels for both drug treatments returned to control values. Moreover, pretreatment of rats with theophylline prevented the action of CPZ, suggesting the two neuroleptics in some way interfere with the central actions of adenosine. In another study, an injection of CPZ (2.5 mg/kg, i.p.) produced a transient rise within 1 hr in the steady-state levels of cyclic AMP in the cerebral cortex of mice which were quickly sacrificed by focused microwave irradiation. This and larger CPZ concentrations plus haloperidol usually lowered the *in vivo* levels of cerebral cyclic AMP from 1–8 hr post-drug injection. However, in the same experiments, using the rapidly fixed mouse cerebellum, CPZ and haloperidol initially reduced cyclic AMP, followed by a rise in a 24 hr post-injection (*Palmer et al.*, 1978b). Acute injections of amphetamine and apomorphine in rats followed by rapid tissue inactivation with focused microwave irradiation caused *in vivo* amounts of cyclic AMP to rise in two brain regions, the striatum and nucleus accumbens. The hypothalamic and cerebellar levels of cyclic AMP were unaffected. Haloperidol prevented the acute elevations in cyclic AMP resulting from apomorphine and amphetamine injections (*Gerhards, Carenzi*, and *Costa.*, 1974; *Carenzi et al.*, 1975; *Costa et al.*, 1975; *Spano* and *Trabucchi*, 1976). In another investigation, the cerebrospinal fluid levels of cyclic AMP were unaffected in the rabbit subsequent to an acute haloperidol (1.5 hr) administration (*Sebens* and *Korf*, 1975).

With subchronic (1–4 days) and chronic (1 week or more) injection schedules of neuroleptics, not all laboratories have been able to establish whether the drugs induce an alteration in the sensitivity of adenylate cyclase. On the positive side, *Palmer* and *Evan* (1974) observed an elevation in daily urinary levels of cyclic AMP in rats treated every 24 hr with CPZ (1 wk). Moreover, daily administration of CPZ, CPZ-SO and haloperidol (1–3 days) led to elevated steady-state levels of cyclic AMP in rapidly inactivated mouse cerebral cortex, cerebellum and diencephalon (*Palmer et al.*, 1977, 1978b). These *in vivo* studies did not identify the specific sensitivities of the enzymes involved, i.e., adenylate and guanylate cyclases and the phosphodiesterases. On the other hand, *Friedhoff, Bonnet*, and *Rosengarten* (1977) correlated the degree of increased striatal binding of ^{3}H-haloperidol to an increased DA stimulation of cyclic AMP in tissue slices of rats treated for 28 days with either haloperidol or trifluperazine. The supersensitivity became manifested 15 days after cessation of neuroleptic administration and was furthermore reversed by L-DOPA injections. The latter precursor of DA and NE was given 5 days after the neuroleptics.

In a recent report *Gnegy, Uzunov*, and *Costa* (1977b) found that after chronic injections of haloperidol or (+)butaclamol to rats, the striatum had a greater sensitivity of DA-activated adenylate cyclase along with increased striatal amounts of the calcium-dependent protein activator of adenylate cyclase and phosphodiesterase. These actions were not observed after injections of either the inactive(—) form of butaclamol or clozapine, which has a low incidence of producing either parkinsonism or tardive dyskinesia. The calcium-dependent activator is especially sensitive to inhibition by neuroleptic agents (see section on the phosphodiesterases). With this in mind, *Fredholm* (1977) showed that the low Km form of phosphodiesterase displayed an increased activity in rats fed haloperidol for 18 days.

On the negative side, four laboratories have been unable to support the contention that chronic administration of phenothiazine-like agents leads to a postsynaptic denervation supersensitivity of central adenylate cyclase receptors to catecholamines. Such studies were conducted by *Rotrosen, Friedman*, and *Gershon* (1975) and *Roufogalis et al.* (1976) who administered CPZ or haloperidol to rats for up to 28 days and saw no change in either DA or apomorphine sensitivity of adenylate cyclase. *Palmer* and *Wagner* (1976) reported similar negative findings in the rat striatum and cerebrum using haloperidol for 16 days. In the latter study the sensitivity of adenylate cyclase was not seen with either NE or DA. However, in parallel experiments reserpine injections did increase enzyme activity in response to NE and DA. Tissue slices of rat cerebral cortex were found to have a subsensitivity to NE-induced cyclic AMP accumulation when the animals had been previously injected from 6–21 days with either CPZ or the tricyclic antidepressant, imipramine (*Schultz*, 1976). In a preliminary experiment *Palmer, Sulser*, and *Robison* (unpublished) were unable to show such a change in regional brain tissue slice responses to NE after 7 days of injection (i.p.) of a low dose of CPZ (2.0 mg/kg/day). Administration of a lower dose of CPZ than described in the above paragraph did not alter the daily pattern of urinary cyclic AMP excretion (*Keatinge, Sinanan,* and *Love,* 1975).

Taken together, these findings do indicate that under relatively specific experimental conditions the receptor components of adenylate cyclase as well as specific enzyme forms of phosphodiesterase may be altered to meet an external challenge such as chronic blockade at adrenergic and dopaminergic receptors. Before this hypothesis can be extended to serve as a molecular model for the phenothiazine-induced syndrome of tardive dyskinesia, a great number of experimental protocols must be examined.

Effect of *In Vivo* Phenothiazines on Cyclic AMP Produced by Trauma

After decapitation of animals, cyclic AMP levels rise rather rapidly throughout the central nervous system (*Breckenridge*, 1964; *Kakiuchi* and *Rall*, 1968a; *Schmidt et al.*, 1970; *Uzunov* and *Weiss*, 1971; for review, also see *Daly*, 1977). In the rat, nucleotide elevations occurred to the greatest extent in the cerebellum, a phenomenon inhibited by pretreatment with trifluperazine, CPZ, haloperidol and chlorprothixene. Neither sulfoxide derivatives of these two phenothiazines nor promethazine were active. The active phenothiazines were not effective in preventing the rather small post-decapitation elevations of cyclic AMP seen in the cerebrum and brain stem (*Uzunov* and *Weiss*, 1971, 1972). Similar findings with phenothiazines in the rabbit and mouse brain were reported by *Kakiuchi* and *Rall* (1968a) and *Lust, Goldberg*, and *Passonneau* (1976). A stab injury resulted in a sevenfold rise in cyclic AMP in the mouse brain within 1 min. Prior treatment of the animals with CPZ and trifluperazine prevented this increase in cyclic AMP (*Watanabe* and *Passonneau*, 1975). Since theophylline was likewise effective, these authors concluded that the rise in cyclic AMP following the injury was due to release of adenosine. However, in many *in vitro* instances phenothiazines did not prevent the cyclic AMP accumulation elicited by adenosine. Clearly other unexplained mechanisms, e.g., massive depolarization, must come into play following cerebral insult.

Clinical Studies with Phenothiazines

Conflicting evidence has been presented to suggest fluctuations in daily urinary, cerebrospinal fluid and plasma levels of cyclic AMP as reflected by changes in mood and behavior. The data indicated a rise in cyclic AMP during mania and psychosis while a decrease was seen in depressive states (*Abdulla* and *Hamadah*, 1970; *Paul et al.*, 1970; *Paul, Cramer*, and *Goodwin*, 1971; *Sinanan et al.*, 1975; *Cramer, Ng,* and *Chase*, 1972). For conflicting data, see *Brown et al.* (1972), *Jenner et al.* (1972), *Biederman et al.* (1977); for review, see *Daly* (1977). Depressed patients undergoing tricyclic antidepressant therapy showed elevations in urinary cyclic AMP which correlated with predrug values (*Abdulla* and *Hamadah*, 1970). No clinical data are presently available with regard to phenothiazines.

In one study administration of epinephrine in man led to a rise in plasma cyclic AMP. Lithium but not haloperidol blocked this effect of epinephrine. The authors concluded that the rise in cyclic AMP was mediated by β-adrenergic receptors with no involvement of DA receptors (*Belmaker et al.*, 1976). Ingestion of phenothiazines and other psychoactive agents in humans had no effect on cyclic AMP in cerebral spinal fluid or urine (*Cramer, Ng*, and *Chase*, 1973).

ACTIONS OF NEUROLEPTICS ON OTHER CENTRAL MECHANISMS WHICH INVOLVE ADENYLATE CYCLASE-CYCLIC AMP

The following series of investigations with regard to phenothiazine actions on cyclic AMP systems reveal that the predominant effects of the drugs are on catecholamine-sensitive receptor sites of adenylate cyclase.

Catalytic Site

Activation of adenylate cyclase by sodium fluoride in cell-free (particulate) preparations from a wide variety of tissues is thought to occur at the so-called catalytic locus of the enzyme. As tissues are subjected to disruption and high-speed centrifugation procedures the action of fluoride progressively becomes more prominent. The ion is not effective in intact (incubated tissue slices or perfused organs) cellular systems (*Robison et al.*, 1968). In a limited number of studies fluoride was used in order to determine with greater specificity the exact location of phenothiazine actions at the adenylate cyclase complex (receptor-transducer-catalytic moieties). *Wolff* and *Jones* (1970), using beef thyroid membranes, showed that CPZ was stimulatory with respect to fluoride activation of adenylate cyclase. As drug concentrations were increased above 1 mM, CPZ antagonized the fluoride-activated enzyme. In homogenates of rat cerebrum, cerebellum and brain stem, trifluperazine at 0.2 mM caused a 50% inhibition of fluoride-induced adenylate cyclase (*Uzunov* and *Weiss*, 1971). We (*Palmer* and *Manian*, 1974a) looked at the action of several phenothiazines on F-activation of adenylate cyclase in either high-speed particulate (100,000 g) fractions of rat cerebrum and hypothalamus or neuronal and glial fractions of rat and rabbit cerebral cortices. In all cases the diOH analogues of CPZ, promazine, perphenazine, prochlorperazine and the 7,8-dioxo derivative of CPZ acted to inhibit both basal- and fluoride-activated adenylate cyclases. The isolated cell preparations were more sensitive to inhibition by the phenothiazines and in some instances the parent compounds and derivatives were active at 0.1 mM.

Transducer Site

Rodbell and coworkers (*Rodbell et al.*, 1975) reported an enhanced elevation in both basal and hormonal but not fluoride-induced adenylate cyclase in liver cells by GTP analogues, notably 5′-guanylyl imidodiphosphate. Kinetic analysis of these data suggested the existence of an "intermediary" mechanism acting between the hormonally sensitive receptor components and the catalytic moiety of the enzyme complex. The transducer apparently is not the site of action of neuroleptic drugs with respect to an antagonism of adenylate cyclase because *Burkard* (1977), *Bockaert et al.* (1977) and ourselves (preliminary findings) were unable to demonstrate an action of fluphenazine on the enhanced basal or DA actions produced with 5′-guanylyl imidodiphosphate in homogenates of rat striatum and frontal cortex. However, 7,8-diOH-CPZ did inhibit these actions of 5′-guanylyl imidodiphosphate in the rat frontal cortex (*Palmer* and *Manian*, unpublished results). As reported above with F-adenylate cyclase, the 7,8-diOH derivatives of pharmacologically active phenothiazines through the generation of free radicals most likely bind nonspecifically to all proteins and inhibit many central enzymes, including ATPase (*Akera et al.*, 1974).

Adenosine

The elevation of cyclic AMP by adenosine via an interaction with a hypothetical receptor is well documented for incubated tissue slices in individual regions from many species (see *Daly*, 1976; *Nathanson*, 1977; *Sattin* and *Rall*, 1970). For the most part, this effect occurs principally in intact cellular preparations (*Premont, Perez*, and *Bockaert*, 1977). At high concentrations CPZ, promethazine and promazine enhanced the release of and prevented the uptake of adenosine into incubated tissue slices of guinea pig cerebral cortex. As a result of these actions, basal cyclic AMP levels were magnified (*Huang* and *Daly*, 1974). Antipsychotic drugs and some analogues (haloperidol, CPZ, thioridazine, pimozide, clozapine, 7-OH-CPZ and 7-OH-CPZ-MeI) at high concentrations (50–100 μM) did not influence the accumulation of cyclic AMP elicited by adenosine in tissue slices of rat limbic forebrain and mouse cerebral cortex (*Blumberg et al.*, 1976; *Cotter et al.*, 1978).

Electrical Stimulation

Electrical pulses applied to brain slices cause a progressive increase in cyclic AMP, an action that may be mediated by adenosine (*Daly*, 1976; *Nathanson*, 1977; *Kakiuchi, Rall*, and *McIlwain*, 1969). Neurohormones, NE and histamine potentiate the action of electric current, while theophylline antagonized the effects of electrical pulses. At 50 μM, CPZ reduced in a significant manner the elevation in cerebral cyclic AMP elicited by electrical stimulation which was associated with a loss of K^+ ions from the tissue (*Kakiuchi et al.*, 1969). At this concentration CPZ has been shown to diminish the displacement of membrane potentials in electrically stimulated, isolated cerebral tissues (*McIlwain*, 1964; Hillman, Campbell and McIlwain, 1963). These actions of CPZ may thus be related to local anesthetic phenomena (*Seeman*, 1977).

Serotonin

Controversy exists as to whether mammalian brain contains an active adenylate cyclase enzyme which is activated by serotonin. On the one hand, preliminary reports showed serotonin to be marginally effective in the stimulation of cyclic AMP in tissue slices of rabbit and guinea pig cerebral cortices (*Kakiuchi* and *Rall*, 1968b; *Huang* and *Daly*, 1972). On the other hand, similar experiments were unable to confirm these findings (*Forn* and *Krishna*, 1971; *Weiss* and *Greenberg*, 1975; *Palmer et al.*, 1973). In addition, serotonin has been shown to inhibit the rise in cyclic AMP induced by NE in the rat brain (*Palmer et al.*, 1973; *Weiss* and *Costa*, 1968). For the most part, cell-free preparations of the rat central nervous system (cerebrum, midbrain, striatum, hippocampus, medulla, hypothalamus and colliculus) do display a sensitivity to serotonin (for reviews, see *Daly*, 1976; *von Hungen* and *Roberts*, 1974; *Nathanson*, 1977). *Nathanson* and *Greengard* (1974), however, were unable to report a serotonin-sensitive enzyme in mammalian brain, while a serotonin-responsive adenylate cyclase was present in homogenates and intact cellular preparations of insect ganglia. Furthermore, the studies of *von Hungen* and *Roberts* (1974) revealed that at birth the serotonin sensitivity of adenylate cyclase in the rat brain was higher than in the adult. In contrast, Pagel and coworkers (*Pagel et al.*, 1976) reported a serotonin-responsive enzyme using an adult rat synaptosomal membrane fraction.

Another paradox relative to serotonin action on brain cyclic nucleotide systems is that lysergic acid diethylamide (LSD) and other serotonin antagonists act at low concentrations to elevate brain adenylate cyclase. At higher concentrations, however, some of the agents inhibit DA- and NE-sensitive adenylate cyclases (see *von Hungen* and *Roberts,* 1974; *von Hungen et al.*, 1975; *Palmer* and *Burks,* 1971; *Nathanson* and *Greengard,* 1974; *Nathanson*, 1977). With regard to an action of phenothiazines on this serotonergic system, *von Hungen et al.* (1975) reported an inhibition of the stimulatory effects of LSD on adenylate cyclase by haloperidol, CPZ, thioridazine and trifluperazine. Preincubation of brain stem tissue slices with trifluperazine abolished the elevation in cyclic AMP elicited by psychotomimetics (LSD, mescaline, dimethyltryptamine, psilocybin and ibotenic acid) (*Weiss* and *Greenberg,* 1975). These investigators were unable to show an elevation in cyclic AMP with serotonin, N-methyl-serotonin or N-methyl-5-methoxytryptamine. In the guinea pig cerebral cortex (tissue slices) *Huang* and *Daly* (1972) found that CPZ reduced the enhanced accumulation of cyclic AMP elicited by a serotonin-KCl combination. In this study chlorpromazine was not active with regard to an antagonism of the stimulatory actions of serotonin-adenosine. On the other hand, the serotonin-adenylate cyclase in insect ganglia was rather potently inhibited with haloperidol (*Nathanson*, 1976). This did not occur in snail ganglia where CPZ and haloperidol did not influence serotonin-adenylate cyclase (*Osborne*, 1977). Conflicting data were also presented by *Pagel et al.* (1976) who demonstrated an insensitivity of fluphenazine toward antagonism of serotonin-adenylate cyclase in rat synaptosomal membranes.

Depolarizing Agents

Extensive investigations conducted by Daly and coworkers (*Daly*, 1976, 1977) utilizing incubated brain slices of guinea pig have shown a high degree of cyclic AMP stimulation in response to added depolarizing agents (KCl, ouabain, batrachotoxin and veratridine). In slices of guinea pig cerebellum the response to veratridine and calcium plus veratridine was reduced by CPZ and promethazine (*Ohga* and *Daly*, 1977). Many of the actions of the depolarizing agents have been attributed to the release of adenosine. Low concentrations of CPZ inhibit the efflux of radioactive adenosine from control and electrically stimulated guinea pig cortical slices, while high concentrations of CPZ enhance adenosine release. In addition, local anesthetic actions of CPZ must be considered, along with actions on ATP levels. Chlorpromazine also antagonized the synergistic action of KCl with either NE, serotonin or histamine (*Huang* and *Daly*, 1972). In a further attempt to delineate the specific characteristics of CPZ actions on systems that generate cyclic AMP in tissue slices of mouse cerebrum, *Cotter et al.* (1978) showed high concentrations (0.1 mM) of CPZ and 7-OH-CPZ to inhibit the stimulatory actions of KCl and ouabain. The 7-OH-CPZ-MeI and 7-OH-CPZ glucuronide derivatives of 7-OH-CPZ were ineffective. Clearly more studies with a wider variety of depolarizing agents are needed in order to determine with any degree of certainty the possible actions of phenothiazine agents on the complex ionic channels mediating the formation of cyclic AMP.

Action of Phenothiazines on Basal Levels of Cyclic AMP

A few investigations with both tissue slices and homogenates of clonal pituitary cells have shown that under some conditions many phenothiazine agents and their corresponding metabolic derivatives elevate adenylate cyclase-cyclic AMP. Chlorpromazine was so active in the guinea pig cortical slice preparation (*Huang* and *Daly*, 1972, 1974). In rat cerebral cortex the 7,8-diOH analogues of CPZ, prochlorperazine, perphenazine, along with 7,8-dioxo-CPZ, 2-Cl-7,8-dioxo-phenothiazine, 3-OH-phenothiazine and 8-OH-7-MeO-CPZ, all significantly raised the basal levels of cyclic AMP (*Palmer* and *Manian*, 1974b). *Clement-Cormier, Heindel*, and *Robison* (1977) showed the following analogues of CPZ, i.e., 7-OH-CPZ, 8-OH-CPZ, 7-MeO-CPZ but not CPZ-5-N-dioxide, acted to elevate adenylate cyclase in cloned pituitary cells. Trifluperazine was also active. In some instances the elevation in cyclic AMP observed in intact cellular preparations was attributed to the release of adenosine by CPZ (*Huang* and *Daly*, 1972, 1974; *Daly*, 1976, 1977). On the other hand, using incubated slices from the rat brain, the following phenothiazines actually inhibited basal levels of cyclic AMP: 3-OH-CPZ, promazine, 8-OH-prochlorperazine, fluphenazine, 7-OH-fluphenazine, 8-OH-fluphenazine and thiothixene (*Palmer* and *Manian*, 1974b). Perhaps a local anesthetic action may explain these latter findings. Most studies, however, did not report any inhibitory actions of phenothiazines on basal cyclic AMP. In an unpublished study (*Palmer* and *Manian*) 3,7-diOH-CPZ and 3,8-diOH-CPZ acted to stimulate basal adenylate cyclase activity in homogenates of rat frontal cortex.

Protein Kinases

Since cyclic AMP-activated protein kinases are responsible for most or all intracellular metabolic actions of the nucleotide (*Greengard* and *Kebabian*, 1974), it is rather surprising that so few neuropharmacological investigations have been carried out with this enzyme system. A cyclic AMP-dependent enzyme from bovine brain was inhibited by 7,8-dioxo- and 7,8-diOH-CPZ (10^{-3} to 10^{-4} M) (*Petzold* and *Greengard,* personal communication). *Maxwell* (1975) found CPZ to be a noncompetitive inhibitor of protein kinase with respect to ATP. The inhibition also occurred in the absence of cyclic AMP. Protein phosphorylation, which may be associated with cyclic AMP-activated protein kinase activity, was increased when either NE or electrical pulses were applied to tissue slices of guinea pig cerebral cortex. Trifluperazine decreased these actions of NE and electrical pulses (*Williams* and *Rodnight*, 1976). It would be of interest to evaluate the level or rate of protein kinase activity following injections of neuroleptics *in vivo*. Possibly correlations and more meaningful conclusions could be drawn to respective changes in cyclic nucleotide levels following such an approach.

STRUCTURE-ACTIVITY RELATIONSHIPS FOR PHENOTHIAZINE ACTIONS ON CENTRAL ADENYLATE CYCLASE

The preceding discussion has dealt with a host of investigations concerning the activity of CPZ and its corresponding derivatives on various aspects of central adenylate cyclase systems. Figure 3 depicts a brief summary of these structure-activity relationships.

ACTIONS OF NEUROLEPTICS ON HORMONALLY SENSITIVE ADENYLATE CYCLASES IN PERIPHERAL ORGANS

As observed with central tissues, the ability of neuroleptic agents to influence adenylate cyclase systems in peripheral organs is usually manifested by a blockade of hormonally sensitive receptor sites on the enzyme. Peripheral tissue has not been studied with the phenothiazines to the same extent as central tissue. At times various investigators use modifications of the same technique, resulting in discrepancies in the data presented.

Lung

In an initial study employing the incubated tissue slice technique, both histamine and epinephrine elevated cyclic AMP levels in rat and guinea pig lung (*Palmer*, 1971). The action of epinephrine in the two species was diminished by CPZ and propranolol, while only in the rat was CPZ capable of inhibiting the histamine-sensitive cyclic nucleotide accumula-

STRUCTURE ACTIVITY RELATIONSHIPS OF CPZ ACTIONS ON ADENYLATE CYCLASE

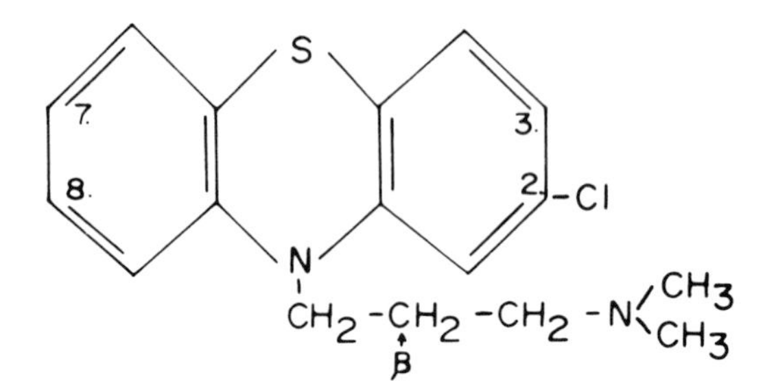

SUBSTITUTIONS	ACTIONS–ADENYLATE CYCLASE
OH at 7	SAME ESPECIALLY NE
OH at 8	LESS EXCEPT DA
MeO at 7	LESS
diMeO at 3,7 or 7,8	LESS
O at S	INACTIVE
diOH at 2,3 or 7,8	POTENT-BROKEN CELLS NONE-WHOLE CELLS
dioxo at 7,8	SAME AS 7,8diOH
diOH at 3,7 or 3,8	ELEVATES-BROKEN CELLS
triOH at 3,7,8	LESS ACTIVE THAN 7,8 diOH
β-OH	SIMILIAR TO CPZ
CH_3 at $N(CH_3)_2$	LESS EXCEPT HISTAMINE
GLUCURONIDE at 7	RELATIVELY INACTIVE
MINUS SIDE CHAIN	INACTIVE EVEN WITH 7,8diOH
MINUS Cl	USUALLY LESS EXCEPT NE
OH at 3	LESS

Fig. 3. Structure-activity relationships for substitutions on the basic phenothiazine molecule with respect to actions on central adenylate cyclase systems.

tion (*Palmer*, 1971). In further studies *Collins et al.* (1973) showed that catecholamines (NE, epinephrine and isoproterenol) and histamine would elevate cyclic AMP in an isolated perfused-intact organ preparation of rat lung. The action of isoproterenol was blocked by equimolar concentrations of CPZ. Adenylate cyclase in tissue slices from mouse lung was also shown to be sensitive to NE, while in a broken cellular-particulate fraction DA as well as NE activated adenylate cyclase. In the tissue slices the action of NE on cyclic AMP levels is more sensitive to inhibition by

CPZ than to haloperidol. With the particulate fractions CPZ but not haloperidol blocked the activation of adenylate cyclase by NE, while the DA-responsive enzyme appears to be about equally inhibited by the two antipsychotic agents (*Palmer et al.*, 1978c). In related broken cellular investigations a variety of phenothiazines (CPZ, trifluperazine, fluphenazine and thioridazine) potently antagonized isoproterenol-sensitive adenylate cyclase in guinea pig lung. To a lesser extent basal enzyme activity was similary affected (*Weinryb et al.*, 1972).

The technique of rapid fixation of pulmonary tissue by high-intensity, focused microwave irradiation has been used to determine the influence of injected (i.p.) neuroleptic agents on steady-state levels of cyclic nucleotides. This method attempts to approximate, as close as possible, actual *in vivo* conditions. When mice are injected (i.p.) on an acute basis (30 min to 8 hr) with either CPZ (2.5, 10, 25 mg/kg) or haloperidol (2.0 mg/kg) and the animals so sacrificed, the levels of both cyclic AMP and GMP are depressed. However, similar subchronic injections of CPZ, haloperidol and the relatively inactive sulfoxide metabolite of CPZ (25 mg/kg) resulted in elevated concentrations of cyclic AMP at periods of and greater than 24 hr post injection. In this case cyclic GMP did not change (*Palmer et al.*, 1977, 1978c). The initial depression of pulmonary cyclic nucleotide levels following *in vivo* administration of neuroleptic agents most likely reflects an ability of the compounds to inhibit hormonally sensitive adenylate cyclase as discussed above, while the lowering of cyclic GMP might be due to an inhibition of guanylate cyclase as reported by *Fujimoto* and *Okabayashi* (1975). In their studies several clinically active phenothiazines, but not haloperidol, inhibited the particulate form of guanylate cyclase in guinea pig trachea. Moreover, this action and the activation of guanylate cyclase by phospholipase were blocked by CPZ. Perhaps the mechanism of the phospholipase effect on the enzyme is to perturbate the structural integrity of the cell membrane, thereby activating guanylate cyclase. The membrane stabilizing role of CPZ prevents this event from occurring. Since subchronic and chronic injections (i.p.) of phenothiazines produce a supersensitivity of postsynaptic adenylate cyclase receptors and perhaps inhibition of phosphodiesterase in central tissue, there is reason to assume that the same events will happen in peripheral organs (*Fertel* and *Weiss*, 1976). Hence, this would account for the elevated level of cyclic AMP observed in rapidly fixed lung tissue following a sequence of neuroleptic administration.

Heart

The adenylate cyclase enzyme in cardiac tissue is readily responsive to a variety of hormonal agents to include NE, epinephrine, isoproterenol, DA, histamine, adenosine, prostaglandins and glucagon. There is considerable conjecture, however, whether these actions of hormones on subsequent cardiac contractile events are mediated by cyclic AMP (for additional references and discussion, see *Martinez* and *McNeil*, 1977). *Langslet* (1971) and *Oye* and *Langslet* (1972), using isolated perfused rat hearts, found CPZ would block both the elevation in cyclic AMP and the activation of phosphorylase that was normally produced with isoproterenol and glucagon. Chlorpromazine did not influence the inotropic response of the heart caused by the two hormones, nor was the activation of phosphorylase induced by dibutyryl cyclic AMP affected. The data indicated that receptor-mediated increases in cyclic AMP were dissociated from the contractile process and that phenothiazines act at the level of the adenylate cyclase receptor. In a more recent study employing a similar preparation *Osnes et al.* (1976) found CPZ to be ineffective toward inhibition of glucagon or isoproterenol stimulation of both cyclic AMP and phosphorylase activation. Using a particulate fraction from another species, rabbit heart, *Wolleman* (1974) found a double effect of CPZ on cardiac adenylate cyclase. The drug inhibited enzyme activation by isoproterenol, NE and histamine but not by glucagon. Moreover, CPZ caused a noncompetitive inhibition of phosphodiesterase and potentiated the action of fluoride on adenylate cyclase. The latter effect was abolished if the samples were solubilized with Lubrol. Recently *Simpson* (1977) injected either CPZ or molindone into rats on a chronic basis and found neither changes in basal adenylate cyclase nor alterations in the mechanical properties of the heart. On the other hand, both drugs antagonized in a noncompetitive manner the elevation of adenylate cyclase by isoproterenol and NE.

Red Blood Cells

Sheppard and *Burghardt* (1971) identified an adrenergic receptor-induced activation of adenylate cyclase in rat red blood cells. Chlorpromazine and haloperidol attenuated the enzyme responsiveness to N-methyldopamine, NE and epinephrine. There were, however, no actions by these neuroleptic agents on the DA-sensitive enzyme. In a related experiment CPZ exerted an antagonism of isoproterenol-induced adenylate cyclase activation in nucleated (turkey) erythrocytes (*Osnes et al.*, 1976).

Adrenal

The synthesis of corticosterone in the adrenal cortex was enhanced via an adrenocorticotrophic hormone (ACTH) action on cyclic AMP. In rat adrenal glands incubated in the presence of CPZ the immediate elevation in ACTH or cyclic AMP-stimulated corticosterone synthesis was blocked. If the cells were preincubated for 30 min with cyclic AMP no inhibition occurred (*Haksar* and *Peron*, 1971). A capability of several phenothiazines to block ACTH actions on steroidogenesis was determined by *Free et al.* (1974). In these studies thioridazine was the most potent, followed by haloperidol > 2-Cl-10-(3 dimethyl amino propyl) phenothiazine > promethazine > CPZ > triflupromazine. Direct investigations with adenylate cyclase (adrenal membranes) revealed an inhibition of the ACTH-sensitive enzyme by CPZ. Furthermore, the agent enhanced fluoride sensitivity of adenylate cyclase (*Wolff* and *Jones*, 1970, 1971).

Other Peripheral Organs

Lipolysis of adipose tissue, an event mediated by an epinephrine-sensitive adenylate cyclase, was antagonized by CPZ, fluphenazine and thioridazine (*Weinryb et al.*, 1972). Thyroid cells elevate cyclic AMP in response to TSH (thyroid stimulating hormone) and prostaglandin E_1. These actions are decreased in the presence of CPZ, trifluperazine and prochlorperazine, which do not affect basal adenylate cyclase but enhance the fluoride activation of the enzyme (*Yamashita et al.*, 1970; *Wolff* and *Jones*, 1970). Cortical tissue of kidney responds similarly to parathyroid hormone but CPZ does not antagonize the actions of this hormone (*Wolff* and *Jones*, 1970). Furthermore, phenothiazines prevented the stimulation of cyclic AMP in the liver by both glucagon and epinephrine but did not affect adenylate cyclase sensitivity to fluoride. However, *Osnes et al.* (1976), using perfused rat livers, could not demonstrate an action by CPZ on cyclic AMP accumulation produced by either epinephrine or glucagon. Dopamine and vasopressin were shown to increase the level of cyclic AMP in the perfusate of rat kidney. The effect of DA but not vasopressin was antagonized by spiroperidol (*Nakajima, Naitoh*, and *Kuruma*, 1977). Urinary excretion of cyclic AMP is influenced by many clinical and hormonal conditions (*Murad*, 1973). When CPZ was injected (i.p.) into rats on a daily basis for 1 wk, the daily amount of cyclic AMP in the urine rose along with the total urine volume (*Palmer* and *Evan*, 1974). Whether these observations reflect a general inhibition of phosphodiesterase in the total body to CPZ or whether selected organs became hypersensitive to adenylate cyclase stimulation is only speculation. The total number of specific organs and cells contributing to the cyclic nucleotide output of course cannot be evaluated in investigations of this type. The possible action of phenothiazines in many of the investigations with peripheral organs has been attributed by *Wolff* and *Jones* (1970) and *Wolleman* (1974) as a membrane-expanding detergent action. This was based on the fact that solubilization of the enzymes with Lubrol or thymol abolishes this action, which could not be restored after addition of phospholipids.

The data from peripheral tissues when taken together show CPZ to have widespread actions on all organ systems in the body. The possibility exists that through these influences on hormonally active adenylate cyclase systems a host of endocrine disturbances could reflect many of the clinically observed side-effects related to phenothiazine actions.

ACTIONS OF NEUROLEPTICS ON CENTRAL CYCLIC GMP SYSTEMS

In Vitro Studies

When added to incubated tissue slices cholinomimetic agents elicit a weak accumulation of cyclic GMP, a process mediated by muscarinic receptors and requiring the presence of calcium (*Lee, Kuo*, and *Greengard*, 1972; *Kinscherf et al.*, 1976; *Palmer* and *Duszynski*, 1975; *Palmer, Manian,* and *Sanborn*, 1976b; for reviews, see *Nathanson*, 1977 and *Daly*, 1977). A culture of mouse neuroblastoma (NIE-115) contains a guanylate cyclase enzyme that is highly sensitive in this manner to cholinomimetics (*Richelson*, 1977; *Richelson* and *Divinetz-Romero*, 1977). In addition, as in the case with cyclic AMP systems, specific depolarizing agents elevate cyclic GMP (*Kinscherf et al.*, 1976; *Ohga* and *Daly*, 1977). *Goldberg, O'Dea*, and *Haddox* (1973) have postulated that under certain conditions the action of neurohumorally induced levels of cyclic AMP and GMP may oppose the physiological effects of one another. By employing the technique of microiontophoresis to single pyramidal tract neurons (rat cerebrum), *Stone, Taylor*, and *Bloom* (1975) found that neuronal firing rates antagonized by NE and cyclic AMP were excited by acetylcholine and cyclic GMP. Moreover, it has been known for some time that many of the side-effects of phenothiazines could be attributed to antimuscarinic actions. *Miller* and *Hiley* (1974) and *Snyder et al.* (1974b) evaluated the antimuscarinic properties of neuroleptics by studying the displacement produced by neuroleptics of radioactive ligands which bind to central muscarinic receptors. These actions were then compared with an ability to inhibit DA adenylate cyclase. Thioridazine and clozapine, which have a low tendency to produce parkinsonism, possessed the greatest degree of antimuscarinic activity but were weaker than trifluperazine and flupenthixol to inhibit DA adenylate cyclase. In contrast, trifluperazine and flupenthixol displayed the weakest antimuscarinic actions. When brain slices from rat cerebral cortex were preincubated with neuroleptic agents, the following drugs inhibited the rise in cyclic GMP elicited by carbamylcholine and choline chloride: CPZ, 7,8-diOH-CPZ and 7,8-diMeO-CPZ (*Palmer et al.*, 1976b). With the highly responsive neuroblastoma cultures the accumulation of cyclic GMP in response to carbamylcholine was inhibited by clozapine > thioridazine, mesoridazine > 8-OH-CPZ > 7-OH-CPZ > promazine,

CPZ, fluphenazine > perphenazine, acetophenazine > haloperidol > trifluphenazine > prochlorperazine. There was a 10,000-fold difference in potency between clozapine and prochlorperazine. Additionally, tertiary tricyclic antidepressants were highly potent (*Richelson,* 1977; *Richelson* and *Divinitz-Romero,* 1977; *Divinitz-Romero,* personal communication). Recently *Ohga* and *Daly* (1977) showed that homogenates of guinea pig cerebella possessed a guanylate cyclase that was inhibited by CPZ and promethazine.

Promethazine and CPZ were found to inhibit the rise in cyclic GMP produced by calcium in incubated guinea pig cerebellar slices. In addition, the agents similarly blocked the action on veratridine with or without added calcium. Local anesthetic effects were attributed to these findings (*Ohga* and *Daly*, 1977). Another interesting finding with regard to neuroleptic actions on cyclic GMP in tissue slices was that several agents, especialy 8-OH-CPZ, 7-OH-CPZ-MeI, thioridazine, promethazine, haloperidol and even CPZ-SO, would elevate basal levels of the cyclic nucleotide. Calcium was required for this action of 8-OH-CPZ (*Palmer et al.*, 1976b). This phenomenon was ascribed to the ability of phenothiazines to release intracellular calcium from brain mitochrondria as described by *Tjioe, Manian*, and *O'Neill* (1974). Inhibition of phosphodiesterases by phenothiazines is another point to be considered (*Levin* and *Weiss,* 1976). However, the elevation in cerebellar cyclic GMP produced by the phosphodiesterase inhibitor, isobutylmethylxanthine, was prevented by preincubation of the tissue slices with CPZ and promethazine (*Ohga* and *Daly*, 1977).

In Vivo Studies Employing Rapid Inactivation of Brain Tissue

Acute (½ to 8 hr) injections (i.p.) of CPZ and haloperidol led to decreased levels of steady-state cyclic GMP in rapidly fixed (focused microwave irradiation) cerebellum and cerebral cortex of mice (*Palmer et al.*, 1978b). Similarly, an acute injection of CPZ was seen to decrease cerebellar cyclic GMP in mice sacrificed by rapid freezing. In addition, the rise in cyclic GMP produced by injected oxotremorine was not blocked by CPZ (*Ferrendelli, Kinscherf*, and *Kipnis*, 1972).

Ferrendelli et al. (1972) first reported an elevation of cyclic GMP in the mouse cerebellum following injection of D-amphetamine. These experiments recently have been extended to the rat cerebellum in an attempt to understand the mechanism in which DA receptor actions produce a rise in cyclic GMP in a brain region devoid of DA nerve endings (*Palmer et al.,* 1978b). Essentially, in experiments from three laboratories injections of apomorphine or LSD induce a rise in steady-state levels of cerebellar cyclic GMP, an event readily antagonized by haloperidol, CPZ and (+)butaclamol, while (−)butaclamol was ineffective. In addition, injections of the neuroleptics alone lowered basal levels of cyclic GMP subsequent to rapid tissue inactivation by microwave irradiation. High doses of neuroleptics (haloperidol and spiroperidol) blocked the stereotyped behavior induced by apomorphine and amphetamine, while lower doses only affected the stereotypy as well as *in vivo* levels of cyclic GMP produced by the two agents that activate DA-mediated receptors. Both doses of neuroleptics alone reduced cerebellar cyclic GMP. These actions were not attributed to actions at muscarinic receptors and the effects of haloperidol and/or apomorphine were not active when administered directly into the cerebellum. Positive effects on cyclic GMP were seen with intraperitoneal or intrastriatal injections of the two neuroleptics. Similar actions were found on *in vivo* levels of cyclic GMP in deep cerebellar nuclei. Furthermore, neither apomorphine nor haloperidol acted *in vitro* to influence guanylate cyclase. The data indicate that striatal DA pathways via polysynaptic connections influence in an indirect manner the cerebellar biochemical mechanisms mediated through cyclic GMP. Electrophysiological evidence suggests such a role of the basal ganglia to modulate Purkinje cell firing rates via the mossy fibers (*Gumulka et al.*, 1976; *Burkard, Pieri*, and *Haefely*, 1976; *Biggio* and *Guidotti*, 1977; *Biggio, Costa*, and *Guidotti*, 1977 a & b; *Fox* and *Williams*, 1970).

Clinical Studies with Neuroleptics and Cyclic GMP

In one reported investigation *Ebstein et al.* (1976) found a tendency for cerebrospinal fluid levels of cyclic GMP to be lower in drug-free schizophrenic patients when compared to those observed in psychiatrically healthy individuals. When cerebrospinal fluid levels of cyclic GMP were measured before and after 2 months of neuroleptic treatment, the mean level of cyclic GMP rose 50% after drug administration. The results of this study indicated a tendency might exist in schizophrenia for a decreased central cholinergic-DA imbalance which may have been restored by the drug regimen. The close anatomical and functional association of dopaminergic and cholinergic neurons in the brain (*Anden* and *Bedard*, 1971; *Wolfarth, Dulska*, and *Lacki*, 1974) has led to an additional hypothesis that alterations in central cholinergic mechanisms may play a role along with DA with regard to a molecular involvement in the pathogenesis of affective disorders (*Davis*, 1970).

ACTIONS OF NEUROLEPTICS ON CENTRAL PHOSPHODIESTERASES

General Background

The first study to examine the influence of major tranquilizers on phosphodiesterase was conducted by *Honda* and *Imamura* (1968). A relatively high concentration of CPZ (5×10^{-5}M) was needed to inhibit an enzyme prepared from the rabbit cerebral cortex. Prochlorperazine, perphenazine, fluphenazine and chlorprothixene were likewise found to be equally active. *Roberts* and *Simonsen* (1970) showed CPZ (10^{-3}–10^{-4} M) to exert a noncompetitive type of inhibition on phosphodiesterase in a mouse brain preparation. They concluded that CPZ did not affect the combination of substrate (cyclic AMP) to the enzyme but instead limited the rate of formation of the free product from the enzyme-substrate complex. A related experiment in the rat cerebrum showed CPZ, triflupromazine and chlorprothixene, but not haloperidol, would act in a noncompetitive fashion to inhibit phosphodiesterase (*Berndt* and *Schwabe*, 1973).

Weinryb and coworkers (*Weinryb et al.*, 1972) used two enzyme preparations (rat brain and cat heart) to evaluate the actions of over one hundred compounds on a soluble form of phosphodiesterase. Of the phenothiazines tested, fluphenazine was the most potent, followed by thioridazine > CPZ > trifluperazine. These investigators then related the actions of agents which reduced anxiety in rats during a "conflict behavioral procedure" to an inhibition of rat brain phosphodiesterase. A relatively high correlation was found between these two parameters, with SQ 20,009 the most potent agent (2 μM), followed by diazepam (33 μM), fluphenazine (48 μM), methylxanthine phosphodiesterase inhibitors (120–150 μM), CPZ (170 μM), thioridazine (180 μM), triflupromazine (300 μM) and haloperidol (> 1000 μM) (*Beer et al.*, 1972).

Using soluble-(100,000 g, 50% ammonium sulfate) enzyme fractions from rat brain, *Uzunov* and *Weiss* (1971) reported that trifluperazine acted to selectively inhibit phosphodiesterase among the various brain regions. The total phosphodiesterase in the cerebral cortex and brain stem was blocked to a greater extent than the enzyme from the cerebellum. Their data suggested regional differences in the sites of action of psychotropic drugs with regard to central enzymes. *Wolleman* (1974) reported a noncompetitive inhibition of rabbit heart phosphodiesterase by CPZ. With homogenates of rat caudate, phosphodiesterase was effectively inhibited only by 7,8-diOH-, 7,8-dioxo-, and 7,8-diMeO-CPZ analogues (*Petzold* and *Greengard*, personal communication). The preliminary findings described above laid the groundwork for a series of recently conducted experiments which have both expanded and contributed significantly to our present knowledge of the mode of interaction between adenylate cyclase-phosphodiesterase and psychotropic drug action.

Actions of Neuroleptics on Multiple Forms of Phosphodiesterases

Several forms of cyclic nucleotide phosphodiesterases are known to exist and to be distributed in specific ratios among the various brain regions of the mammalian central nervous system. Different technical approaches have been employed to determine phosphodiesterase enzymes which differ in molecular weight, substrate specificity (either cyclic AMP or GMP), stability, isoelectric points, pH optima, kinetic properties, response to inhibitors and response to activators (*Kakiuchi, Yamazaki*, and *Teshima*, 1972; *Thompson* and *Appleman*, 1971; *Uzunov*, *Shein*, and *Weiss*, 1974; *Weiss et al.*, 1974; *Weiss* and *Greenberg*, 1975; *Levin* and *Weiss*, 1976, 1977; *Brostrom et al.*, 1975; *Cheung*, 1971). The accumulation of findings to date indicates that the cellular mechanisms of controlling the levels of cyclic nucleotides within the brain are highly complex. Most likely cellular organelles, in order to carry out selective metabolic functions, contain highly specific forms of the phosphodiesterases to control precisely the cyclic nucleotide ratios within designated intracellular microenvironments. Recent work, especially with the phenothiazines, has focused attention on the endogenous calcium-dependent, heat-stable activator (Ca^{++} activator) protein which enhances the catalytic activity of only one form of phosphodiesterase. This protein presumably resides within the synaptic membrane and is necessary for catecholamine-induced activation of adenylate cyclase to occur. After interaction of the receptor site of adenylate cyclase with a catecholamine, an event that coincides with or follows release of calcium, the Ca^{++} activator through a phosphorylation reaction is released from the synaptic membrane into the cytosol. Therein, calcium interacts with the activator and changes the protein to a more helical, stable structure and the Ca^{++} activator complex formed then enhances the activity of a high Km phosphodiesterase, thereby lowering intracellular cyclic AMP levels. Thus, the Ca^{++} activator regulates specific types of both phosphodiesterase and adenylate cyclase, because when the activator is released from the synaptic membrane the stimulation of membrane-bound adenylate cyclase by catecholamines is inhibited. Furthermore, the catecholamine-induced release of Ca^{++} activator from synaptic membranes is subject to regional differences. For example, NE acts more selectively than DA to release the Ca^{++} activator in rat cerebellar tissue which contains only an NE-sensitive adenylate cyclase, while DA is more potent to effect Ca^{++} activator release in the striatum. (For discussion and theories, see *Gnegy, Nathanson*, and *Uzunov*, 1977a; *Gnegy*, *Uzunov*, and *Costa*, 1976; *Revuelta, Uzunov*, and *Costa*, 1976; *Brostrom*, *et al.*, 1975; *Liu* and *Cheung*, 1976; *Levin* and *Weiss*, 1976, 1977.) In further support of these hypotheses *Brostrom* (in press) recently reported the existence of two forms of adenylate cyclase in brain. One form is dependent upon the Ca^{++} activator protein, its stimulation by fluoride is inhibited by the calcium chelator, ethyleneglycol-bis-(β-aminoethyl-ether)-N,N′-tetraacetic acid (EGTA), and low concentrations of CPZ inhibit basal enzyme activity. The Ca^{++} activator independent form of adenylate cyclase appears unaffected by EGTA or CPZ.

Weiss and coworkers have conducted extensive investigations with regard to phenothiazine-induced inhibiton of the enzyme form of phosphodiesterase subject to regulation by the Ca^{++} activator. As discussed above, using a crude phosphodiesterase from rat brain, *Uzunov* and *Weiss* (1971, 1972) reported that trifluperazine exhibited greater enzyme antagonism in the cerebral cortex and the brain stem than in the cerebellum. Subsequently, when these regional enzymes were separated by acrylamide gel electrophoresis four peaks of phosphodiesterase activity were present. In the cerebral cortex the major peak was a high Km form of peak II which was highly sensitive to activation by both calcium and the activator protein. Trifluperazine potently inhibited this enzyme, while its sulfoxide metabolite and promethazine produced a lesser antagonism. The pattern of enzyme inhibition by trifluperazine was highly selective because known phosphodiesterase inhibitors, theophylline and papaverine, were barely active. In addition, much lower concentrations of trifluperazine (40 μM) were 25-fold more potent than theophylline and papaverine in blocking the activation of the peak II enzyme by the Ca^{++} activator. Inhibition by theophylline and papaverine was competitive, while that of trifluperazine was neither competitive nor noncompetitive. This was evident whether the activator was present or not. Therefore, the phenothiazine did not prevent the access of the substrate to the enzyme. However, if the concentration of Ca^{++} activator was increased, only the inhibition of low concentrations of trifluperazine was overcome. Another major enzyme form in the cerebrum, peak III phosphodiesterase, was potently inhibited by theophylline and papaverine, while trifluperazine was 6-fold less active than with the peak II enzyme. Moreover, the peak III phosphodiesterase was the major form in the cerebellum and this accounted for the reason stated above in that trifluperazine was not highly active in this brain region (*Weiss et al.*, 1974; *Weiss* and *Greenberg*, 1975).

In extension of the previous findings with the Ca^{++} activator, it was shown that trifluperazine and CPZ effectively blocked peak II enzymes from other central sources, namely the cerebellum, C-6 astrocytoma cells and the bovine brain (*Uzunov et al.*, 1974; *Weiss* and *Greenberg*, 1975). In the bovine brain a peak I form of phosphodiesterase was blocked only 30% with high concentrations of the phenothiazine. In this study CPZ inhibited the activation (Ca^{++} activator) of peak II phosphodiesterase at concentrations below which affected the unactivated enzyme. In addition, the following agents (in descending order of potency) blocked the action of Ca^{++} activator on the peak II enzyme: pimozide, trifluperazine, chlorprothixene, thioridazine, CPZ and benperidol. Chlorpromazine sulfoxide and promethazine were considerably weaker. Further studies with the bovine brain peak II enzyme revealed that the activation by calcium could be diminished by EGTA, excess calcium could overcome this action and phenothiazines did not influence the combination of calcium with the activator. This was in contrast to phenothiazines which blocked only the activator protein, an event overcome by excess activator but not calcium (*Weiss* and *Greenberg*, 1975; *Levin* and *Weiss*, 1976). Using binding studies of trifluperazine to various proteins *Levin* and *Weiss* (1977) and *Weiss* (in press) recently showed that trifluperazine and other neuroleptics possessed both high-affinity, calcium-dependent and low-affinity, calcium-independent binding properties. Only the Ca^{++} activator protein displayed a high-affinity Ca^{++}-dependent site and only neuroleptic agents were specifically bound to and inhibited this protein. In one peripheral study in the rat lung, a fraction was separated from the supernatant that displayed a Ki for trifluperazine of 50 μM, while another fraction had a Ki of 1000 μM. No activated form of phosphodiesterase was present (*Fertel* and *Weiss*, 1976).

These findings concerning phenothiazine inhibition of specific forms of phosphodiesterase indicate the selective inhibition of the Ca^{++} activator is due to a highly specific binding of the drug to the activator protein. Since the activator is identical to the activator of adenylate cyclase, this might account for differences in phenothiazine actions in different brain regions. For example, in the rat striatum the Ki for trifluperazine inhibition of DA-sensitive adenylate cyclase is 11 nM (*Clement-Cormier et al.*, 1975) versus a Ki for inhibition of the predominant peak III phosphodiesterase of 250 μM. Thus, phenothiazines act in a more potent manner to decrease cyclic AMP in this region. On the other hand, relatively high concentrations (10^{-4} M) of phenothiazine were needed in the rat cerebral cortex to block the accumulation of cyclic AMP via NE stimulation (*Uzunov* and *Weiss*, 1971; *Palmer* and *Manian*, 1974b), whereas the trifluperazine inhibited the Ca^{++}-activated peak II phosphodiesterase at 10 μM. Therefore, in some regions phenothiazines might lower cyclic AMP levels and in others raise nucleotides. Thus, the diverse clinical effects seen by the neuroleptics might be explained on the basis of a selective drug action on a particular form of adenylate cyclase or phosphodiesterase present in specific cell types in discrete brain regions (*Weiss* and *Greenberg*, 1975).

Involvement of Phosphodiesterase in Neuroleptic-Induced Cellular Mechanisms of Supersensitivity

Recent findings indicate that cellular levels and activities of the phosphodiesterases exist in a dynamic state. For example, chronic stimulation of central adenylate cyclase by catecholamines is under a form of feedback control in that specific forms of phosphodiesterase are increased to compensate for the elevated levels of intracellular cyclic AMP (*Oleshansky* and *Neff*, 1975; *Schwartz, Morris*, and *Breckenridge*, 1973). With regard to the phenothiazines, *Gnegy et al.* (1976, 1977b) studied the role of the Ca^{++} activator in the DA-receptor supersensitivity of the rat caudate nucleus. Rats received injections of haloperidol or clozapine for 4 weeks. After sacrifice by microwave irradiation the amount of the membrane-bound Ca^{++} activator was increased by 40%. The peak II levels of phosphodiesterase were not, however, measured in this study. In one other related investigation *Fredholm* (1977) treated rats for 18 days with haloperidol and observed a reduction in a low K_m soluble phosphodiesterase from the striatum. The chromatographic behavior of the enzyme on DEAE-cellulose was altered in that the enzyme was eluted at lower ionic strength. In control rats the phosphodiesterase activity was enhanced by calcium, while in the haloperidol-treated animals addition of calcium ion caused a reduction in phosphodiesterase activity. This observation was not related to any changes in the levels of Ca^{++} activator protein in either control or experimental animals. Furthermore, the drug regimen did not change the activities of high K_m cyclic AMP phosphodiesterase, the particulate cyclic AMP phosphodiesterase or the cyclic GMP phosphodiesterase. These preliminary data reveal that the molecular substrates subserving the phenomenon of adenylate cyclase receptor supersensitivity following administration of neuroleptic agents may in part include the phosphodiesterases.

CONCLUSIONS

Based on the extensive data presented, the two major actions of neuroleptic drugs on cyclic nucleotide systems appear to be 1) a highly selective inhibition of the receptor site mediating both the DA- and to a lesser extent the NE-induced increase in cyclic AMP and 2) a highly specific interaction of the calcium-activated form (calcium-dependent regulator protein) of phosphodiesterase. With the wide variety of tissues and different experimental techniques utilized, all hormonally activated but not depolarization-activated adenylate cyclases are inhibited to varying degrees by pharmacologically active neuroleptics. Even though there is general agreement as to how phenothiazines affect *in vitro* adenylate cyclase preparations, these findings are at times difficult to extrapolate to *in vivo* conditions. For example, in most instances receptor activation of adenylate cyclase is inhibited under *in vitro* conditions, while, on the other hand, subchronic and chronic injections of phenothiazines produce rises in steady-state levels of cyclic AMP. However, acute administration of neuroleptics to animals may yield findings somewhat identical to *in vitro* experiments, i.e., an inhibition of adenylate cyclase. Chronic administration of phenothiazines most likely causes two actions: 1) an

Table 4. Summary of Neuroleptic Actions on Cyclic Nucleotide Systems

System	Action	References
In vitro Neurohormones-Receptors		
DA-adenylate cyclase	Profound block	6, 8, 15, 19, 24
NE-adenylate cyclase	Block	2, 8, 16, 21, 23, 30
Histamine-adenylate cyclase	Block	16, 29
5-HT-adenylate cyclase	No effect	20
Peripheral hormones-adenylate cyclase	Block	7, 13, 33
In vitro Depolarizing Agents		
Basal cyclic AMP	Elevate or no effect	2, 8, 14, 16, 21, 22, 30
KCl cyclic AMP	Slight block	8
Ouabain cyclic AMP	Slight block	8
Adenosine cyclic AMP	No effect	2, 8
In vitro Adenylate Cyclase		
Fluoride catalytic site	Slight block	23, 30
5′Guanylyl imidodiphosphate transducer	No effect	unpublished
CA^{++} dependent activator	Block	3
CA^{++} independent activator	Slight block	3
Protein kinase phosphorylation	Block	18, 32
Phosphodiesterase calcium activator	Profound block	17
In vitro Cyclic GMP Systems		
Basal cyclic GMP	Elevate	25
Cholinergic cyclic GMP	Block	25, 28
In vivo Injections of Neuroleptics		
Acute cyclic AMP	Lower	27
Subchronic cyclic AMP	Elevate	26, 27
Acute adenylate cyclase	Supersensitivity	15
Chronic adenylate cyclase	Supersensitivity	12
Chronic urinary cyclic AMP	Elevate	22
Acute amphetamine-induced cyclic AMP	Block	5
Postdecapitation cyclic AMP	Block	30
Trauma cyclic AMP	Block	31
Chronic phosphodiesterase	Decrease	11
Acute cyclic GMP	Lower	1, 4, 10, 27
Chronic cyclic GMP	Elevate or no effect	9, 27

Acute injections are less than 24 hr; subchronic injections are 1–4 days; chronic injections are greater than 4 days.

References: 1) *Biggio et al.*, 1977a; 2) *Blumberg et al.*, 1976; 3) *Brostrom*, in press; 4) *Burkard, Pieri*, and *Haefely*, 1976; 5) *Carenzi et al.*, 1975; 6) *Clement-Cormier et al.*, 1974; 7) *Collins et al.*, 1973; 8) *Cotter et al.*, 1978; 9) *Ebstein et al.*, 1976; 10) *Ferrendelli, Kinscherf*, and *Kipnis*, 1972; 11) *Fredholm*, 1977; 12) *Friedhoff, Bonnet*, and *Rosengarten*, 1977; 13) *Haksar* and *Peron*, 1971; 14) *Huang* and *Daly*, 1972; 15) *Iwatsubo* and *Clouet*, 1975; 16) *Kakiuchi* and *Rall*, 1968b; 17) *Levin* and *Weiss*, 1976; 18) *Maxwell*, 1975; 19) *Miller, Horn*, and *Iversen*, 1974; 20) *Pagel et al.*, 1976; 21) *Palmer, Robison*, and *Sulser*, 1971; 22) *Palmer* and *Evan*, 1974; 23) *Palmer* and *Manian*, 1974a & b; 24) *Palmer* and *Manian*, 1976; 25) *Palmer, Manian*, and *Sanborn*, 1976b; 26) *Palmer et al.*, 1977; 27) *Palmer et al.*, 1978b; 28) *Richelson* and *Divinitz-Romero*, 1977; 29) *Spiker, Palmer*, and *Manian*, 1976; 30) *Uzunov* and *Weiss*, 1971; 31) *Watanabe* and *Passonneau*, 1975; 32) *Williams* and *Rodnight*, 1976; 33) *Wolff* and *Jones*, 1970.

increase in the sensitivity of the postsynaptic receptor to activation by catecholamines, and 2) inhibition of specific molecular forms of phosphodiesterase.

Neuroleptic agents most likely inhibit cholinomimetic-induced cyclic GMP accumulation in both *in vitro* and *in vivo* central preparations via their well-known anticholinergic side-effects. There are two possible explanations for the neuroleptic-induced rise in basal cyclic GMP under *in vitro* conditions: 1) the agents inhibit phosphodiesterase, and 2) the agents release calcium from brain mitochrondria, which in turn activates guanylate cyclase. These elevated *in vitro* levels of cyclic GMP may in turn contribute in a minor way to the modulation of cyclic AMP.

A summary of phenothiazine actions on cyclic nucleotides is presented in Table 4.

One final point which should be made is that the total therapeutic outcome of any pharmacologically active parent phenothiazine is additionally reflected through the presence of one or more active metabolites. Likewise, these active metabolites may contribute to the adverse effects occurring with phenothiazine therapy.

REFERENCES

Abdulla, Y.H., and Hamadah, K. 1970. 3′,5′-cyclic adenosine mono-phosphate in depression and mania. Lancet 1:378–381

Ahn, H.S., Mishra, R.K., Demirjian, C., and Makman, M.H. 1976. Catecholamine-sensitive adenylate cyclase in frontal cortex of primate brain. Brain Res 116:437–454

Akera, T., Baskin, S.I., Tobin, T., Brody, T.M., and Manian, A.A. 1974. 7,8-Dihydroxychlorpromazine: (Na^{+}-K^{+})-ATPase inhibition and positive inotropic effect. In *Phenothiazines and Structurally Related Drugs,* Eds. I.S. Forrest, C.J. Carr, and E. Usdin, pp. 633–640. New York: Raven Press

Anden, N.E., and Bedard, P. 1971. Influences of cholinergic mechanisms on the function and turnover of brain dopamine. J Pharm Pharmacol 23:460–462

Baca, G.M., and Palmer, G.C. 1978. Presence of hormonally-sensitive adenylate cyclase receptors in capillary-enriched fractions from rat cerebral cortex. Blood Vessels 15:286–298

Baldessarini, R.J. 1977. Schizophrenia. New Eng J Med 297:988–995

Beer, B., Chasin, M., Clody, D.E., Vogel, J.R., and Horovitz, Z.P. 1972. Cyclic adenosine monophosphate phosphodiesterase in brain. Effect on anxiety. Science 176: 428–430

Belmaker, R.H., Ebstein, R.P., Schoenfeld, H., and Rimon, R. 1976. The effect of haloperidol on epinephrine-stimulated adenylate cyclase in humans. Psychopharmacologia 49:215–217

Berndt, S., and Schwabe, U. 1973. Effect of psychotropic drugs on phosphodiesterase and cyclic AMP levels in rat brain *in vivo.* Brain Res 63:303

Biederman, J., Rimon, R., Ebstein, R., Belmaker, R.H., and Davidson, J.T. 1977. Cyclic AMP in the CSF of patients with schizophrenia. Brit J Psychiat 130:64–67

Biggio, G., Costa, E., and Guidotti, A. 1977a. Pharmacologically induced changes in the 3′,5′-cyclic guanosine monophosphate content of rat cerebellar cortex. Difference between apomorphine, haloperidol and harmaline. J Pharmacol Exp Ther 200: 207–215

Biggio, G., Costa, E., and Guidotti, A. 1977b. Regulation of 3′,5′-cyclic guanosine monophosphate content in deep cerebellar nuclei. Neurosciences 2:49-52

Biggio, G., and Guidotti, A. 1977. Regulation of cyclic GMP in cerebellum by a striatal dopaminergic mechanism. Nature 265:240-242

Blumberg, J.B., Taylor, R.E., and Sulser, F. 1975. Blockade by pimozide of a noradrenaline sensitive adenylate cyclase in the limbic forebrain: Possible role of limbic noradrenergic mechanisms in the mode of action of antipsychotics. J Pharm Pharmacol 27:125-128

Blumberg, J.B., Vetulani, J., Stawarz, R.J., and Sulser, F. 1976. The noradrenergic cyclic AMP generating system in the limbic forebrain. Pharmacological characterization *in vitro* and possible role of limbic noradrenergic mechanisms in the mode of action of antipsychotics. Europ J Pharmacol 37:357-366

Bockaert, J., Tassin, J.P., Thierry, A.M., Glowinski, J., and Premont, J. 1977. Characteristics of dopamine and β-adrenergic sensitive adenylate cyclases in the frontal cerebral cortex of the rat. Comparative effects of neuroleptics on frontal cortex and striatal dopamine sensitive adenylate cyclases. Brain Res 122:71-86

Breckenridge, B.M. 1964. The measurement of cyclic adenylate in tissues. Proc Nat Acad Sci USA 52:1580-1586

Brostrom, C.O., Huang, Y.C., Breckenridge, B.McL., and Wolff, D.J. 1975. Identification of a calcium-dependent regulator of brain adenylate cyclase. Proc Nat Acad Sci USA 72:64-68

Brostrom, M. 1978. Calcium-dependent regulation of brain adenylate cyclase. In *Advances in Cyclic Nucleotide Research*, Vol. 9, New York: Raven Press, in press

Brown, B.L., Salway, J.G., Albano, J.D.M., Hullin, R.P., and Ekins, R.P. 1972. Urinary excretion of cyclic AMP and manic-depressive psychosis. Brit J Psychiat 120: 405-408

Brown, J.H., and Makman, M.H. 1972. Stimulation by dopamine of adenylate cyclase in retinal homogenates and of adenosine 3′,5′-cyclic monophosphate formation in intact retina. Proc Nat Acad Sci USA 69:539-543

Bucher, M-B., and Schorderet, M. 1974a. Selective stimulation by dopamine of adenylate cyclase in homogenates of rabbit retina. Experientia 30:694

Bucher, M-B., and Schorderet, M. 1974b. Apomorphine-induced accumulation of cyclic AMP in isolated retinas of the rabbit. Biochem Pharmacol 23:3079-3082

Bucher, M-B., and Schorderet, M. 1975. Dopamine- and apomorphine-sensitive adenylate cyclase in homogenates of rabbit retina. Naunyn-Schmiedeberg's Arch Pharmacol 288:103-107

Burkard, W.P., Pieri, L., and Haefely, W. 1976. *In vivo* changes of guanosine 3′,5′-cyclic phosphate in rat cerebellum by dopaminergic mechanisms. J Neurochem 27: 297-298

Burkard, W.P. 1977. Different activation of striatal adenylate cyclase by dopamine and GTP-analogues. Experientia 33:788

Calcutt, C.R. 1976. The role of histamine in the brain. Gen Pharmacol 7:15-25

Carenzi, A., Cheney, D.L., Costa, E., Guidotti, A., and Racagni, G. 1975. Action of opiates, antipsychotics, amphetamine and apomorphine on dopamine receptors in rat striatum. *In vivo* changes of 3′,5′-cyclic AMP content and acetylcholine turnover rate. Neuropharmacology 14:927-940

Carlsson, A., and Lindqvist, M. 1963. Effect of chlorpromazine and haloperidol on formation of 3-methoxytyramine and normetanephrine in mouse brain. Acta Pharmacol Toxicol 20:140-144

Chand, N., and Eyre, P. 1975. Classification and biological distribution of histamine receptor sub-types. Agents and Actions 5:277-295

Cheung, W.Y. 1971. Cyclic 3′,5′-nucleotide phosphodiesterase. Evidence for and properties of a protein activator. J Biol Chem 246:2859-2869

Christensen, A.V., Fjalland, B., and Moller-Nielsen, I. 1976. On the supersensitivity of dopamine receptors, induced by neuroleptics. Psychopharmacology 48:1-6

Clement-Cormier, Y.C., Kebabian, J.W., Petzold, G.L., and Greengard, P. 1974. Dopamine-sensitive adenylate cyclase in mammalian brain: A possible site of action of antipsychotic drugs. Proc Nat Acad Sci USA 71:1113-1117

Clement-Cormier, Y.C., Parrish, R.G., Petzold, G.L., Kebabian, J.W., and Greengard, P. 1975. Characterization of a dopamine-sensitive adenylate cyclase in the rat caudate nucleus. J Neurochem 25:143-149

Clement-Cormier, Y.C., and Robison, G.A. 1977. Adenylate cyclase from various dopaminergic areas of the brain and the action of antipsychotic drugs. Biochem Pharmacol 26:1719-1722

Clement-Cormier, Y. 1977. Phenothiazine side effects: DA receptor site a clue? by B.J. Montgomery. J Amer Med Assoc 238:2113

Clement-Cormier, Y.C., Heindel, J.J., and Robison, G.A. 1977. Adenylyl cyclase from a prolactin producing tumor cell-effect of phenothiazines. Life Sci 21:1357-1364

Collins, M., Palmer, G.C., Baca, G., and Scott, H.R. 1973. Stimulation of cyclic AMP in the isolated perfused rat lung. Res Comm Chem Path Pharmacol 6:805-812

Costa, E., Cheney, D.L., Racagni, G., and Zsilla, G. 1975. An analysis at synaptic level of the morphine action in striatum and N. accumbens. Dopamine and acetylcholine interactions. Life Sci 17:1-8

Cotter, G.W., Palmer, G.C., Palmer, S.J., and Manian, A.A. 1978. Modification of adenylate cyclase systems in mouse cerebral cortex by chlorpromazine and respective 7-hydroxy analogs. Comm Psychopharmacol 2:51-58

Coupet, J., Szucs, V.A., and Greenblatt, E.N. 1976. The effects of 2-chloro-11-(4-methyl-1-piperazinyl)-dibenz (b,f)(1,4) oxazepine (Loxapine) and its derivatives on the dopamine-sensitive adenylate cyclase of rat striatal homogenates. Brain Res 116:177-180

Cramer, H., Ng, L.K.Y., and Chase, T.N. 1972. Effect of probenecid on levels of cyclic AMP in human cerebrospinal fluid. J Neurochem 19:1601-1602

Cramer, H., Ng, L.K.Y., and Chase, T.N. 1973. Adenosine 3′,5′-monophosphate in cerebrospinal fluid. Effect of drugs and neurologic disease. Arch Neurol 29:197-199

Creese, I., Manian, A.A., Prosser, T.D., and Snyder, S.H. 1978. ^{3}H-Haloperidol binding to dopamine receptors in rat corpus striatum. Influence of chlorpromazine metabolites and derivatives. Europ J Pharmacol 47:291-296

Daly, J.W., and Manian, A.A. 1967. The metabolism of hydroxychlorpromazines by rabbit liver microsomes. Biochem Pharmacol 16:2131-2136

Daly, J. 1976. Role of cyclic nucleotides in the nervous system. In *Handbook of Psychopharmacology,* Vol. 5, eds. L.L. Iverson, D.S. Iversen, S.H. Snyder, pp. 47-130. New York: Plenum Publishing Corp.

Daly, J. 1977. *Cyclic Nucleotides in the Nervous System.* New York: Plenum Press

Davis, J.M. 1970. Theories of biological etiology of affective disorders. Int Rev Neurobiol 12:145-175

Dismukes, R.K., and Daly, J.W. 1976. Adaptive responses of brain cyclic AMP-generating systems to alterations in synaptic input. J Cyclic Nucleotide Res 2:321-336

Dunstan, R., and Jackson, D.M. 1976. The demonstration of a change in adrenergic receptor sensitivity in the central nervous system of mice after withdrawal from long-term treatment with haloperidol. Psychopharmacology 48:105-114

Ebstein, R.P., Biederman, J., Rimon, R., Zohar, J., and Belmaker, R.H. 1976. Cyclic GMP in the CSF of patients with schizophrenia before and after neuroleptic treatment. Psychopharmacology 51:71-74

Feinberg, A.P., and Snyder, S.H. 1975. Phenothiazine drugs. Structure-activity relationships explained by a conformation that mimics dopamine. Proc Nat Acad Sci USA 72:1899-1903

Ferrendelli, J.A., Kinscherf, D.A., and Kipnis, D.M. 1972. Effects of amphetamine, chlorpromazine and reserpine on cyclic GMP and cyclic AMP levels in mouse cerebellum. Biochem Biophys Res Comm 46:2114-2120

Fertel, R., and Weiss, B. 1976. Properties and drug responsiveness of cyclic nucleotide phosphodiesterases of rat lung. Molec Pharmacol 12:678-687

Forn, J., and Krishna, G. 1971. Effect of norepinephrine, histamine and other drugs on cyclic 3′,5′-AMP formation in brain slices of various animal species. Pharmacology 5:193-204

Forn, J., Krueger, B.K., and Greengard, P. 1974. Adenosine 3′,5′-monophosphate content in rat caudate nucleus. Demonstration of dopaminergic and adrenergic receptors. Science 186:1118-1120

Fox, M., and Williams, T.D. 1970. The caudate nucleus-cerebellar pathways. An electrophysiological study of their route through the midbrain. Brain Res 20:140–144
Fredholm, B.B. 1977. Decreased adenosine cyclic 3′,5′-monophosphate phosphodiesterase activity in rat striatum following chronic haloperidol treatment. Med Biol 55:61–65
Free, C.A., Paik, V.S., and Shada, D.J. 1974. Inhibition by phenothiazines of adenylate cyclase in adrenal and brain tissue. In *The Phenothiazines and Structurally Related Drugs,* eds. I.S. Forrest, C.J. Carr, E. Usdin, pp. 739–748. New York: Raven Press
Freedman, R., and Hoffer, B.J. 1974. Phenothiazine antagonism of the noradrenergic inhibition of cerebellar Purkinje neurons. J Neurobiol 6:277–288
Freedman, R., Hoffer, B.J., and Siggins, G.R. 1974. Neuroleptic antagonism of catecholamine inhibition in rat cerebellum and caudate. Soc Neurosci 4:213
Freedman, R. 1977. Interactions of antipsychotic drugs with norepinephrine and cerebellar neuronal circuitry. Implications for the psychobiology of psychosis. Biol Psychiat 12:181–197
Friedhoff, A.J., Bonnet, K., and Rosengarten, H. 1977. Reversal of two manifestations of dopamine receptor supersensitivity by administration of L-DOPA. Res Comm Chem Path Pharmacol 16:411–423
Fujimoto, M., and Okabayashi, T. 1975. Proposed mechanisms of stimulation and inhibition of guanylate cyclase with reference to the actions of chlorpromazine, phospholipase and triton X-100. Biochem Biophys Res Comm 67:1332–1336
Garbarg, M., Barbin, G., Feger, J., and Schwartz, J-C. 1974. Histaminergic pathway in rat brain evidenced by lesions of the median forebrain bundle. Science 186:833–835
Gerhards, H.J., Carenzi, A., and Costa, E. 1974. Effect of nomifensine on motor activity, dopamine turnover rate and cyclic adenosine 3′,5′-monophosphate concentrations of rat striatum Naunyn-Schmiedeberg's Arch Pharmacol 286:49–64
Gianutsos, G., Drawbaugh, R.B., Hynes, M.D., and Lal, H. 1974. Behavioral evidence for dopamine supersensitivity after chronic haloperidol. Life Sci 14:887–898
Gnegy, M., Uzunov, P., and Costa, E. 1976. Drug-induced supersensitivity of dopamine receptors and membrane bound adenylate cyclase activator. Pharmacologist 18:185
Gnegy, M.E., Nathanson, J.A., and Uzunov, P. 1977a. Release of the phosphodiesterase activator by cyclic AMP-dependent ATP: Protein phosphotransferase from subcellular fractions of rat brain. Biochim Biophys Acta 497:75–85
Gnegy, M., Uzunov, P., and Costa, E. 1977b. Participation of an endogenous Ca^{++}-binding protein activator in the development of drug-induced supersensitivity of striatal dopamine receptors. J Pharmacol Exp Ther 202:558–564
Goldberg, N.D., O'Dea, R.F., and Haddox, M.K. 1973. Cyclic GMP. In *Advances in Cyclic Nucleotide Research,* Vol. 3, eds. P. Greengard, G.A. Robison, pp. 155–224. New York: Raven Press
Greengard, P., and Kebabian, J.W. 1974. Role of cyclic AMP in synaptic transmission in the mammalian peripheral nervous system. Fed Proc 33:1059–1067
Gumulka, S.W., Dinnendahl, V., Schonhofer, P.S., and Stock, K. 1976. Dopaminergic stimulants and cyclic nucleotides in mouse brain. Naunyn-Schmiedeberg's Arch Pharmacol 295:21–26
Haksar, A., and Peron, F.G. 1971. Chlorpromazine. Inhibition of ACTH and cyclic 3′,5′-AMP stimulated corticosterone synthesis. Biochem Biophys Res Comm 44:1376–1380
Harris, J.E. 1976. *Beta* adrenergic receptor-mediated adenosine cyclic 3′,5′-monophosphate accumulation in the rat corpus striatum. Mol Pharmacol 12:546–558
Henn, F.A., Anderson, D.J., and Sellstrom, A. 1977. Possible relationship between glial cells, dopamine and the effects of antipsychotic drugs. Nature 266:237–238
Hillman, H.H., Campbell, J.R., and McIlwain, H. 1963. Membrane potentials in isolated and electrically stimulated mammalian cerebral cortex, effects of chlorpromazine, cocaine, phenobarbitone and protamine on the tissue's electrical and chemical responses to stimulation. J Neurochem 10:325–339
Hoffer, B.J., Freedman, R., Woodward, D.J., Daly, J.W., and Skolnick, P. 1976. β-adrenergic-control of cyclic AMP-generating systems in cerebellum: Pharmacological heterogeneity confirmed by destruction of interneurons. Exp Neurol 51:653–667

Honda, F., and Imamura, H. 1968. Inhibition of cyclic 3′,5′-nucleotide phosphodiesterase by phenothiazine and reserpine derivatives. Biochim Biophys Acta 161: 267–269

Horn, A.S., Cuello, A.C., and Miller, R.J. 1974. Dopamine in the mesolimbic system of the rat brain: Endogenous levels and the effects of drugs on the uptake mechanism and stimulation of adenylate cyclase activity. J Neurochem 22:265–270

Huang, M., and Daly, J.W. 1972. Accumulation of cyclic adenosine monophosphate in incubated slices of brain tissue. 1. Structure-activity relationships of agonists and antagonists of biogenic amines and of tricyclic tranquilizers and antidepressants. J Med Chem 15:458–462

Huang, M., and Daly, J.W. 1974. Interrelationships among the levels of ATP, adenosine and cyclic AMP in incubated slices of guinea-pig cerebral cortex. Effects of depolarizing agents, psychotropic drugs and metabolic inhibitors. J Neurochem 23:393–404

Iversen, L.L. 1975. Dopamine receptors in the brain. Science 188:1084–1089

Iwatsubo, K., and Clouet, D.H. 1975. Dopamine-sensitive adenylate cyclase of the caudate nucleus of rats treated with morphine or haloperidol. Biochem Pharmacol 24:1499–1503

Jenner, F.A., Sampson, G.A., Thompson, E.A., Somerville, A.R., Beard, N.A., and Smith, A.A. 1972. Manic-depressive psychosis and urinary excretion of cyclic AMP. Brit J Psychiat 121:236–237

Kakiuchi, S., and Rall, T.W. 1968a. The influence of chemical agents on the accumulation of adenosine 3′,5′-phosphate in slices of rabbit cerebellum. Mol Pharmacol 4:367–378

Kakiuchi, S., and Rall, T.W. 1968b. Studies on adenosine 3′,5′-phosphate in rabbit cerebral cortex. Mol Pharmacol 4:379–388

Kakiuchi, S., Rall, T.W., and McIlwain, H. 1969. The effect of electrical stimulation upon the accumulation of adenosine 3′,5′-phosphate in isolated cerebral tissue. J Neurochem 16:485–491

Kakiuchi, S., Yamazaki, R., and Teshima, Y. 1972. Regulation of brain phosphodiesterase activity: $Ca^2 +$ plus $Mg^2 + -$ dependent phosphodiesterase and its activating factor from rat brain. In *Advances in Cyclic Nucleotide Research,* Vol. 1, eds. P. Greengard, R. Paoletti, G.A. Robison, pp. 455–477. New York: Raven Press

Kalisker, A., Rutledge, C.O., and Perkins, J.P. 1973. Effect of nerve degeneration by 6-hydroxydopamine on catecholamine-stimulated adenosine 3′,5′-monophosphate formation in rat cerebral cortex. Mol Pharmacol 9:619–629

Karobath, M., and Leitich, H. 1974. Antipsychotic drugs and dopamine-stimulated adenylate cyclase prepared from corpus striatum of rat brain. Proc Nat Acad Sci USA 71:2915–2918

Karobath, M.E. 1975a. Tricyclic antidepressive drugs and dopamine-sensitive adenylate cyclase from rat brain striatum. Europ J Pharmacol 30:159–163

Karobath, M.E. 1975b. Dopamin-receptor-blockade, ein möglicher Wirkungsmechanismus antipsychotisch wirksamer Pharmaka. Pharmakopsych 8:151–161

Kaufman, J.J., and Kerman, E. 1974. Quantum chemical and other theoretical techniques for the understanding of the psychoactive action of the phenothiazines. In *The Phenothiazines and Structurally Related Drugs,* eds. I.S. Forrest, C.J. Carr, E. Usdin, pp. 55–75. New York: Raven Press

Keatinge, A.M.B., Sinanan, K., and Love, W.C. 1975. Effects of reserpine, chlorpromazine, imipramine and amitriptyline on urinary excretion of adenosine 3′,5′-cyclic monophosphate in rats. Irish J Med Sci 144:249–254

Kebabian, J.W., Petzold, G.L., and Greengard, P. 1972. Dopamine-sensitive adenylate cyclase in caudate nucleus of rat brain, and its similarity to the "dopamine receptor." Proc Nat Acad Sci USA 69:2145–2149

Kebabian, J.W., and Saavedra, J.M. 1976. Dopamine-sensitive adenylate cyclase occurs in a region of substantia nigra containing dopaminergic dendrites. Science 193:683–685

Kinscherf, D.A., Chang, M.M., Rubin, E.H., Schneider, D.R., and Ferrendelli, J.A. 1976. Comparison of the effects of depolarizing agents and neurotransmitters on regional CNS cyclic GMP levels in various animals. J Neurochem 26:527–530

Laduron, P. 1976. Limiting factors in the antagonism of neuroleptics on dopamine-sensitive adenylate cyclase. J Pharm Pharmacol 28:250–251
Laduron, P., Verwimp, M., Janssen, P.F.M., and Leysen, J. 1976. Subcellular localization of dopamine-sensitive adenylate cyclase in rat brain striatum. Life Sci 18:433–440
Langer, G., Ahn, H.S., Perel, J.M., Makman, M.H., and Sachar, E.J. 1977. No effects of quaternary neuroleptics on human prolactin and adenylcyclase. Lancet i:493
Langslet, A. 1971. Effects of chlorpromazine, d,l-propranolol, and d-propranolol on the isolated rat heart. Modification of the response to isoprenaline and glucagon. Europ J Pharmacol 15:164–170
Lee, T-P., Kuo, J.F., and Greengard, P. 1972. Role of muscarinic cholinergic receptors in regulation of guanosine 3′,5′-cyclic monophosphate content in mammalian brain, heart, muscle, and intestinal smooth muscle. Proc Nat Acad Sci USA 69:3287–3291
Levin, R.M., and Weiss, B. 1976. Mechanism by which psychotropic drugs inhibit adenosine cyclic 3′,5′-monophosphate phosphodiesterase of brain. Molec Pharmacol 12:581–589
Levin, R.M., and Weiss, B. 1977. Binding of trifluoperazine to the calcium-dependent activator of cyclic nucleotide phosphodiesterase. Molec Pharmacol 13:690–697
Leysen, J., and Laduron, P. 1977. Differential distribution of opiate and neuroleptic receptors and dopamine-sensitive adenylate cyclase in rat brain. Life Sci 20:281–288
Liu, Y.P., and Cheung, W.Y. 1976. Cyclic 3′,5′-nucleotide phosphodiesterase. J Biol Chem 251:4193–4198
Lust, W.D., Goldberg, N.D., and Passonneau, J.V. 1976. Cyclic nucleotides in murine brain. The temporal relationship of changes induced in adenosine 3′,5′-monophosphate and guanosine 3′,5′-monophosphate following maximal electroshock or decapitation. J Neurochem 26:5–10
Martinez, T.T., and McNeill, J.H. 1977. Cyclic AMP and the positive inotropic effect of norepinephrine and phenylephrine. Can J Physiol Pharmacol 55:279–287
Matthysse, S. 1974. Schizophrenia: Relationships to dopamine transmission, motor control, and feature extraction. In *The Neurosciences*: Third Study Program, eds. F.O. Schmitt, F.G. Worden, pp. 721–732. Cambridge, MA: M.I.T.
Matthysse, S., and Lipinski, J. 1975. Biochemical Aspects of Schizophrenia. Ann Rev Med 26:551–565
Maxwell, D.M. 1975. Inhibition of cyclic adenosine 3′,5′-monophosphate-dependent protein kinase by chlorpromazine. Edgewood Arsenal Technical Report EB-TR-75068:1–9
McDowell, J.J.H. 1974. The Molecular Structures of Phenothiazine Derivatives. In *The Phenothiazines and Structurally Related Drugs*, eds. I.S. Forest, C.J. Carr, E. Usdin, pp. 33–54. New York: Raven Press
McIlwain, H. 1964. Actions of haloperidol, meperidine and related compounds on the excitability and ion content of isolated cerebral tissue. Biochem Pharmacol 13:523–529
Mendels, J., Stern, S., and Frazer, A. 1976. Biochemistry of depression. Dis Nerv System 37:3–9
Miller, R.J., and Hiley, C.R. 1974. Anti-muscarinic properties of neuroleptics and drug-induced parkinsonism. Nature 248:596–597
Miller, R.J., Horn, A.S., and Iversen, L.L. 1974. The action of neuroleptic drugs on dopamine-stimulated adenosine cyclic 3′,5′-monophosphate production in rat neostriatum and limbic forebrain. Molec Pharmacol 10:759–766
Miller, R.J., and Iversen, L.L. 1974. Effect of chlorpromazine and some of its metabolites on the dopamine-sensitive adenylate cyclase of rat brain striatum. J Pharmacol 26:142–144
Mishra, R.K., Demirjian, C., Katzman, R., and Makman, M.H. 1975. A dopamine-sensitive adenylate cyclase in anterior limbic cortex and mesolimbic region of primate brain. Brain Res 96:395–399
Mishra, R.K. 1977. Pre and postsynaptic effects of sulpiride and other substituted benzamides. Fed Proc 36:319
Murad, F. 1973. Clinical studies and applications of cyclic nucleotides. In *Advances in Cyclic Nucleotide Research*, Vol. 3, eds. P. Greengard, G. A. Robison, pp. 355–383. New York: Raven Press

Nakajima, T., Naitoh, F., and Kuruma, I. 1977. Elevation of adenosine 3′,5′-monophosphate in the perfusate of rat kidney after addition of dopamine. Europ J Pharmacol 45:195-197

Nathanson, J.A., and Greengard, P. 1974. Serotonin-sensitive adenylate cyclase in neural tissue and its similarity to the serotonin receptor. A possible site of action of lysergic acid diethylamide. Proc Nat Acad Sci USA 71:797-801

Nathanson, J.A. 1976. Octopamine-sensitive adenylate cyclase and its possible relationship to the octopamine receptor. In *Trace Amines and the Brain*, Vol. 1, eds. E. Usdin and M. Sandler, pp. 161-190. New York: Marcel Dekker, Inc.

Nathanson, J.A. 1977. Cyclic nucleotides and nervous system function. Physiol Rev 57:157-256

Ohga, Y., and Daly, J.W. 1977. Calcium ion-elicited accumulations of cyclic GMP in guinea pig cerebellar slices. Biochim Biophys Acta 498:61-75

Oleshansky, M.A., and Neff, N.H. 1975. Rat pineal adenosine cyclic 3′,5′-monophosphate phosphodiesterase activity: modulation *In Vivo* by a β-adrenergic receptor. Molec Pharmacol 11:552-557

Osborne, N.N. 1977. Adenosine 3′,5′-monophosphate in snail (*Helix pomatia*) nervous system. Experientia 33:917-918

Osnes, J-B., Christoffersen, T., Morland, J., and Oye, I. 1976. Chlorpromazine and hormonal elevation of cyclic AMP contents in turkey erythrocytes and in perfused rat heart and liver. Acta Pharmacol Toxicol 38:195-208

Oye, I., and Langslet, A. 1972. The Role of Cyclic AMP in the Inotropic Response to Isoprenaline and Glucagon. In *Advances in Cyclic Nucleotide Research*, Vol. 1, eds. P. Greengard, R. Paoletti, G.A. Robison, pp. 291-300. New York: Raven Press

Pagel, J. Christian, S.T., Quayle, E.S., and Monti, J.A. 1976. A serotonin sensitive adenylate cyclase in mature rat brain synaptic membranes. Life Sci 19:819-824

Palmer, G.C., Robison, G.A., and Sulser, F. 1971. Modification by psychotropic drugs of the cyclic AMP response to norepinephrine in rat brain. Biochem Pharmacol 20:236-239

Palmer, G.C. 1971. Characteristics of the hormonal induced cyclic adenosine 3′,5′-monophosphate response to the rat and guinea pig lung *in vitro*. Biochim Biophys Acta 252:561-566

Palmer, G.C., and Burks, T.F. 1971. Central and peripheral adrenergic blocking actions of LSD and BOL. Europ J Pharmacol 16:113-116

Palmer, G.C. 1972. Increased cyclic AMP response to norepinephrine in the rat brain following 6-hydroxydopamine. Neuropharmacology 11:145-149

Palmer, G.C., Robison, G.A., Manian, A.A., and Sulser, F. 1972. Modification by psychotropic drugs of the cyclic AMP response to norepinephrine in the rat brain *in vitro*. Psychopharmacologia 23:201-211

Palmer, G.C., Sulser, F., and Robison, G.A. 1973. Effects of neurohumoral and adrenergic agents on cyclic AMP levels in various areas of the rat brain *in vitro*. Neuropharmacology 12:327-337

Palmer, G.C., and Evan, A.P. 1974. Effect of psychotropic drugs on the urinary excretion of cyclic AMP in the rat. Proc West Pharmacol Soc 17:204-209

Palmer, G.C., and Manian, A.A. 1974a. Inhibition of the catalytic site of adenylate cyclase in the central nervous system by phenothiazine derivatives. Neuropharmacology 13:651-664

Palmer, G.C., and Manian, A.A. 1974b. Modification of the receptor component of adenylate cyclase in the rat brain by phenothiazine derivatives. Neuropharmacology 13:851-866

Palmer, G.C., and Duszynski, C.R. 1975. Regional cyclic GMP content in incubated tissue slices of rat brain. Europ J Pharmacol 32:375-379

Palmer, G.C., and Manian, A.A. 1975. Modification of DA-sensitive adenylate cyclase in neuronal and glial-enriched fractions by phenothiazine derivatives. In *Advances in Cyclic Nucleotide Research*, eds. G.I. Drummond, P. Greengard, G.A. Robison, p. 814. New York: Raven Press

Palmer, G.C., and Manian, A.A. 1976. Actions of phenothiazine analogues on dopamine-sensitive adenylate cyclase in neuronal and glial-enriched fractions from rat brain. Biochem Pharmacol 25:63-71

Palmer, G.C., Wagner, H.R., and Putnam, R.W. 1976a. Neuronal localization of the enhanced adenylate cyclase responsiveness to catecholamines in the rat cerebral cortex following reserpine injections. Neuropharmacology 15:695–702
Palmer, G.C., Manian, A.A., and Sanborn, C.R. 1976b. Effects of neuroleptic agents on cyclic GMP in rat cerebral cortex. Europ J Pharmacol 38:205–209
Palmer, G.C., and Wagner, H.R. 1976. Supersensitivity of striatal and cortical adenylate cyclase following reserpine: Lack of effect of chronic haloperidol. Res Comm Psychol Psychiat Behav 1:567–570
Palmer, G.C., Jones, D.J., Medina, M.A., and Stavinoha, W.B. 1977. Influence of injected psychoactive drugs on cyclic AMP levels in mouse brain and lung following microwave irradiation. Neuropharmacology 16:435–443
Palmer, G.C., Wagner, H.R., Palmer, S.J., and Manian, A.A. 1978a. Histamine-, norepinephrine-, and dopamine-sensitive central adenylate cyclases: Effects of chlorpromazine derivatives and butaclamol. Arch Int Pharmacodyn Ther 233:101–112
Palmer, G.C., Jones, D.J., Medina, M.A., Palmer, S.J., and Stavinoha, W.B. 1978b. Actions *In Vitro* and *In Vivo* of chlorpromazine and haloperidol on cyclic nucleotide systems in mouse cerebral cortex and cerebellum. Neuropharmacology 17:491–498
Palmer, G.C., Jones, D.J., Medina, M.A., Palmer, S.J., and Stavinoha, W.B. 1978c. Role of neuroleptic agents on mouse pulmonary cyclic nucleotide systems. Pharmacology 17:280–287
Paul, M.I., Ditzion, B.R., Pauk, G.L., and Janowsky, D.S. 1970. Urinary adenosine 3′,5′-monophosphate excretion in affective disorders. Amer J Psychiat 126:1493–1497
Paul, M.I., Cramer, H., and Goodwin, F.K. 1971. Urinary cyclic AMP excretion in depression and mania. Effects of levodopa and lithium carbonate. Arch Gen Psychiat 24:327–333
Phillipson, O.T., and Horn, A.S. 1976. Substantia nigra of the rat contains a dopamine sensitive adenylate cyclase. Nature 261:418–420
Phillis, J.W. 1977. The role of cyclic nucleotides in the CNS. J Can Sci Neurobiol 4:152–195
Premont, J., Perez, M., and Bockaert, J. 1977. Adenosine-sensitive adenylate cyclase in rat striatal homogenates. FEBS Letters 75:209–212
Pugsley, T.A., Merker, J., and Lippmann, W. 1976. Effect of structural analogs of butaclamol (a new antipsychotic drug) on striatal homovanillic acid and adenyl cyclase of olfactory tubercle in rats. Can J Physiol Pharmacol 54:510–515
Revuelta, A., Uzunov, P., and Costa, E. 1976. Release of phosphodiesterase activator from particulate fractions of cerebellum and striatum by putative neurotransmitters. Neurochem Res 1:217–227
Richelson, E. 1977. Antipsychotics block muscarinic acetylcholine receptor-mediated cyclic GMP formation in cultured mouse neuroblastoma cells. Nature 266:371–373
Richelson, E., and Divinetz-Romero, S. 1977. Blockade by psychotropic drugs of the muscarinic acetylcholine receptor in cultured nerve cells. Biol Psychiat 12:771–785
Roberts, E., and Simonsen, D.G. 1970. Some properties of cyclic 3′,5′-nucleotide phosphodiesterase of mouse brain. Effects of imidazole-4-acetic acid, chlorpromazine, cyclic 3′,5′-GMP, and other substances. Brain Res 24:91–111
Robison, G.A., Butcher, R.W., and Sutherland, E.W. 1968. Cyclic AMP. Ann Rev Biochem 37:149–173
Rodbell, M., Lin, M.C., Salomon, Y., Londos, C., Harwood, J.P., Martin, B.R., Rendell, M., and Berman, M. 1975. Role of adenine and guanine nucleotides in the activity and response of adenylate cyclase systems to hormones. Evidence for multisite transition states. In *Advances in Cyclic Nucleotide Research.* Vol. 5, eds. G.I. Drummond, P. Greengard, G.A. Robison, pp. 3–29. New York: Raven Press
Rotrosen, J., Friedman, E., and Gershon, S. 1975. Striatal adenylate cyclase activity following reserpine and chronic chlorpromazine administration in rats. Life Sci 17: 563–568
Roufogalis, B.D., Thornton, M., and Wade, D.N. 1976. Specificity of the dopamine sensitive adenylate cyclase for antipsychotic antagonists. Life Sci 19:927–934
Sattin, A., and Rall, T.W. 1970. The effect of adenosine and adenine nucleotides on the cyclic adenosine 3′,5′-phosphate content of guinea pig cerebral cortex slices. Molec Pharmacol 6:13–23
Sawaya, M.C.B., Dolphin, A., Jenner, P., Marsden, D.C., and Meldrum, B.S. 1977.

Noradrenergic-sensitive adenylate cyclase in slices of mouse limbic forebrain: Characterization and effect of dopaminergic agonists. Biochem Pharmacol 26:1877–1884
Sayers, A.C., Burki, H.R., Ruch, W., and Asper, H. 1975. Neuroleptic-induced hypersensitivity of striatal dopamine receptors in the rat as a model of tardive dyskinesias. Effects of clozapine, haloperidol, loxapine and chlorpromazine. Psychopharmacology 41:97–104
Scatton, B., Theirry, A.M., Glowinski, J., and Julou, L. 1975. Effects of thioproperazine and apomorphine on dopamine synthesis in the mesocortical dopaminergic systems. Brain Res 88:389–393
Scatton, B., Bischoff, S., Dedek, J., and Korf, J. 1977. Regional effects of neuroleptics on dopamine metabolism and dopamine-sensitive adenylate cyclase activity. Eur J Pharmacol 44:287–292
Schildkraut, J.J., and Kety, S.S. 1967. Biogenic amines and emotion. Science 156: 21–30
Schmidt, M.J., Palmer, E.C., Dettbarn, W-D., and Robison, G.A. 1970. Cyclic AMP and adenyl cyclase in the developing rat brain. Develop Psychobiol 3:53–67
Schorderet, M. 1976. Direct evidence for the stimulation of rabbit retina dopamine receptors by ergot alkaloids. Neurosci Letters 2:87–91
Schorderet, M., and Frangaki, A. 1976. Effects of antiparkinsonian drugs on cAMP accumulation in rabbit retina *in vitro*. Experientia 32:782.
Schubert, D., Tarikas, H., and LaCorbiere, M. 1976. Neurotransmitter regulation of adenosine 3′,5′-monophosphate in clonal nerve, glia, and muscle cell lines. Science 192:471–472
Schultz, J. 1976. Psychoactive drug effects on a system which generates cyclic AMP in brain. Nature 261:417–418
Schwarcz, R., and Coyle, J.T. 1976. Adenylate cyclase activity in chick retina. Gen Pharmacol 7:349–354
Schwartz, J-C. 1975. Histamine as a transmitter in brain. Life Sci 17:503–518
Schwartz, J.P., Morris, N.R., and Breckenridge, B. McL. 1973. Adenosine 3′,5′-monophosphate in glial tumor cells. J Biol Chem 248:2699–2704
Sebens, J.B., and Korf, J. 1975. Cyclic AMP in cerebrospinal fluid. Accumulation following probenecid and biogenic amines. Exp Neurol 46:333–344
Seeber, U., and Kuschinsky, K. 1976. Effects of penfluridol on dopamine-sensitive adenylate cyclase in corpus striatum and substantia nigra of rats. Experientia 32: 1558–1559
Seeman, P., Staiman, A., Lee, T., and Chau-Wong, M. 1974. The membrane actions of tranquilizers in relation to neuroleptic-induced parkinsonism and tardive dyskinesia. In *The Phenothiazines and Structurally Related Drugs,* eds. I.S. Forrest, C.J. Carr, E. Usdin, pp. 137–148. New York: Raven Press
Seeman, P., Chau-Wong, M., Tedesco, J., and Wong, K. 1975. Brain receptors for antipsychotic drugs and dopamine: Direct binding assays. Proc Nat Acad Sci USA 72:4376–4380
Seeman, P., and Lee, T. 1975. Antipsychotic drugs. Direct correlation between clinical potency and presynaptic action on dopamine neurons. Science 188:1217–1219
Seeman, P. 1977. Anti-schizophrenic drugs—Membrane receptor sites of action. Biochem Pharmacol 26:1741–1748
Sheppard, H., and Burghardt, C.R. 1971. The effect of alpha, beta, and dopamine receptor-blocking agents on the stimulation of rat erythrocyte adenyl cyclase by dihydroxyphenelthyamines and their β-hydroxylated derivatives. Molec Pharmacol 7:1–7
Shimizu, H., Daly, J.W., and Creveling, C.R. 1969. A radioisotopic method for measuring the formation of adenosine 3′,5′-monophosphate in incubated slices of brain. J Neurochem 16:1609–1619
Siggins, G.R., Battenberg, E.F., Hoffer, B.J., Bloom, F.E., and Steiner, A.L. 1973. Noradrenergic stimulation of cyclic adenosine monophosphate in rat Purkinje neurons: An immunocytochemical study. Science 179:585–588
Siggins, G.R., Hoffer, B.J., and Ungerstedt, U. 1974. Electrophysiological evidence for involvement of cyclic adenosine monophosphate in dopamine responses of caudate neurons. Life Sci 15:779–792
Simpson, L.L. 1977. The effects of behaviorally relevant doses of chlorpromazine and

molindone on cardiac adenylate cyclase and on myocardial contractility. Biochem Pharmacol 26:1315–1319
Sinanan, K., Keatinge, A.M.B., Beckett, P.G.W., and Love, W.C. 1975. Urinary cyclic AMP in endogenous and neurotic depression. Brit J Psychiat 126:49–55
Skolnick, P., Daly, J.W., Freedman, R., and Hoffer, B.J. 1976. Interrelationship between catecholamine-stimulated formation of adenosine 3′,5′-monophosphate in cerebellar slices and inhibitory effects on cerebellar Purkinje cells: Antagonism by neuroleptic compounds. J Pharmacol Exp Therap 197:280–292
Smith, R.C., and Davis, J.M. 1976. Behavioral evidence for supersensitivity after chronic administration of haloperidol, clozapine, and thioridazine. Life Sci 19:725–732
Snyder, S.H., Banerjee, S.P., Yamamura, H.I., and Greenberg, D. 1974a. Drugs, neurotransmitters, and schizophrenia. Science 184:1243–1253
Snyder, S., Greenberg, D., and Yamamura, H.I. 1974b. Antischizophrenic drugs and brain cholinergic receptors. Arch Gen Psychiat 31:58–61
Snyder, S.H. 1976. The dopamine hypothesis of schizophrenia. Focus on the dopamine receptor. Am J Psychiat 133:197–202
Soudijn, W., and van Wijngaarden, I. 1972. Localization of (^{3}H) pimozide in the rat brain in relation to its antiamphetamine potency. J Pharm Pharmacol 24:773–780
Spano, P.F., and Trabucchi, M. 1976. On the cyclic nucleotides involvement in rat striatal function. Pharmacol Res Comm 8:143–148
Spano, P.F., Dichiara, G., Tonon, G.C., and Trabucchi, M. 1976. A dopamine-stimulated adenylate cyclase in rat substantia nigra. J Neurochem 27:1565–1568
Spiker, M.D., Palmer, G.C., and Manian, A.A. 1976. Action of neuroleptic agents on histamine-sensitive adenylate cyclase in rabbit cerebral cortex. Brain Res 104:401–406
Stone, T.W., Taylor, D.A., and Bloom, F.E. 1975. Cyclic AMP and cyclic GMP may mediate opposite neuronal responses in the rat cerebral cortex. Science 187:845–847
Suran, A.A. 1977. CAMP production in rat cerebral cortex: Effects of histaminergic and alpha adrenergic reagents. Pharmacologist 19:202
Sutherland, E.W., and Robison, G.A. 1966. The role of cyclic 3′,5′-AMP in responses to catecholamines and other hormones. Pharmacol Rev 18:145–162
Tang, L.C., and Cotzias, G.C. 1977. Opposing effects of dopaminergic to cholinergic compounds on a cerebral dopamine-activated adenylate cyclase. Proc Nat Acad Sci USA 74:769–773
Thompson, W.J., and Appleman, M.M. 1971. Multiple cyclic nucleotide phosphodiesterase activities from rat brain. Biochemistry 10:311–316
Tjioe, S.A., Manian, A.A., and O'Neill, J.J. 1974. Effects of hydroxylated phenothiazine metabolites on rat brain mitochondria. In *Phenothiazines and Structurally Related Drugs,* eds. I.S. Forrest, C.J. Carr, E. Usdin, pp. 603–616. New York: Raven Press
Traficante, L.J., Friedman, E., Oleshansky, M.A., and Gershon, S. 1976. Dopamine-sensitive adenylate cyclase and cAMP phosphodiesterase in substantia nigra and corpus striatum of rat brain. Life Sci 19:1061–1066
Ungerstedt, U. 1971. Stereotaxic mapping of the monoamine pathways in the rat brain. Acta Physiol Scand Suppl 367:49–68
Uzunov, P., and Weiss, B. 1971. Effects of phenothiazine tranquilizers on the cyclic 3′,5′-adenosine monophosphate system of rat brain. Neuropharmacology 10:697–708
Uzunov, P., and Weiss, B. 1972. Psychopharmacological agents and the cyclic AMP system of rat brain. In *Advances in Cyclic Nucleotide Research,* Vol. 1, eds. P. Greengard, R. Paoletti, G.A. Robison, pp. 435–453. New York: Raven Press
Uzunov, P., Shein, M.H., and Weiss, B. 1974. Multiple forms of cyclic 3′,5′-AMP phosphodiesterase of rat cerebrum and cloned astrocytoma and neuroblastoma cells. Neuropharmacology 13:377–391
von Hungen, K., and Roberts, S. 1974. Neurotransmitter-sensitive adenylate cyclase systems in the brain. Rev Neurosci 1:232–281
von Hungen, K., Roberts, S., and Hill, D.F. 1974. LSD as an agonist and antagonist at central dopamine receptors. Nature 252:588–589
von Hungen, K., Roberts, S., and Hill, D.F. 1975. Interactions between lysergic acid diethylamide and dopamine-sensitive adenylate cyclase systems in rat brain. Brain Res 94:57–66

Walker, J.B., and Walker, J.P. 1973. Neurohumoral regulation of adenylate cyclase activity in rat striatum. Brain Res 54:386–390

Watanabe, H., and Passonneau, J.V. 1975. Cyclic adenosine monophosphate in cerebral cortex. Arch Neurol 32:181–184

Weinryb, I., Chasin, M., Free, C.A., Harris, D.N., Goldenberg, H., Michel, I.M., Paik, V.S., Phillips, M., Samaniego, S., and Hess, S.M. 1972. Effects of therapeutic agents on cyclic AMP metabolism *in vitro*. J Pharm Sci 61-1556–1567

Weinryb, I., and Michel, I. 1976. Amine-responsive adenylate cyclase activity from brain. Comparisons between rat and rhesus monkey and demonstration of dopamine-stimulated adenylate cyclase in monkey neocortex. Psychopharmacol Comm 2:39–48

Weiss, B., and Costa, E. 1967. Adenyl cyclase activity in rat pineal gland. Effects of chronic denervation and norepinephrine. Science 156:1750–1752

Weiss, B., and Costa, E. 1968. Selective stimulation of adenyl cyclase of rat pineal gland by pharmacologically-active catecholamines. J Pharmacol Exp Ther 161:310–391

Weiss, B., Fertel, R., Figlin, R., and Uzunov, P. 1974. Selective alteration of the activity of the multiple forms of adenosine 3′,5′-monophosphate. Molec Pharmacol 10:615–625

Weiss, B., and Greenberg, L.H. 1975. Cyclic AMP and brain function: Effects of psychopharmacologic agents on the cyclic AMP system. In *Cyclic Nucleotides in Disease,* ed. B. Weiss, pp. 269–320. Baltimore: University Park Press

Weiss, B. 1978. Mechanisms for selectively altering the multiple forms of cyclic nucleotide phosphodiesterase. In *Advances in Cyclic Nucleotide Research,* Vol. 9. New York: Raven Press

Williams, M., and Rodnight, R. 1976. Protein phosphorylation in respiring slices of guinea-pig cerebral cortex. Biochem J 154:163–170

Wolfarth, S., Dulska, E., and Lacki, M. 1974. Comparison of the effects of intranigral injections of cholinomimetics with systemic injections of the dopamine receptor stimulating and blocking agents in the rabbit. Neuropharmacology 13:867–875

Wolfe, B.B., Harden, T.K., and Molinoff, P.B. 1977. *In Vitro* study of β-adrenergic receptors. Ann Rev Pharmacol Toxicol 17:575–604

Wolff, J., and Jones, A.B. 1970. Inhibition of hormone-sensitive adenyl cyclase by phenothiazines. Proc Nat Acad Sci USA 65:454–459

Wolff, J., and Jones, A.B. 1971. The purification of bovine thyroid plasma membranes and the properties of membrane-bound adenyl cyclase. J Biol Chem 246:3939–3947

Wollenman, M. 1974. The action of chlorpromazine on the metabolism of cyclic adenosine 3′,5′-monophosphate. In *The Phenothiazines and Structurally Related Drugs,* eds. I.S. Forrest, C.J. Carr, E. Usdin, pp. 731–738. New York: Raven Press

Yamashita, K., Bloom, G., Rainard, B., Zor, U., and Field, J.B. 1970. Effects of chlorpromazine, propranolol and phospholipase C on thyrotropin and prostaglandin stimulation of adenyl cyclase-cyclic AMP system in dog thyroid slices. Metabolism 19:1109–1118

Yarbrough, G.G. 1975. Supersensitivity of caudate neurons after repeated administration of haloperidol. Europ J Pharmacol 31:367–369

Cyclic AMP and Adenylate Cyclase in Psychiatric Illness

Ghanshyam N. Pandey and John M. Davis
Illinois State Psychiatric Institute
Chicago, Illinois 60612

INTRODUCTION

It has been hypothesized that affective illness and schizophrenia are associated with a dysfunction of biogenic amines. Whereas a functional deficiency of norepinephrine (NE) or serotonin (5-HT) may be involved in certain types of depressive illness, an excess of these amines may be present in mania. Schizophrenia, on the other hand, has been related to the hyperactivity of the dopaminergic system. An abnormality of amine function may be caused by altered levels of the amines as a result of changes in their biosynthesis, release, reuptake or metabolism or due to an imbalance in the transmitter systems. It is also possible that amine dysfunction results from altered sensitivity of the amine receptors.

Several strategies have been used in studies of the abnormality of amine function in psychiatric illness. Studies have been performed on amine and amine metabolites in urine, plasma and cerebrospinal fluid (CSF); on the enzymes associated with the biosynthesis of amines, e.g., tyrosine hydroxylase (TH) and dopamine β-hydroxylase (DBH), and on the enzymes associated with amine metabolism, e.g., monoamine oxidase (MAO) and catechol-*O*-methyl transferase (COMT); in peripheral tissues such as platelets, plasma and red cells; and in postmortem human brain specimens. Indirect studies of biogenic amines have been carried out by means of neuroendocrine strategies, because release of several of the pituitary hormones, e.g., growth hormone (GH) and prolactin, appears to be regulated by the central biogenic amine system. Another indirect strategy for the study of biogenic amine function in psychiatric illness involves the use of the cyclic nucleotide system.

There are several lines of evidence which suggest that cyclic nucleotides play an important role in brain function and in the regulation of mood states. Cyclic AMP, which is formed by the catalytic action of the

---

Some of the work reported was carried out in collaboration with Drs. A. Frazer, J. Mendels and Y. Wang at the Veterans Administration Hospital, Philadelphia, and with Drs. R Casper, D. Garver, W. Heinze and F. Jones at the Illinois State Psychiatric Institute. The authors wish to thank B. Brown and L. Kujovic for technical assistance and P. Hilligoss for help in preparation of the manuscript.

enzyme adenylate cyclase on adenosine triphosphate (ATP), has been shown to mediate the cellular effects of several hormones and neurotransmitters in both peripheral and brain tissues (*Sutherland* and *Robison*, 1966; *Sutherland, Robison,* and *Butcher,* 1968). Neurotransmitters such as NE, whose dysfunction has been hypothesized in affective illness, have been shown to increase the cyclic AMP concentration in brain (*Kakiuchi* and *Rall,* 1968; *Kakiuchi, Rall,* and *McIlwain,* 1969). Cyclic AMP, which has been implicated in synaptic transmission (*Siggins, Hoffer,* and *Bloom,* 1969), has been reported to be decreased in the urine of patients with affective illness. The responsiveness of the enzyme adenylate cyclase to NE, which appears to be closely coupled to and is an integral part of the adrenergic receptors (*Robison, Butcher,* and *Sutherland,* 1969b), is altered by treatment with psychopharmacological agents. In fact, it is observed that the clinical potency of the neuroleptic drugs is directly correlated with their ability to block the dopamine receptors, suggesting that their therapeutic effects are mediated as a result of the blockage of DA-sensitive adenylate cyclase receptors in the brain (*Seeman, Lee,* and *Chau-Wong,* 1976; *Creese, Burt,* and *Snyder,* 1976).

These observations thus suggest that studies of adenylate cyclase and its responsiveness to catecholamines could be used as an index for adrenergic receptor sensitivity. Since the accumulation of cyclic AMP is related to the effective concentration of the amines and to the sensitivity of amine receptors coupled to adenylate cyclase, studies of cyclic AMP may provide useful information about the amine dysfunction in these illnesses.

An absolute or relative deficiency of the amines or a reduced sensitivity of adrenergic receptors would thus result in decreased intracellular levels of cyclic AMP. Initial reports of decreased excretion of cyclic AMP in patients with depression led *Abdullah* and *Hamadah* (1970) to suggest that depressive illness is associated with reduced levels of cyclic AMP in all tissues, including those of the nervous system.

As an explanatory aid for the relationship between catecholamines and the cyclic nucleotide system in the central nervous system, the diagram in Figure 1 of an adrenergic neuron shows the processes mediated by catecholamines and the cyclic nucleotide system and their interaction.

Catecholamines are released from the presynaptic neuron in the synaptic cleft and then bind to the postsynaptic membrane receptors, which are intimately associated with the catalytic subunit of the enzyme adenylate cyclase. It has been shown by direct binding studies with the adrenergic agonist and antagonist, e.g., ^{3}H-dihydroalprenolol (DHA) and hydroxybenzylpindalol (HBP), in brain and peripheral tissues and by stimulation studies of adenylate cyclase that adrenergic receptors are coupled to adenylate cyclase and are, in fact, an integral part of it

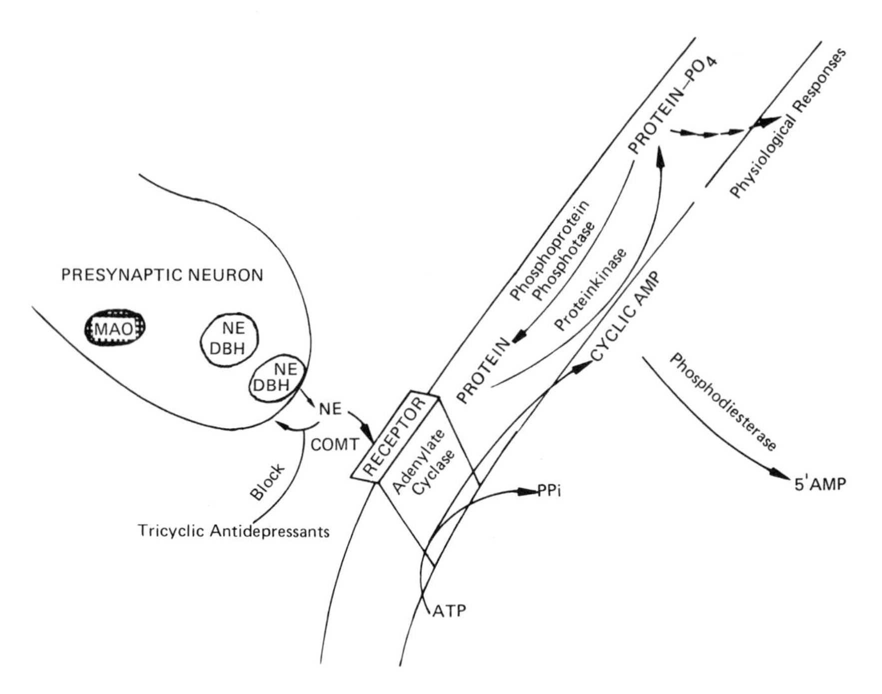

Fig. 1. Schematic representation of a catecholaminergic neuron showing the reactions mediated by catecholamines and cyclic AMP and their interaction.

(see *Nathanson,* 1977; *Lefkowitz,* 1976; *Bylund* and *Snyder,* 1976; *Robison, Butcher,* and *Sutherland,* 1969b). Binding of catecholamines to receptor sites causes stimulation of the enzyme adenylate cyclase, which in turn results in increased synthesis of cyclic AMP from ATP. The cyclic AMP thus formed may then interact with a cyclic AMP-dependent protein kinase, resulting in the catalytic transfer of a phosphate group from ATP to a protein substrate. Dephosphorylation of the phosphoprotein is achieved by the enzyme phosphoprotein phosphatase. Cyclic AMP is ultimately inactivated by its conversion to AMP due to the action of the enzyme phosphodiesterase.

The inactivation of the catecholamines in the synaptic cleft is achieved by their metabolism by the enzyme COMT or by reuptake in the neuron. Several psychotropic agents affect the catecholamine concentrations in the synaptic cleft. Thus, tricyclic antidepressants such as imipramine (Imi) increase the concentration of catecholamines in the synaptic cleft by blocking the reuptake of amines. MAO inhibitors increase the levels of amines by inhibiting their metabolism in the presynaptic neuron.

Chronic treatment with drugs such as 6-hydroxy dopamine and reserpine decreases the catecholamines at the synaptic cleft as a result of the depletion of amines and also causes an increase in adrenergic receptor sensitivity (*Palmer*, 1972; *Iversen*, 1977). *Deguchi* and *Axelrod* (1973) have shown that over- or underexposure of the adrenergic receptors to their agonists causes an alteration in their sensitivity. Since reserpine and 6-OH-DA deplete the amines, underexposure of receptors to the amines thus results in increased sensitivity, as shown by adenylate cyclase and binding studies.

The above description of the events occurring in the neuron, mediated by the catecholamines and the cyclic nucleotides, suggests a close relationship between the two systems. It also suggests that the cyclic nucleotide system could be used as a tool for studies of catecholamine dysfunction in psychiatric disorders. A presynaptic abnormality in the catecholamine system caused by altered synthesis, metabolism, release or reuptake would result in abnormal concentrations of the catecholamines in the synaptic cleft. This, in turn, would create an altered response of the postsynaptic adenylate cyclase-coupled receptor because of changes in catecholamine concentrations in the synaptic cleft, resulting in an altered synthesis of cyclic AMP. On the other hand, it is possible, as suggested by some investigators, that an abnormality in the synaptic receptor sensitivity may be involved in these illnesses, namely, hypersensitive DA receptors in schizophrenia and subsensitive adrenergic receptors in depression. Since these catecholamine receptors are an integral part of adenylate cyclase, it would mean an altered response to catecholamine stimulation by adenylate cyclase, even though the levels of these amines remain unchanged. Thus, a postsynaptic receptor abnormality could be studied by determination of the responses produced by catecholamines on the enzyme adenylate cyclase, or by determination of turnover rates of cyclic AMP.

Initial studies of cyclic nucleotide in affective illness involved the determination of cyclic AMP in the urine of patients. Cyclic AMP levels in plasma and CSF were subsequently studied. Even though earlier results with urinary cyclic AMP studies were encouraging, their interpretation in relation to psychiatric illness remains complicated. In recent years adenylate cyclase and its responsiveness to catecholamines and other hormones have been studied in such illnesses.

For the past several years we have been engaged in investigating the possibility that an abnormality of catecholamine receptors may be associated with schizophrenia and depression. Using a neuroendocrine strategy, we recently demonstrated that hypersensitive DA receptors may be associated with a subgroup of patients diagnosed as acute schizophrenics (*Pandey et al.*, 1977). We have extended the studies by determining the platelet and leukocyte adenylate cyclase levels in patients with these illnesses. Some of our results and related reports on the cyclic nucleotide studies are discussed in this chapter.

Another approach for elucidating the role of cyclic nucleotides in psychiatric disorders is to study the basic underlying mechanisms by which psychotropic drugs and psychopharmacological agents modify the cyclic nucleotide system. Since many such agents may alter the cyclic nucleotide system, either directly by binding on the adrenergic receptor sites or indirectly by altering the availability of the amines to the receptor, the effects of treatment with antidepressants on human tissue and rat brain have been included here.

ANTIDEPRESSANT DRUGS AND RAT BRAIN ADENYLATE CYCLASE

The therapeutic effects of antidepressants have been attributed to their capacity to alter the catecholamine systems. Thus, the tricyclic antidepressants Imi and amitriptyline and some of their demethylated metabolites have been shown to inhibit the reuptake of the biogenic amines in the neuron. MAO inhibitors, on the other hand, increase the levels of the biogenic amines by inhibiting their metabolism. Electroconvulsive therapy (ECT) increases the availability of NE by increasing its turnover and reuptake (*Bliss, Ailion,* and *Zwanziger,* 1968; *Hendley* and *Welch,* 1975). Since a close relationship between the catecholamines and the cyclic nucleotide system has been suggested, the earlier approach for elucidating the association of cyclic nucleotides was directed toward studies of the effects of treatment with psychotropic agents on the brain adenylate cyclase system in experimental laboratory animals *in vitro* and *in vivo*. The significance of some of these studies carried out in our laboratories and in others is described in the following section.

Effect of Imipramine and Triiodothyronine on Adenylate Cyclase of Rat Cerebral Cortex

The tricyclic antidepressant Imi is used in the treatment of certain types of depression. It was reported by *Prange et al.* (1970) that the antidepressant effect of Imi was potentiated by the addition of small amounts of triiodothyronine (T_3) in a group of depressed females. Based on these observations, they suggested that the potentiation of the antidepressant effects of Imi could be caused by alterations produced in the adrenergic receptor sensitivity by T_3. To examine whether treatment with antidepressants such as Imi produced alterations in adrenergic receptor sensitivity and whether these were potentiated by T_3, we studied (in collaboration with Drs. A. Frazer and J. Mendels, 1974) the *in vivo* effects of Imi alone or in conjunction with T_3 on the adenylate cyclase activity and its responsiveness to NE in rat cerebral cortex.

Under *in vitro* conditions, Imi has been reported to increase the basal adenylate cyclase activity of guinea pig cortical slices (*Kodama et al.*, 1971). In rat cerebral cortex, however, although we did not observe that Imi had a significant effect on basal adenylate cyclase activity, it caused a significant reduction in NE-stimulated adenylate cyclase activity (Table 1). Palmer and co-workers (*Palmer*, 1976; *Palmer, Robison,* and *Sulser*, 1971; *Palmer et al.*, 1972) reported that antidepressants such as Imi and amitryptyline have similar effects on rat cortical slices. T_3 had no significant effect on either basal or NE-stimulated activity under *in vitro* conditions. For *in vivo* studies, rats were injected with Imi (20 mg/kg, i.p.), T_3 (15 μg, i.m.), or a combination once a day for 5 successive days. Rats that were injected only with the vehicle were used as controls. Adenylate cyclase activity and its responsiveness to NE and isoproterenol (ISO) were determined in the cortical slices essentially according to the procedure of *Shimizu, Daly,* and *Creveling* (1969) with the pulse-labeling technique. The results were expressed as percent conversion of total [^{3}H] nucleotides incorporated in the slices to [^{3}H] cyclic AMP, which was used as an index of adenylate cyclase activity.

Under *in vitro* conditions, additions of NE to the cortical slices produced a dose-dependent increase in adenylate cyclase activity (Fig. 2). NE-stimulated adenylate cyclase activity was significantly lower in the cortical slices of rats treated with Imi than in those of control rats. Conversely, treatment with T_3 alone did not produce significant changes in the NE stimulation of the adenylate cyclase activity as compared to

Table 1. Effect of Imipramine *In Vitro* on NE- and Isoproterenol-induced Stimulation of [^{3}H]-Cyclic AMP Net Synthesis

Drug	n	Percentage Production of [^{3}H]-cyclic AMP	*P** (compared with NE)
None	4	0.24 ± 0.03†	< 0.01
Imipramine‡	4	0.23 ± 0.04	< 0.005
Norepinephrine‡	4	0.94 ± 0.12	—
Isoproterenol§	4	0.60 ± 0.09	< 0.005
Imipramine and Norepinephrine	4	0.70 ± 0.10	< 0.005
Imipramine and Isoproterenol	4	0.59 ± 0.09	< 0.01 > 0.1§§

*Paired *t* test (two-tailed)
†Mean ± SEM
‡2 × 10^{-4} M
§4 × 10^{-4} M (±) isoproterenol
§§Compared with isoproterenol value
Reproduced from *Frazer et al.* (1974)

the control rats. The NE-stimulated activity in the cortical slices of rats treated with Imi plus T_3 also was lower than that in control rats. However, this decrease was slightly lower than that produced by Imi alone (Fig. 2). These results suggested that activity with T_3 plus Imi was slightly lower than that produced by Imi alone (Fig. 2), and that T_3 did not potentiate the inhibitory effect of Imi on NE-stimulated cortical adenylate cyclase activity. Treatment with Imi or T_3, alone or in combination, did not alter basal adenylate cyclase activity.

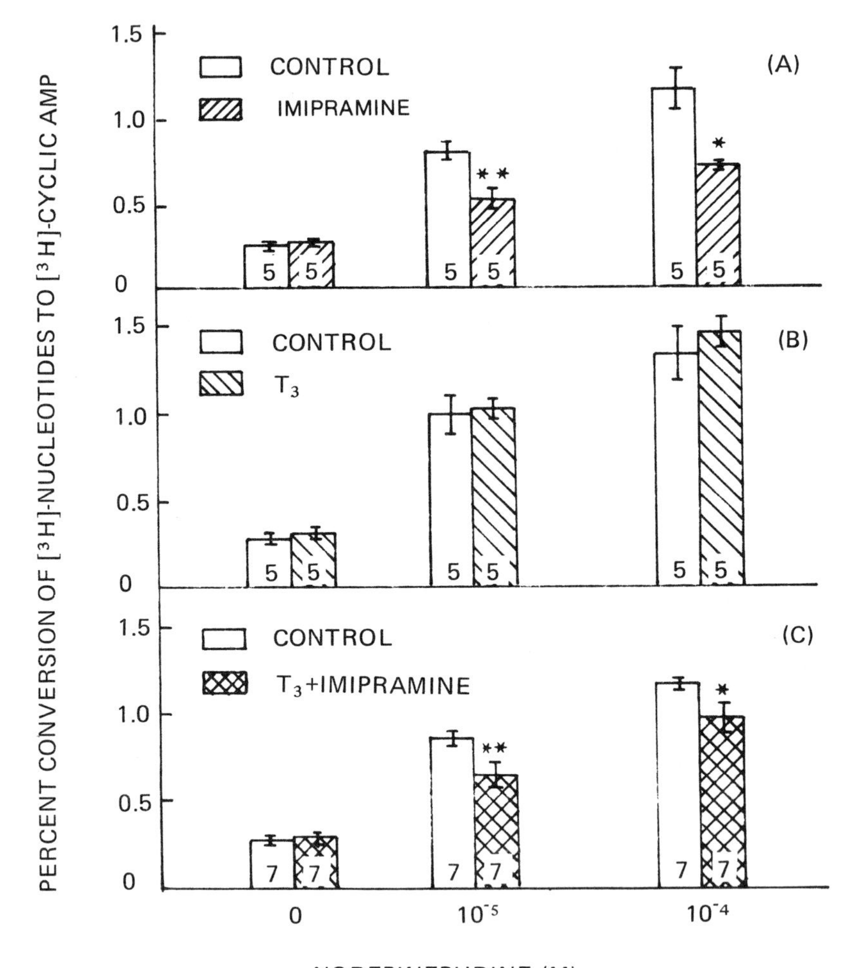

Fig. 2. The effect of imipramine (20 mg/kg) for 5 days (A), triiodothyronine (15 g) for 5 days (B), or imipramine plus triiodothyronine for 5 days (C) on NE-stimulated production of [^{3}H]-cyclic AMP. The bars and brackets represent the mean and SEM, respectively; the number of observations is indicated in each bar. *$P < 0.05$; ***$P < 0.01$, paired t test. Reproduced from *Frazer et al.* (1974).

The important finding in this investigation was that treatment with the antidepressant Imi causes a reduction in the NE-stimulated adenylated cyclase activity in the rat cerebral cortex. Treatment with T_3, which has been suggested to potentiate the antidepressant effect of Imi, did not produce significant changes in the basal adenylate cyclase activity or in its responsiveness to NE in control or Imi-treated rats.

Effect of Electroconvulsive Therapy on Rat Brain Adenylate Cyclase

Electroconvulsive therapy has been widely used in the treatment of depression. Since we observed a significant reduction in NE-sensitive adenylate cyclase in rat cortical slices after treatment with Imi, it was of interest to examine the effect of treatment with ECT on the adenylate cyclase system. If clinical response to treatment with antidepressants is mediated by adenylate cyclase, it would be expected that ECT, which is not an antidepressant drug but is useful in the treatment of depression, would have similar effects.

We recently completed a study in which the effect of ECT on adenylate cyclase activity and its responsiveness to NE were examined (*Pandey et al.*, 1976). For acute studies, male Fisher rats were shocked for 0.5 sec at 25 mamp through corneal electrodes and were killed 1 hr after the last shock. Control rats were treated similarly, but did not receive the shock. For chronic studies, the rats were shocked as above, once a day for 12 days, and were killed either one-half or 24 hr after the last shock. Adenylate cyclase in the cortical slices was assayed as described previously.

Acute treatment with ECT caused a significant increase in the basal adenylate cyclase activity, as shown in Table 2. The NE- and isoproterenol-stimulated adenylate cyclase activity in the slices remained unchanged, however, after an acute shock of ECT. When the rats were treated

Table 2. Effect of acute ECT treatment on rat brain adenylate cyclase

Treatment	Basal †	Adenylate Cyclase* NE (10 μM)-stimulated ‡	ISOP (10 μM)-stimulated ‡
Control	0.89 ± 0.10	1.07 ± 0.08	0.45 ± 0.07
Shocked	1.32 ± 0.13	1.04 ± 0.10	0.56 ± 0.05

*Adenylate cyclase activity expressed as percent conversion of total ^3H-incorporated nucleotides in the cortical slices to ^3H-cyclic AMP.
†$P > 0.001$
‡Not significant
Derived from data in *Pandey et al.*, 1976.

chronically for 12 days with ECT and killed one-half hr after the last shock, the basal adenylate cyclase activity in the shocked rats was significantly lower than in the control rats (Table 3). This reduction in basal adenylate cyclase activity was observed as much as 24 hr after the last shock in chronically treated rats (Table 3). The NE-stimulated adenylate cyclase activity in the chronically treated rats was significantly lower than that in control rats, although the reduction in the NE-stimulated adenylate cyclase was much more significant in rats killed one-half hr after the last shock. On the other hand, the isoproterenol-stimulated adenylate cyclase activity in chronically treated rats killed either one-half hr or 1 day after the last shock was not significantly different from that in control rats (Table 3).

The results obtained on NE-stimulated adenylate cyclase activity of rat cerebral cortex after treatment with ECT are similar to those observed after treatment with Imi; however, the results on basal adenylate cyclase activity differ from those observed with Imi. The increase in basal adenylate cyclase in acutely shocked rats probably is caused by either increased release of NE or increased synthesis of adenosine, which in turn results in increased synthesis of cyclic AMP. These results appear to be in agreement with those obtained by *Lust, Goldberg,* and *Passonneau* (1976) who observed increased levels of cyclic AMP in murine brain after acute ECT. The decrease in basal adenylate cyclase activity in chronically treated rats probably is caused by depletion of NE and adenosine in the slices.

The results described here indicate that two different forms of treatment for depression have similar effects on NE-sensitive adenylate cyclase. These observations are very similar to those reported by Vetulani and co-workers (*Vetulani* and *Sulser,* 1975; *Vetulani, Stawarz,* and *Sulser,* 1976b; *Vetulani et al.,* 1976a), who studied the effects of treatment with ECT, several antidepressant drugs (e.g., desmethylimipramine and iprindole) and MAO inhibitors (e.g., pargyline and nialamide). They observed that in slices obtained from rat limbic forebrain, NE-

Table 3. Effect of Chronic Treatment with ECT on Rat Brain Adenylate Cyclase

Treatment	Basal P	NE (10 μM)-stimulated	ISOP-stimulated
Control (14)	1.3	1.37	0.63
Shocked ½ hr (14)	1.0*	0.74†	0.46
Shocked 24 hr (14)	0.83†	1.06	0.70

Numbers in parentheses represent number of rats studied in each group.
P values are those obtained as compared to control rats.
*$P < 0.025$
†$P < 0.001$

sensitive adenylate cyclase was significantly reduced after chronic treatment with desmethylimipramine, iprindole (Tables 4 and 5) and MAO inhibitors. Chronic treatment with ECT caused similar changes in NE-sensitive adenylate cyclase (Table 6). In acute experiments, treatment with MAO inhibitors resulted in increased synthesis of cyclic AMP in response to NE (Table 7).

These reports suggest that chronic treatment with antidepressant drugs has a common effect of reducing the NE-stimulated increase in cyclic AMP synthesis, implying a decreased response to NE of the brain

Table 4. Effect of Short-term (1-2 wk) Treatment with Desipramine or Iprindole on Response of the Cyclic AMP Generating System in Rat Limbic Forebrain to NE

	n	Time of Sacrifice (hr)[a]	Basal Level of Cyclic AMP (pmoles/mg protein ± SEM)	Cyclic AMP Response to NE (pmoles/mg protein ± SEM)[b]	Control Response (%)
Control	23	1 or 24	18.0 ± 1.7	19.4 ± 3.2	100
Desipramine	8	1	14.7 ± 1.7	14.6 ± 3.3	75
Desipramine	28	24	14.5 ± 1.4	16.6 ± 3.2	85
Iprindole	8	1	12.8 ± 1.8	15.6 ± 5.6	80
Iprindole	17	24	18.6 ± 2.0	21.1 ± 4.3	108

[a]Time after last injection.
[b]Difference in the level of cyclic AMP between the preparation exposed to 5 μM NE and that of the control preparation (corresponding hemisection).
Reproduced from *Vetulani et al.* (1976a)

Table 5. Effect of Long-term (4-8 wk) Treatment with Desipramine or Iprindole on Response of the Cyclic AMP Generating System in Rat Limbic Forebrain to NE

	n	Time of Sacrifice (hr)[a]	Basal Level of Cyclic AMP (pmoles/mg protein ± SEM)	Cyclic AMP Response to NE (pmoles/mg protein ± SEM)[b]	Percent of Control Response
Control	15	1 or 24	17.8 ± 2.6	20.4 ± 2.7	100
Desipramine	12	1	20.5 ± 2.7	9.9 ± 3.5*	49
Desipramine	14	24	16.6 ± 1.6	6.9 ± 2.1***	34
Iprindole	13	1	22.3 ± 3.6	9.4 ± 4.7*	46
Iprindole	15	24	16.9 ± 1.5	7.9 ± 2.4**	38

[a]Time after last injection.
[b]Difference in the level of cyclic AMP between the preparation exposed to 5 μM NE and that of the control preparation (corresponding hemisection).
*$P < 0.05$; **$P < 0.01$; ***$P < 0.001$ (difference from control response; Student's *t* test).
Reproduced from *Vetulani et al.* (1976a)

Table 6. Effect of ECT on Response of the Cyclic AMP Generating System in Rat Limbic Forebrain to NE

	n	Time of Sacrifice (days)[a]	Basal Level of Cyclic AMP (pmoles/mg protein + SEM)		Cyclic AMP Response to NE (pmoles/mg protein + SEM)[b]		Percent of Control Response
Handling	20	1	35.4	1.9	57.9	5.5	100
ECT	20	1	34.1	1.2	40.3	3.7***	70
Handling	5	2	30.9	3.3	88.4	6.2	100
ECT	5	2	32.6	1.3	44.3	7.8	50
Handling	10	3	22.6	1.3	50.4	4.4	100
ECT	7	3	32.4	2.8	26.7	5.8	53
Handling	6	8	20.1	1.2	37.7	3.4	100
ECT	6	5	18.6	1.4	27.5	2.7*	73
ECT	7	8	22.4	2.7	26.2	3.1*	69

[a]Time after ECT or handling.
[b]Difference in the level of cyclic AMP between the preparation exposed to 10 μM NE and that of the control preparation (corresponding hemisection).
*$P < 0.005$; **$P < 0.01$; ***$P < 0.001$ (difference from control response; Student's t test).
Data from *Vetulani et al.* (1976a)

Table 7. Effect of Treatment with MAO Inhibitors on the Cyclic AMP Response to NE in the Limbic Forebrain and on the Level of NE in the Brain

Daily Doses		Cyclic AMP pmol/mg Protein ± SEM		Reactivity Index[1]	NE level (μg/g)	Percent of Control
		Basal Level	5 μM—NE			
1	Saline	25.4 ± 2.9(5)	74.6 ± 4.6(12)	100	0.37 ± 0.03(8)	100
	Pargyline	25.0 ± 3.0(5)	114.6 ± 10.4(12)†	182	0.66 ± 0.04(6)‡	178
	Nialamide	28.0 ± 6.3(5)	141.8 ± 24.9(12)*	231	0.50 ± 0.05(6)*	135
3	Saline	28.4 ± 4.8(4)	53.6 ± 5.4(9)	100	0.30 ± 0.02(12)	100
	Pargyline	28.6 ± 4.7(4)	53.6 ± 4.7(9)	99	0.52 ± 0.05(7)‡	173
	Nialamide	28.0 ± 2.8(4)	54.3 ± 3.8(8)	104	0.55 ± 0.02(7)‡	183
7	Saline	26.9 ± 5.5(4)	50.3 ± 4.21(8)	100	0.30 ± 0.02(12)	100
	Pargyline	18.8 ± 1.8(4)	51.1 ± 9.0(9)	138(N.S.)	0.65 ± 0.06(5)‡	217
	Nialamide	24.6 ± 0.9(4)	40.9 ± 3.6(9)	70	0.53 ± 0.03(7)‡	177
14	Saline	33.6 ± 2.9(4)	51.3 ± 4.6(10)	100	0.32 ± 0.02(7)	100
	Pargyline	32.4 ± 4.2(4)	57.6 ± 6.1(9)	142(N.S.)	0.58 ± 0.05(7)‡	181
	Nialamide	51.0 ± 1.4(3)†	64.4 ± 6.9(9)	76	0.61 ± 0.02(7)‡	191
21	Saline	33.1 ± 5.0(5)	78.1 ± 5.3(18)	100	0.31 ± 0.03(8)	100
	Pargyline	26.9 ± 4.1(6)	57.2 ± 6.4(12)*	67	0.67 ± 0.03(6)‡	216
	Nialamide	32.7 ± 4.1(6)	42.9 ± 3.0(12)‡	23	0.66 ± 0.04(9)‡	212

Animals were killed 18 hr after the last injection. Nialamide was given at a dose of 100 mg/kg i.p. (primer dose) followed by daily doses of 40 mg/kg; the doses of pargyline were 75 and 25 mg/kg, respectively.

[1]Calculated with the formula: $\frac{\text{Stimulated level (Drug)—Basal level (Drug)}}{\text{Stimulated level (Control)—Basal level (Control)}} \times 100$

*P < 0.05
†P < 0.01
‡P < 0.001
N.S. = not significant

Reproduced from *Vetulani et al.* (1976a)

adenylate cyclase. Since the stimulation produced by NE in cyclic AMP synthesis has been shown to be primarily a β-adrenergic response, antidepressant treatment seems to cause a reduction in the sensitivity of β-adrenergic receptors in the rat brain.

In recent years, radioligand binding techniques have been developed for studies of catecholamine receptors in several tissues. Using the radiolabeled β-adrenergic antagonist ^{3}H-dihydroalprenolol (^{3}H-DHA), *Lefkowitz* (1976) and *Bylund* and *Snyder* (1976) studied the characteristics of β-adrenergic receptors in several tissues. Radiolabeled hydroxybenzylpindolol has also been used for such studies (*Brown et al.*, 1976). By means of these techniques, it has been possible to study the affinity and density of β-adrenergic receptors, and *Banerjee et al.* (1977) reported recently that prolonged treatment with desmethyl-imipramine causes a significant reduction in the density of β-adrenergic receptors (binding sites of ^{3}H-DHA) in the rat brain without causing significant changes in the affinity of these receptors (Table 8). This reduction in the density of β-adrenergic receptors is also observed in chronic treatment

Table 8. Effect of Chronic Administration of Antidepressants on Specific Dihydroalprenolol Binding

Drug Treatment	Specific [^{3}H]-Dihydroalprenolol binding (pmoles/g protein)	
None	70.0 ± 3.5	
Desipramine	24.2 ± 1.5	(P < 0.001)
Iprindole	40.4 ± 2.8	(P < 0.001)
Doxepin	48.6 ± 2.4	(P < 0.001)

Desipramine, iprindole and doxepin were injected intraperitoneally in male Sprague-Dawley rats weighing 180–220 g at a dose of 10 mg per kg body weight daily for 6 wk. The rats were killed by decapitation 1 hr after the last injection and the brains were rapidly removed and chilled in ice-cold 0.9% NaCl. The whole brain minus cerebellum was homogenized using a Brinkman Polytron in 20 volumes of ice-cold 0.05 M Tris buffer (pH 8.0 at 25 °C), and centrifuged at 49,000 g for 15 min. The pellet was rehomogenized in the same buffer and centrifuged as before. The pellet was finally resuspended in 50 volumes of Tris buffer (pH 8.0 at 25 °C) and used fresh for the binding assay. ^{3}H-dihydroalprenolol (32 Ci mmol^{-1}) binding was determined at 23 °C by combining 0.97 ml of the above particulate suspension, 4.0 nM ^{3}H-dihydroalprenolol and sufficient distilled water to make 1 ml as the final volume of the reaction mixture. After 20 min incubation, the reaction mixtures were filtered under reduced pressure through Whatman glass fibers (GF/B). The filters were rinsed four times with 4 ml ice-cold Tris buffer to remove most of the unbound radioactive ligand. The filter papers were dried and placed in a counting vial containing 10 ml of Scintiverse (Fisher) and counted in a Packard tri-carb liquid scintillation spectrometer (Model 3380) at 30% efficiency. Corrections were made for nonspecific accumulation of radioactivity by assaying parallel incubations which contained a large excess (100 μM) of noradrenaline. Specific binding, defined as the difference between total and nonspecific radioactivity, was 60–70% of the total binding. Values are means ± SEM. Reproduced from *Banerjee et al.* (1977).

with iprindole and doxepin. Sarai and his associates (*Sarai et al.*, 1978) have reported a significant decrease in the density of β-adrenergic receptors in rat brain after chronic treatment with desmethylimipramine (desipramine or DMI). The reduced number of β-adrenergic receptors observed after chronic treatment with the antidepressants reported in the above studies is thus consistent with earlier reports of diminished accumulation of [^{3}H]-cyclic AMP by addition of NE in brain slices obtained from rats chronically treated with tricyclics and other antidepressants. The interaction of phenothiazine and other neuroleptics with the adenylate cyclase system is discussed in a separate chapter of this volume (*Palmer* and *Manian,* Actions of Neuroleptic Agents on Central Cyclic Nucleotide Systems).

In summary, animal experiments suggest that repeated administration of antidepressants results in the subsensitivity of β-adrenergic receptors in the brain. Whether or not a relationship between the development of subsensitivity of β-adrenergic receptors and clinical response exists is a matter of speculation. The therapeutic actions of the antidepressants presumably are mediated by their pharmacological actions on the catecholamine systems. Thus, acute administration of tricyclics significantly inhibits the uptake of NE in the synaptosome and slows the turnover of catecholamines. Prolonged administration of tricyclic drugs causes a reduction of tyrosine hydroxylase in the brain. MAO inhibitors cause an increase in the levels of NE in the brain, and ECT results in an increased release of NE. Thus, different forms of antidepressant treatment have different pharmacological effects on the catecholamine system but have a common effect on the adenylate cyclase system. With prolonged administration, all decrease the number of β-adrenergic receptor sites. The decrease in the β-adrenergic receptor sites and the reduction in NE-sensitive cyclase are probably caused by overexposure of the receptors to NE. It has been reported that exposure of the receptors to elevated concentrations of agonists results in subsensitivity of the receptor to the agonist (*Deguchi* and *Axelrod*, 1973). There may be some association between the development of subsensitivity of adrenergic receptors and clinical response, but the existence of such a relationship is yet to be demonstrated.

CLINICAL STUDIES

Platelet Adenylate Cyclase

In recent years there have been suggestions, reinforced by *Ashcroft et al.* (1972), of a receptor defect in affective illness. Because the enzyme adenylate cyclase has been shown to be closely related to the adrenergic and dopaminergic receptors, and because cyclic AMP is associated with the intracellular transmission of a receptor stimulus, it is logical to study the metabolism of cyclic AMP in patients with psychiatric disorders. Because of the inaccessibility of the human brain, peripheral tissues and fluids have been used for such studies. Strategies for study of cyclic nucleotides in psychiatric disorders involve the enzyme adenylate cyclase and its responsiveness to catecholamines and other hormones in platelets, leukocytes and postmortem brain, and cyclic AMP levels in urine, plasma and CSF. Some of our studies, and others which are related, are discussed in this section.

The presence of adenylate cyclase in human platelets has been reported by several investigators. Platelet adenylate cyclase can be stimulated by prostaglandin E_1 (PGE_1) (*Robison, Arnold,* and *Hartmann*, 1969a; *Wang et al.*, 1974a), prostacyclin (*Gorman*, *Bunting*, and *Miller*, 1977a & b), and several other agents (*Deykin* and *Parris,* 1972). Several prostaglandin analogues also have been shown to stimulate platelet adenylate cyclase activity, although PGE_1 appears to be the most potent. The PGE_1 stimulation of platelet adenylate cyclase can be partially inhibited by NE (Fig. 3). The inhibition produced by NE on PGE_1 stimulation of platelet adenylate cyclase is blocked by the α-adrenergic antagonist phentolamine (*Robison, Arnold,* and *Hartmann,* 1969a; *Wang et al.*, 1974a). The NE-induced inhibition of PGE_1-stimulated platelet adenylate cyclase, which is blocked by phentolamine, suggests that the responses produced by NE in human platelets are primarily α-adrenergic responses. The presence of α-adrenergic receptors in human platelets also has been confirmed by direct binding studies with radiolabeled antagonists. *Kafka*, *Tallman*, and *Smith* (1978) recently reported that the binding characteristics of human platelet fragments to ^3H-dihydroergocryptive suggest the presence of α-adrenergic receptors in human platelets.

These observations suggest that platelets, which are easily available and homogeneous, could be used as a model for the study of adrenergic receptor sensitivity in a variety of etiological conditions such as affective illness and schizophrenia. While the platelet obviously does not provide a measure of neuronal receptor sensitivity, it has been used widely in studies of biogenic amine function, including biogenic amine uptake and release and the catecholamine-metabolizing enzyme MAO.

Three studies on platelet adenylate cyclase in psychiatric disorders have been reported. In an initial study reported by us (*Wang et al.*, 1974b), *in vitro* measurements of basal adenylate cyclase and its responsiveness to PGE_1 and NE were made on 11 patients with affective disorders and on 8 normal control subjects. The pulse-labeling technique described by *Kuo* and *De Renzo* (1969) in fat cells and by *Shimizu*, *Daly*, and *Creveling* (1969) in brain slices was used as an index of

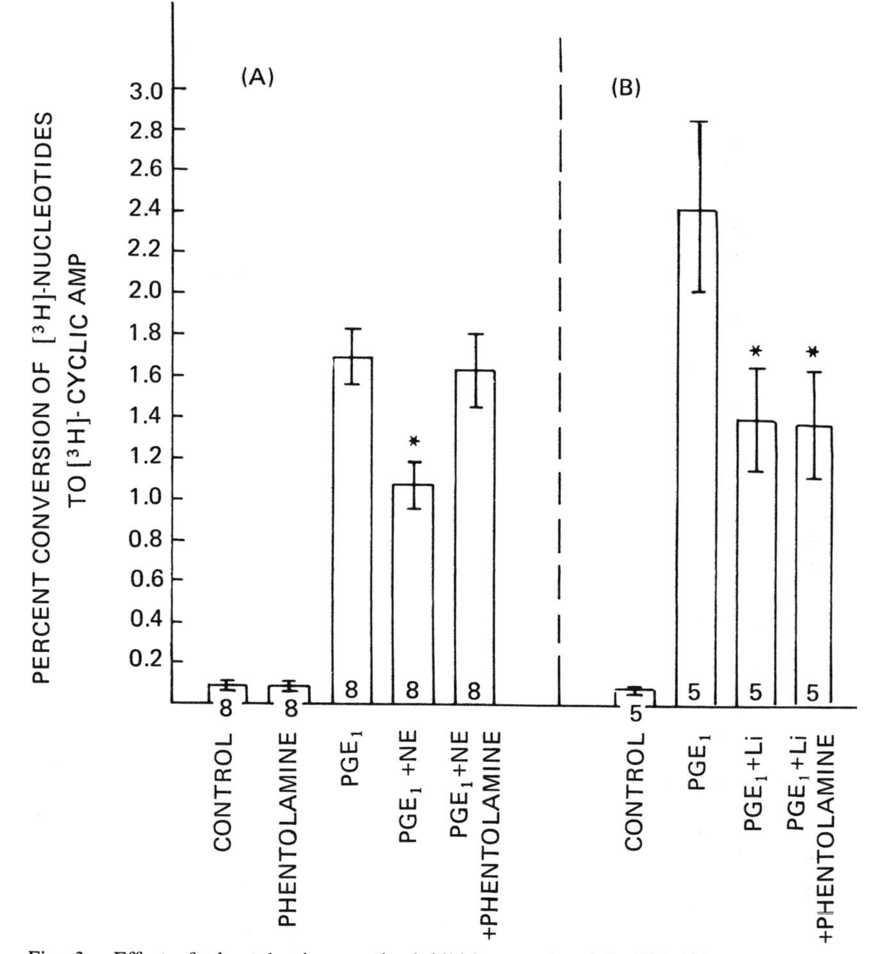

Fig. 3. Effect of phentolamine on the inhibition produced by NE (A) and Li (B) on PGE_1-induced accumulation of [^{3}H]-cyclic AMP in intact platelets. The bars and brackets represent the mean and SEM, respectively; the number of experiments is indicated in each bar. Drugs were used in the following concentrations: PGE_1, 2×10^{-5}M; Li, 36 mM; NE_1, 2×10^{-5}M; phentolamine, 2×10^{-4}M. Li was added during the incubation with [^{3}H] adenine; the other drugs were added during the 2 min reaction. Asterisks indicate values significantly different from the corresponding PGE_1 mean value ($P < 0.01$, paired t test. Reproduced from *Wang et al.* (1974a) with permission.

adenylate cyclase activity. The activity is expressed as percent conversion of total ^{3}H nucleotides to ^{3}H cyclic AMP (for details, see *Wang et al.*, (1974b).

The basal platelet adenylate cyclase activity determined in 11 patients with depression did not differ significantly from that of 8 normal control volunteers. *In vitro* addition of PGE_1 caused a significant, dose-dependent increase in platelet adenylate cyclase, as shown in Figure 4. However,

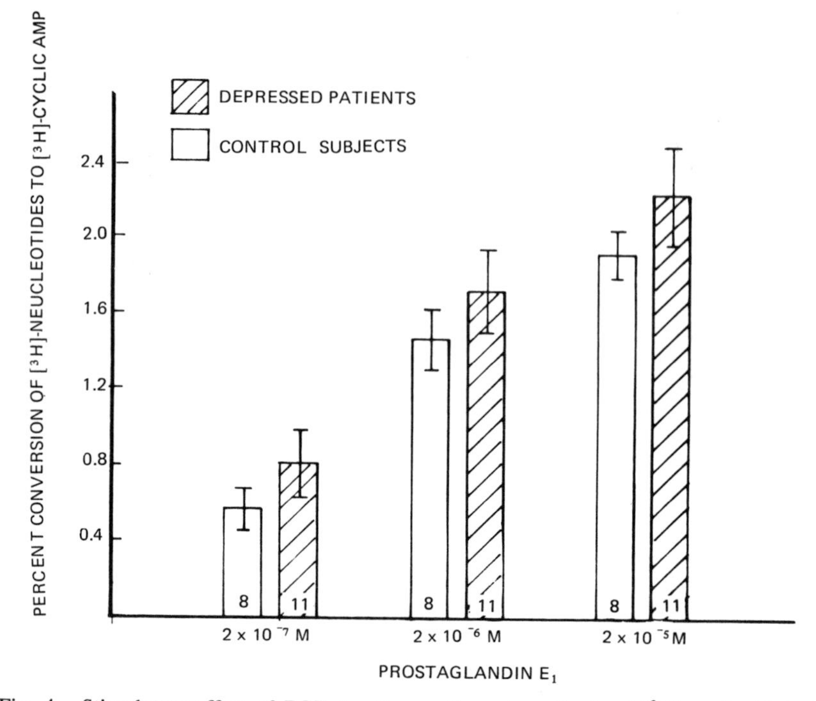

Fig. 4. Stimulatory effect of PGE *in vitro* on the net synthesis of [^{3}H]-cyclic AMP in intact platelets obtained from depressed patients and control subjects. The bars and brackets represent the mean and SEM, respectively; the number of subjects is indicated in each bar. Reproduced from *Wang et al.* (1974b).

the PGE_1-stimulated adenylate cyclase activity (determined under three different conditions) in depressed patients was not significantly different from that in normal controls. Addition of NE produced a significant, dose-dependent decrease in PGE_1-stimulated adenylate cyclase activity (Fig. 5). However, the degree of inhibition produced by NE in the depressed patients was not significantly different from that in normal controls.

A similar study has been reported by *Murphy*, *Donnelly*, and *Moskowitz* (1974) on unipolar and bipolar depressives and normal control subjects. In contrast to the study of *Wang et al.* (1974b), most of the subjects in this study were women. Stimulation of platelet adenylate cyclase (^{3}H-cyclic AMP accumulation) by PGE_1 and inhibition by NE were studied. As shown in Table 9, no significant differences were observed between the patients and normal controls, either in the PGE_1-stimulated increase in cyclic AMP synthesis or in the NE inhibition of PGE_1-stimulated increase in cyclic AMP synthesis.

In these reports, only patients with depression were studied. Since the

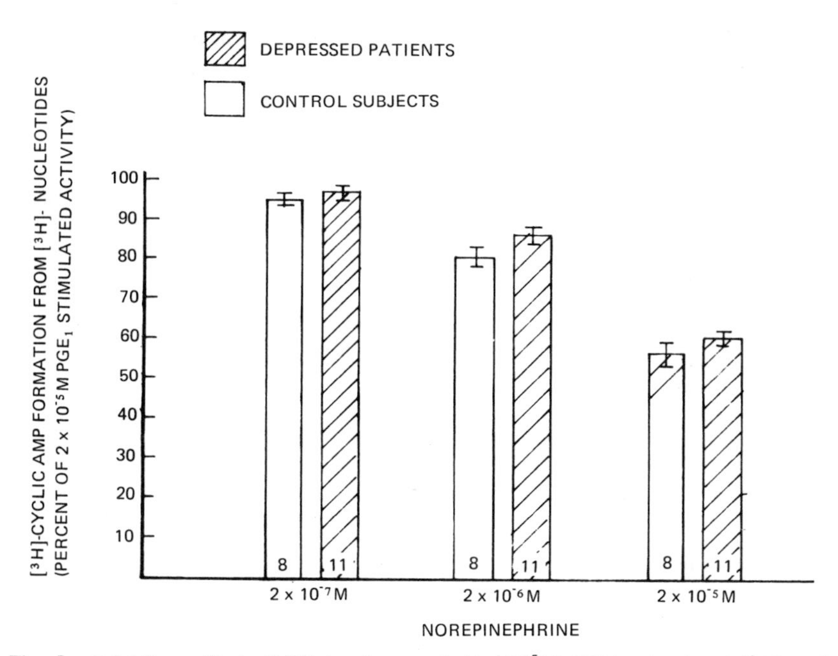

Fig. 5. Inhibitory effect of NE *in vitro* on 2×10^{-5}M PGE_1-stimulated [^{3}H]-cyclic AMP formation in intact platelets obtained from depressed patients and control subjects. The response to PGE_1 has been taken as 100%. In the control subjects, the net synthesis of [^{3}H] cyclic AMP was $1.92 \pm 0.13\%$ in the presence of PGE_1, whereas it was 2.24 $\pm$ 0.27 in the depressed patients ($P < 0.3$). This represents approximately a twentyfold stimulation over basal values. The bars and brackets represent the mean and SEM, respectively; the number of subjects is indicated in each bar.

results were negative, it seemed useful to study patients with schizophrenia and bipolar mania. We recently examined platelet adenylate cyclase and its responsiveness to PGE_1 and NE in such patients in a study conducted in the research department of the Illinois State Psychiatric Institute (*Pandey et al.*, 1978a).

The experimental design was similar to that reported by *Wang et al.* (1974b). *In vitro* determinations of platelet adenylate cyclase and its stimulation by PGE_1 and by PGE_1 plus NE were carried out in intact platelets obtained from patients during the drug-free baseline period. All patients were hospitalized, diagnosed according to the Research Diagnostic Criteria of *Spitzer et al.* (1975) and grouped as having schizophrenia, major depressive illness or bipolar mania. The normal control population consisted of volunteers who were free of any major medical or psychiatric disorders and were drug-free at the time of the study. Statistical analysis was done by one-way analysis of variance.

The results are shown in Figure 6. The PGE_1-stimulated increase in cyclic AMP synthesis was not significantly different among the subject

Table 9. [^{3}H]-Cyclic AMP Formation in Platelets in Response to Prostaglandin E_1 and Norepinephrine Administration

Item	Depressed Patients (n = 17) Unipolar (n = 9)	Bipolar (n = 8)	Normal Subjects (n = 11)
[^{3}H]-cyclic AMP formation			
Prostaglandin E_1 (2 μM)*	674 ± 82	646 ± 91	663 ± 77
Prostaglandin E_1 plus norepinephrine (10^{-4} M)*	263 ± 36	233 ± 39	279 ± 28
Percent of inhibition by norepinephrine	61	63	59
^{3}H-adenine incorporation**	4.3 ± 0.6	4.4 ± 0.8	4.3 ± 0.8

*Results expressed in mean counts/min/mg platelet protein, incubated for 5 min, ± SEM.
**Results expressed in mean counts/10^5/min/mg platelet protein, incubated for 30 min, ± SEM.
Reproduced from *Murphy, Donnelly,* and *Moskowitz* (1974).

groups. Similarly, the inhibition produced by NE (10^{-4}M) on the PGE_1-stimulated increase of ^{3}H-cyclic AMP synthesis did not differ significantly among the various groups. These results were thus similar to previous results. We have reported (*Pandey et al.*, 1977) in a preliminary study, however, that the PGE_1-stimulated increase in cyclic AMP synthesis in a subgroup of patients diagnosed as acute or subacute schizophrenics was significantly higher than that of normal controls or chronic schizophrenics (Fig. 7).

Effect of Lithium on Platelet Adenylate Cyclase

Lithium has been shown to alter the hormone-stimulated adenylate cyclase in several tissues. Effects of lithium on platelet adenylate cyclase also have been reported. In an *in vitro* study, *Wang et al.* (1974a) reported that Li^+ at higher concentrations inhibits PGE_1-stimulated adenylate cyclase both in intact platelets and in platelet homogenates. This antagonism of Li^+ to PGE_1-stimulated platelet adenylate cyclase appears to be independent of the antagonism produced by NE, thus suggesting that Li^+ probably acts on PGE_1-sensitive adenylate cyclase sites that are not sensitive to NE, i.e., Li^+ does not interact with the α-adrenergic receptors of the platelets. *Murphy*, *Donnelly*, and *Moskowitz* (1973) have reported that PGE_1-stimulated cyclic AMP production in human platelets is inhibited by treatment with lithium carbonate. They observed that the mean PGE_1-stimulated ^{3}H-cyclic AMP formation determined in 16 individuals during Li^+ therapy was significantly re-

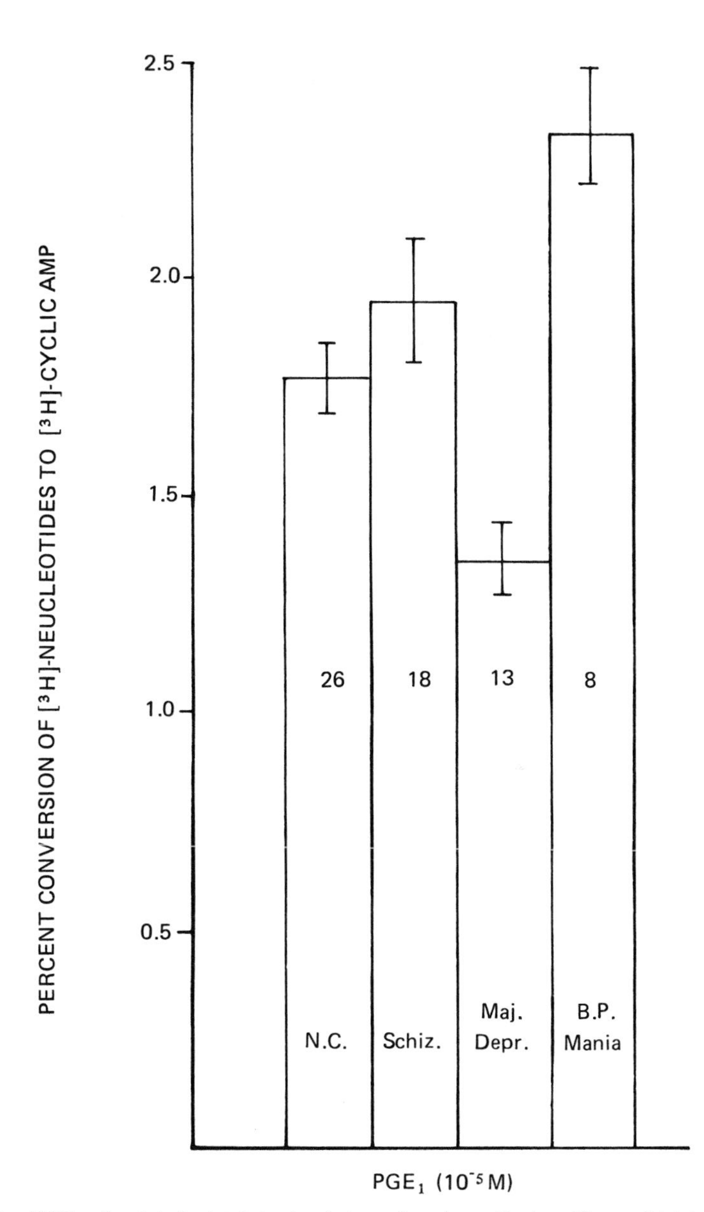

Fig. 6. PGE_1-stimulated platelet adenylate cyclase in patients with psychiatric illness and normal controls (NC). The number of subjects in each group is indicated in the bars. Derived from data in *Pandey et al.* (1978a).

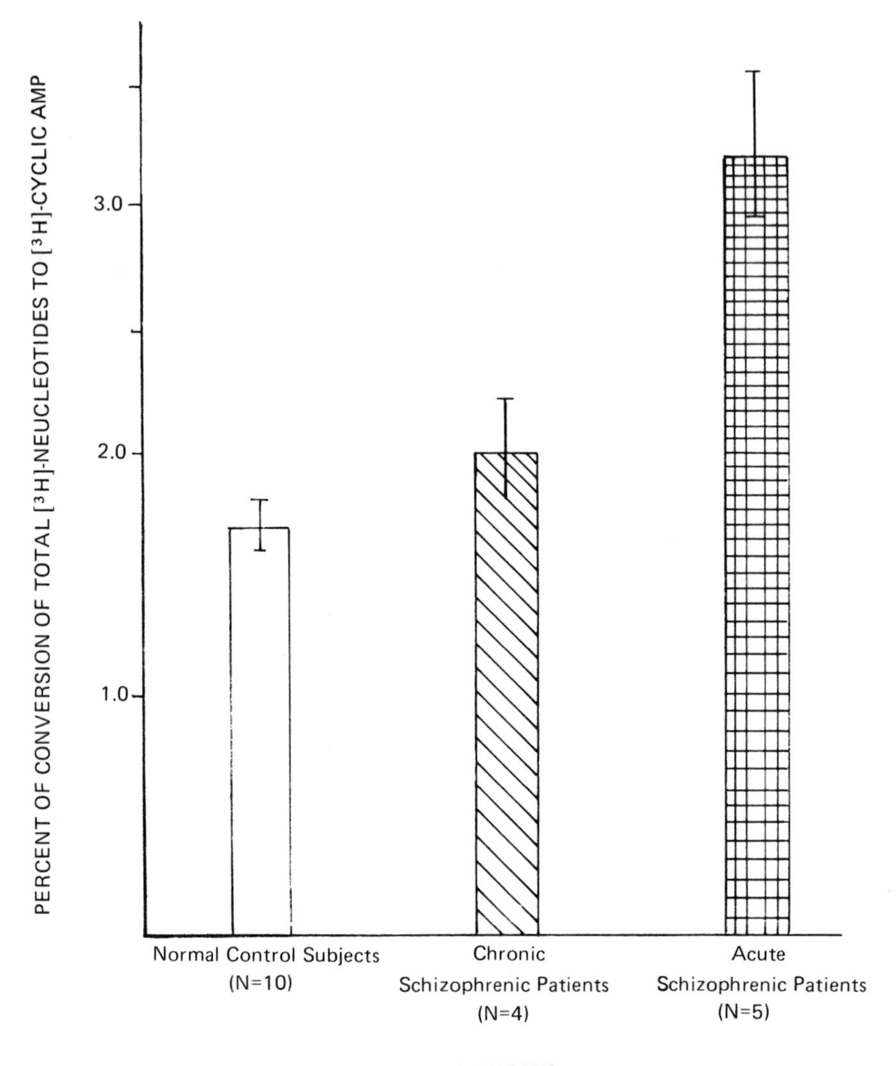

Fig. 7. PGE_1-stimulated adenylate cyclase platelet activity in schizophrenic patients and normal controls. Reproduced from *Pandey et al.* (1977), copyright American Psychiatric Association, reprinted with permission.

duced (about 50%) as compared to 19 controls. The *in vitro* inhibition produced by Li^+ in the PGE_1-stimulated accumulation of ^{3}H-cyclic AMP in the platelets obtained from patients on Li^+ therapy was also reduced as compared to the platelets obtained from controls. Four patients who were studied before and during Li^+ therapy showed similar changes. The results obtained by *Murphy et al.* (1973) and *Wang et al.* (1974a) thus indicate that Li^+, under both *in vitro* and *in vivo* conditions, decreases *in vitro* stimulation of platelet adenylate cyclase by PGE_1.

Leukocyte Adenylate Cyclase

The platelet adenylate cyclase studies discussed above indicate that the enzyme and its sensitivity to PGE_1 and NE appear to be normal in patients with affective illness. Although a subgroup of patients diagnosed as acute schizophrenics appeared in a preliminary study to exhibit hypersensitivity to PGE_1, sensitivity to NE was indistinguishable from that of the normal control population. The results with platelet adenylate cyclase are thus generally negative but should be interpreted with caution. Alpha-adrenergic sensitivity in patients has been studied by indirect methods, because it involves stimulation of platelet adenylate cyclase and subsequent inhibition of the PGE_1-stimulated activity by NE; NE by itself does not appreciably inhibit platelet cyclase. Because of this indirect measurement, there may be methodological limitations in the quantitation of the α-adrenergic receptors in platelets. Recently, *Kafka, Tallman,* and *Smith* (1978) described the binding characteristics of human platelets to the α-adrenergic antagonist [^{3}H]-dihydroergocryptine (DHE) and confirmed the presence of α-adrenergic receptors. It is now possible to quantitate α-adrenergic receptors in human platelets directly by their binding to ^{3}H-DHE, and this technique should be used in future studies of α-adrenergic receptor sensitivity in affective illness.

It is also possible that adenylate cyclase responses in platelets are not related to adenylate cyclase responses in the brain. Elevation of cyclic AMP synthesis in brain preparations has been reported to occur as a result of the responses produced by β-adrenergic receptors, although α-receptors may also be involved (*Kakiuchi* and *Rall,* 1968; *Chasin et al.,* 1971; *Huang* and *Daly,* 1972). Whereas studies of α-adrenergic receptor response in human platelets for evaluating a generalized abnormality of this response may not be illogical, in light of the negative results obtained in studies of platelet adenylate cyclase it may be pertinent to perform such studies in a peripheral tissue in which alterations in cyclic AMP synthesis are produced by responses of the β-adrenergic receptors. Human leukocytes may thus be a useful model for the study of β-adrenergic receptor sensitivity.

Several investigators have reported the presence of adenylate cyclase in human leukocytes (*Scott,* 1970; *Bourne* and *Melmon,* 1971; *Logsdon et al.,* 1973; *Logsdon, Middleton,* and *Coffey,* 1972; *Mue et al.,* 1975; *Alston, Patel,* and *Kerr,* 1974). Leukocyte adenylate cyclase can be stimulated by NE, IP, PGE_1 and NaF. Increases in cyclic AMP synthesis produced as a result of interaction of leukocyte adenylate cyclase and IP or NE are inhibited by the β-adrenergic antagonist propranolol but not by α-adrenergic antagonists such as phentolamine and phenoxybenzamine, thus indicating that the increase in cyclic AMP synthesis by NE and IP in human leukocytes is not due to an α-adrenergic receptor response.

The presence of a β-adrenergic receptor in human leukocytes has also been demonstrated by direct binding studies with the β-adrenergic antagonist [^{3}H]-DHA (*Williams, Snyderman,* and *Lefkowitz,* 1976). Because human leukocytes are easily available, they provide a useful peripheral model for the study of β-adrenergic receptor sensitivity in a variety of pathological conditions. Leukocyte adenylate cyclase and its responsiveness to NE and IP have been studied in subjects with bronchial asthma (*Logsdon et al.,* 1973; *Logsdon, Middleton,* and *Coffey,* 1972; *Mue et al.,* 1975; *Alston, Patel,* and *Kerr,* 1974). However, studies of leukocyte adrenergic receptor sensitivity have not been reported in patients with psychiatric disorders. To evaluate β-adrenergic receptor sensitivity in affective illness, we have done preliminary studies of leukocyte adenylate cyclase and its responsiveness to NE and isoproterenol in such patients (*Pandey et al.,* unpublished results).

Leukocyte adenylate cyclase studies were performed on patients admitted to the research ward of the Illinois State Psychiatric Institute during the end of a three-week, drug-free baseline period. Normal controls were nonhospitalized volunteers who were free of any known major medical disorders, and patients were diagnosed according to the Research Diagnostic Criteria of *Spitzer et al.* (1975). Intact leukocytes were isolated according to the method of *Bourne et al.* (1971), and evaluation of adenylate cyclase and its responsiveness to NE, IP and PGE_1 in the intact leukocytes, carried out by the pulse-labeling technique, was similar to that described by *Bourne et al.* (1971). As an index of adenylate cyclase activity, percent conversion of total [^{3}H] incorporated in nucleotides to [^{3}H] cyclic AMP was determined and evaluated by one-way analysis of variance.

Adenylate cyclase activity and the stimulation produced by *in vitro* additions of PGE_1 (10^{-5} M) and IP (10^{-4} M) were determined in 26 normal control volunteers, 21 patients with major unipolar depression, 9 with bipolar mania and 29 with schizophrenia. The mean basal activity observed in the patient population was not significantly different from that in the normal control group (Fig. 8). *In vitro* addition of PGE_1 (10^{-5} M) produced approximately a sixfold increase in the net [^{3}H] cyclic AMP synthesis in the intact leukocytes over the basal activity. One-way analysis of variance showed that the mean activity in the normal controls was not significantly different from that in the patient groups (Fig. 8).

In vitro addition of NE produced approximately a 1.5-fold increase in net [^{3}H] cyclic AMP synthesis in normal controls. The mean activity

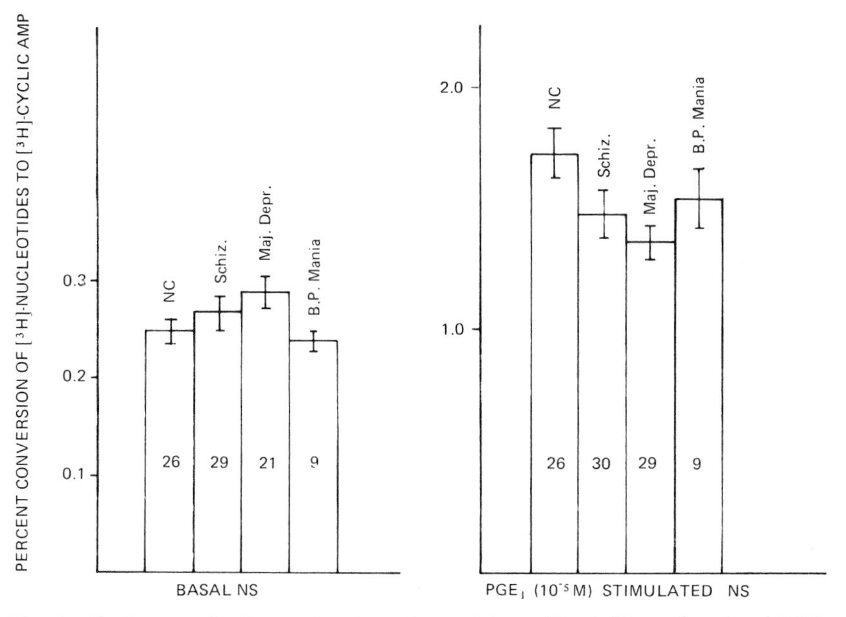

Fig. 8. Leukocyte adenylate cyclase in patients with psychiatric illness (basal and PGE_1-stimulated). The number of subjects in each group is indicated in the bars. Derived from data in *Pandey et al.* (1978b).

observed in the patient groups was, however, significantly lower than in the normal controls, as determined by one-way analysis of variance (Fig. 9). This suggested that the [^{3}H]cyclic AMP synthesis produced by NE, although lower in the patient population, could not distinguish patients with affective illness from schizophrenic patients. Synthesis of [^{3}H]cyclic AMP by human leukocytes also was measured after the *in vitro* addition of the β-adrenergic agonist isoproterenol (10^{-4} M). Addition of IP produced about a twofold increase in net [^{3}H]cyclic AMP synthesis above the baseline. One-way analysis of variance indicated significant differences between the various subject groups (Fig. 9). Further analysis by Scheef's test indicated that mean activity in patients with unipolar major depressive disorders was significantly lower than normal controls ($P < 0.001$) and the schizophrenic population ($P < 0.05$). Patients with bipolar mania were significantly different from normal controls but did not differ from the unipolar depressed patients. Schizophrenics did not differ significantly from normal controls.

In summary, the leukocyte studies indicated that there was no difference in either the basal activity or the PGE_1-stimulated activity between the population groups studied. NE-sensitive leukocyte adenylate cyclase was significantly lower in the patient population than in normal controls, and IP-stimulated activity in the depressed group was signifi-

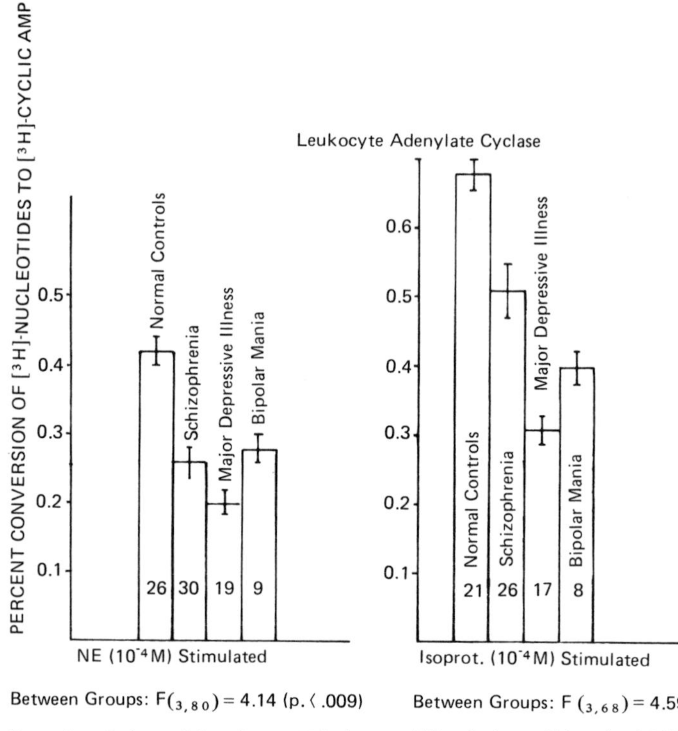

Between Groups: $F_{(3,80)} = 4.14$ (p. $<$ .009)

Controls vs Patients: $F_{(1,80)} = 10.08$ (p $<$.002)

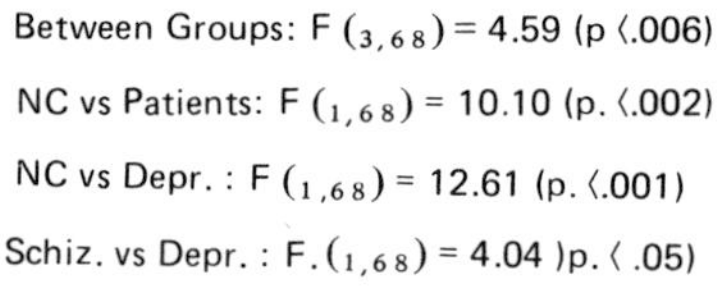

Fig. 9. Leukocyte adenylate cyclase in patients with psychiatric illness (NE- and ISOP-stimulated).

cantly lower than in normal controls and schizophrenic groups. Although these results in general were suggestive of a reduced sensitivity of β-adrenergic receptors in the leukocytes of patients with affective illness, the observation that the schizophrenic group also was significantly lower than normal controls in NE-sensitive leukocyte adenylate cyclase was contrary to our expectation.

Several explanations could be offered for this observation. NE is both an α- and a β-adrenergic agonist, and the net increase in cyclic AMP synthesis in human leukocytes produced by NE is a cumulative response of both α- and β-adrenergic receptors. Because some α-receptors may be present in human leukocytes, it is quite possible that the decreased response produced by NE in the schizophrenic population reflects hypersensitivity of α-adrenergic receptors in schizophrenic patients. Further observation that [^{3}H]cyclic AMP synthesis produced by IP, a much cleaner β-adrenergic agonist, is not significantly different in schizophrenic patients compared to normal controls may tend to support the above assumption. Thus, whereas increased α-adrenergic receptor sensitivity may be associated with schizophrenic illness, a reduction in β-adrenergic receptor sensitivity may be involved in affective illness. It is also possible that a common biological abnormality of β-adrenergic receptor deficiency may be associated with both schizophrenic and affective illness, i.e., this abnormality, which may be associated with altered behavior, may not be specific to either disease.

The preliminary results, however, suggest a generalized receptor defect in affective illness and should be pursued further, possibly by means of alternative strategies.

Postmortem Human Brain Adenylate Cyclase in Schizophrenia

Because of the inaccessibility of the human brain for biological studies, direct evaluation of the action of adenylate cyclase in the brain is difficult. Attempts have been made, however, to derive indirect information about human brain adenylate cyclase, either by studies of CSF cyclic AMP, which will be discussed later, or by determination of adenylate cyclase activity in the postmortem brain. A recent report by *Carenzi et al.* (1975) describes a study of the basal adenylate cyclase activity in human postmortem caudate, determined in eight psychiatrically normal subjects and in seven patients diagnosed as schizophrenic. They reported the presence of a dopamine-sensitive adenylate cyclase in human caudate and observed that the stimulation produced by DA was greater than that observed with NE. The investigators, however, observed no significant differences between the normal controls and the schizophrenics, either in basal caudate adenylate cyclase activity or in the stimulation produced by the addition of dopamine (10^{-4} M).

These results were clearly negative in terms of the adenylate cyclase activity and its sensitivity to dopamine in the caudate of human postmortem brain. However, they should be taken with caution since most of the brains had been preserved for prolonged periods of time. Even though some parallels between rat brain cyclase activity and the human brain were reported, it is not clear what the effects, if any, of prolonged storage of brain might be. Because of the inherent difficulties discussed before, there do not appear to be many investigations of adenylate cyclase activity and its responsiveness to biogenic amines in the human brain.

Summary of Human Adenylate Cyclase Studies

The studies of adenylate cyclase in human peripheral tissues such as platelets and leukocytes thus provide a useful tool for studies of adrenergic receptor sensitivity under *in vitro* conditions. Although these peripheral tissues do not provide an index of neuronal receptor sensitivity, they could at least be used to provide information about whether a generalized abnormality of adrenergic receptor sensitivity is involved in psychiatric illness. Leukocytes have been used successfully by some investigators for studies of abnormal adrenergic receptor sensitivity in other pathological conditions such as asthma and leukemia; the results reported thus far are encouraging in that they clearly indicate an abnormality of adrenergic receptor sensitivity in bronchial asthma. The question often raised about the use of such peripheral systems in psychiatric disorders is whether or not the results of these *in vitro* measurements are related to the neuronal receptor sensitivity that might be involved in various other behavioral disorders. It is possible that an abnormality in the neuronal adenylate cyclase is reflected in other peripheral systems, since the properties of adenylate cyclase in peripheral tissues such as leukocytes are very similar to those observed in the brain. Recent progress made in the understanding of the nucleotide system, especially in animal systems, and the hypothesis of *Aschroft et al.* (1972) have given further impetus to studies of the role of cyclic nucleotides in general, and adenylate cyclase β-receptors in particular, in psychiatric illness.

Because of the complexity of human studies in psychiatric illness, little progress has been made in this area. To date, only three or four studies of platelet adenylate cyclase in depression are available (in two of which the author was involved), and probably only one study of leukocyte adenylate cyclase in affective illness and schizophrenia exists. The results presented here are generally negative as far as platelet adenylate cyclase in affective illness is concerned. Two encouraging aspects of these studies are apparent, however. The studies on leukocyte adenylate cyclase reported here and carried out in our laboratories suggest that either β-adrenergic receptor activity or α-adrenergic receptor sensitivity, or both, may be altered in affective illness and schizophrenia. Our initial studies, although not conclusive, suggest that other, more careful studies need to be undertaken for a better understanding of the nucleotide system, which may be very closely related to the adrenergic receptors in affective illnesses. Further studies could be undertaken on biopsy tissue such as muscle.

Urinary Cyclic AMP

Some of the early attempts at evaluating the role of cyclic nucleotides in affective illness were directed toward the determination of cyclic AMP in urine. This approach was very similar to those used for the study of biogenic amine metabolism in affective illness, which consisted of the quantitation of catecholamines and their metabolites such as NE, normetanephrine (NM), 3-methoxy-4-hydroxy-phenylethylene glycol (MHPG) and homovanillic acid (HVA), and serotonin (5-HT) and its metabolite 5-hydroxy indoleacetic acid (5-HIAA).

Initial studies of urinary cyclic AMP were reported by *Paul et al.* (1970a & b) and *Abdullah* and *Hamadah* (1970). *Paul et al.* (1970a & b) found that patients with psychotic depression excreted significantly lower amounts of cyclic AMP than normal controls and patients with neurotic depression. *Abdullah* and *Hamadah* (1970) also reported that urinary cyclic AMP excretion in depressed patients (females only) was lower than in normal controls, and that clinical improvement resulted in a change in cyclic AMP toward the normal value. They also reported increased cyclic AMP excretion in patients with mania and hypomania. Clinical changes from hypomania to depression, or from depression to mania, were associated with changes in cyclic AMP excretion. In a subsequent study *Paul, Cramer,* and *Bunney* (1971) reported that in patients rapidly cycling between depression and mania, an increase in urinary cyclic AMP was observed on the day of change from depression to mania. *Jenner et al.* (1972) reported a correlation between mood changes and cyclic AMP excretion in a patient with cyclical mood disturbances. These observations thus suggested, for urinary cyclic AMP, (1) decreased excretion in patients with depressive illness and increased excretion in mania, (2) changes toward normal with clinical improvement and (3) changes with mood in patients rapidly cycling between depressive and manic phases. *Abdullah* and *Hamadah* (1970) proposed on the basis of these conclusions that depressive illness is associated with a decrease in intracellular levels of cyclic AMP in all tissues, including the central nervous system.

Since the initial reports of Paul and co-workers (*Paul, Kramer,* and *Bunney,* 1971; *Paul, Ditzion,* and *Janowsky,* 1970b; *Paul et al.* 1970a) and *Abdullah* and *Hamadah* (1970), several other investigators have studied urinary cyclic AMP in affective illness and schizophrenia (Table 10). Most of the studies listed in the table report a generalized decrease in urinary cyclic AMP excretion in subgroups of patients with affective illness. In some of these studies, however, the decrease is not statistically significant. *Brown et al.* (1972) did not find significant correlations between cyclic AMP excretion and changes in mood, and *Hullin et al.* (1974) reported that there was no significant change in urinary cyclic AMP in a patient rapidly shifting from depression to mania.

The effect of treatment with psychotropic drugs and ECT on levels of urinary cyclic AMP has been reported by some investigators. Whereas many reports suggest that some psychotropic drugs alter adenylate cyclase activity and its responsiveness to NE in several tissues, only a

Table 10. Urinary Cyclic AMP Excretion in Patients with Affective Illness and Normal Controls

Study	Normal	Psychotic Depression	Neurotic Depression	Recovered Depression
Paul, Ditzion, and *Janowsky* (1970b)	5.64 ± 0.68	3.64 ± 0.19	6.70 ± 0.37	—
n	10	7	25	
Abdullah and *Hamadah* (1970)	2.28 ± 0.09	0.52 ± 0.08	—	1.28 ± 0.02
n	18	21		15
Paul, Cramer, and *Bunney* (1971)	5.64 ± 0.68	3.14 ± 0.36	6.70 ± 0.36	—
n	10	9	25	
Brown et al. (1972)	3.58 ± 0.37	3.00	—	—
n	16	3		
Somerville (1973)	4.46 ± 1.0	5.16 ± 0.76	—	2.31
n	6	8		4
Naylor et al. (1974)	—	4.50 ± 0.44	—	6.03 ± 0.59
n		12		12
Sinanan et al. (1975)	3.98 ± 1.55	2.88 ± 1.55	2.27 ± 1.77	3.33 ± 1.78
n	25	27	15	18
Jarrett et al. (1977)	4.36 ± 0.17	2.95 ± 0.22	—	4.99 ± 0.60
n	14	19		6

Reproduced from *Jarrett et al.* (1977)

few studies were done on human subjects. *Paul et al.* (1971) reported a reduction in cyclic AMP excretion by hypomanic patients during treatment with Li^+ and a rise in depressed patients on treatment with Li^+. Furthermore, *Brown et al.* (1972) reported a decrease in cyclic AMP excretion in a manic patient during treatment with Li^+. However, no significant effect of treatment on cyclic AMP levels was observed with amitriptyline, haloperidol or chlorpromazine.

There are essentially two reports on the effect of ECT on urinary cyclic AMP excretion. *Hamadah et al.* (1972) noted a significant increase in urinary cyclic AMP excretion on the day on which ECT was given to patients. During the period after ECT administration, the levels remained elevated. The effect of ECT on one of these patients is shown in Figure 10. The levels of cyclic AMP excretion rose immediately after each administration of ECT and then gradually dropped. These findings have not been confirmed by Moyes and co-workers (*Moyes* and *Moyes,* 1976a & b; *Moyes,* 1972), who observed either no change or a decrease in urinary cyclic AMP excretion in 15 patients undergoing ECT. They reported no significant correlation between the effect of ECT and urinary cyclic AMP excretion.

Although not all investigators observe a significant correlation between mood changes and cyclic AMP excretion, the majority suggest

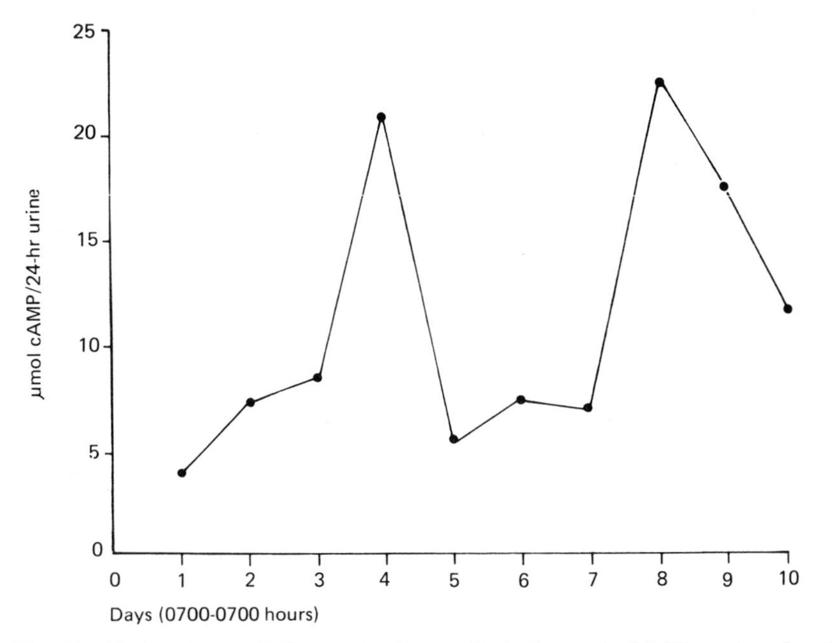

Fig. 10. Daily urinary cAMP excretion in a patient who received ECT on days 4 and 8. Reproduced from *Hamadah et al.* (1972).

a decreased excretion of cyclic AMP in depression. The inconsistency may be due in part to the selection of patients. However, the significance of the findings in relation to affective illness is unclear.

Some investigators have suggested that factors such as exercise, age, diet (see *Moyes* and *Moyes*, 1976b) and urinary volume (*Owen* and *Moffat*, 1973) could alter cyclic AMP excretion. Lowered physical activity in depressed patients may be a cause of lowered cyclic AMP excretion. *Eccleston et al.* (1970) found that exercise in normal subjects caused changes in the urinary cyclic AMP levels. Conversely, *Paul et al.* (1971) observed no changes in urinary cyclic AMP due to physical activity in normal children and adults and in hyperkinetic children. They also reported that agitated as well as retarded depressed patients excreted lowered amounts of cyclic AMP. Age and diet thus appear to have no significant effects on urinary cyclic AMP.

Another controversy with regard to the significance of cyclic AMP levels in urine and its relation to depressive illness is the source of cyclic AMP. The major source of cyclic AMP in urine is believed to be the kidney. *Broadus et al.* (1970a) have shown that up to 50% of urinary cyclic AMP is derived from the renal tubules and the rest from plasma by glomerular filtration. Furthermore, stimulation by parathyroid hormone results in a considerable increase in cyclic AMP levels in the kidney (*Broadus et al.*, 1970b). It is not certain how much of the cyclic AMP in plasma, and therefore in urine, is derived from the brain; therefore, it is not known whether alterations observed in patients with clinical changes reflect cyclic AMP in the brain or whether they are an index of physical and hence peripheral metabolic activity. In summary, the results from urinary cyclic AMP studies in depression, though encouraging, need further substantiation by alternate methods.

CSF Cyclic AMP

For further examination of the results of urinary cyclic AMP studies in depressed patients, attempts have been made to determine cyclic AMP levels in CSF, because CSF cyclic AMP changes may provide a better, more direct indication of changes in brain cyclic AMP. However, to date only a few clinical studies of CSF cyclic AMP in psychiatric illness have been carried out (Table 11). Several investigators have reported values for normal subjects which range from 5–31 mM.

Robison et al. (1970) measured the CSF cyclic AMP concentration in depressed patients and in patients with mania. Patients undergoing neurological examinations were used as controls. Samples from each group of subjects were pooled prior to the determination of cyclic AMP. The results (Table 10) did not suggest significant differences between groups.

Table 11. CSF Cyclic AMP in Psychiatric Patients

	Cyclic AMP pmoles/ml CSF			
Study	Controls	Depression	Mania	Schizo-phrenia
	Without Probenecid			
Robison et al. (1970)	16.2 (Neurological controls)	15.5	14.2	—
Geisler et al. (1976)	16.0 (Character disorders) 10.0 (Neurological disorders)	8.8	11.2	—
Biederman et al. (1977)	28.5	—	—	32.8
Cramer et al. (1972)	11.0	14.1	10.5	—
	With Probenecid			
Cramer et al. (1972)	28.0	27.9	47.5	

Another study of CSF cyclic AMP was reported by *Geisler et al.* (1976). The patient group consisted of melancholic patients subdivided as bipolar, unipolar, unclassified and manic. The control group was made up of patients with reactive depression, paranoid psychosis or character disorder, and of neurologic and orthopedic patients. No significant differences were observed either between melancholic patients and other diagnostic groups or among the subgroups. No significant correlations with age, height, motor activity or sex were observed. Treatment with ECT, tricyclic antidepressants, lithium or neuroleptics did not alter CSF cyclic AMP concentrations. *Biederman et al.* (1977), who reported CSF cyclic AMP levels in schizophrenic patients and psychiatrically normal controls, observed no significant differences between the two groups.

Cramer et al. (1972), in a study of the CSF cyclic AMP in a group of 12 depressed patients, found that they had significantly higher levels than the mean values obtained in 15 neurologic controls and in 6 manic patients (Table 10). The mean CSF cyclic AMP levels in the manic patients were not significantly different from those in normal controls. Four hours of intense psychomotor activity did not alter cyclic AMP levels, and the levels in the depressed group were not related to age, sex or type of illness. Probenecid caused a significant increase in CSF cyclic AMP levels. In manic patients the increase was greater than in the depressed and control subjects; this suggests a higher turnover rate of cyclic AMP in manic patients.

Post, Cramer, and *Goodwin* (1977) recently reported a study of CSF cyclic AMP levels in depressed and manic patients and in neurologic controls. CSF cyclic AMP levels in depressed patients (n = 29) were slightly but significantly higher than normal controls, but were not significantly different from manic patients. Administration of probenecid caused a significant increase in CSF of both depressed and manic patients, but the levels were not significantly different from each other. CSF levels of cyclic AMP (basal or probenecid-induced) obtained during treatment with tricyclics, lithium, methadone or ECT were not significantly different from the pretreatment levels or from medication-free depressed patients.

The results of the CSF cyclic AMP studies thus do not confirm the earlier hypothesis of *Abdullah* and *Hamadah* (1970), based on urinary cyclic AMP data during a depressed state, that cyclic AMP levels are decreased in all tissues, including the CNS.

Plasma Cyclic AMP

It is estimated that as much as 40–50% of urinary cyclic AMP is derived from plasma, through glomerular filtration (*Broadus et al.,* 1970a). If the observed decrease in urinary cyclic AMP does not reflect changes in the production of cyclic AMP in the kidney, one would expect that plasma cyclic AMP would be low in patients with depression. However, not many studies of plasma cyclic AMP have been published.

Hansen (1972) reported the nucleoside and nucleotide content in whole blood of schizophrenic patients. The mean whole-blood cyclic AMP content (0.26 ± 0.04 mg/100 ml blood) in these patients (n = 24) was not significantly different from normal controls (0.23 ± 0.02). However, patients diagnosed as psychotic depressives (n = 11) had lowered levels of whole-blood cyclic AMP (0.12 ± 0.02) compared to levels in normal controls. On the other hand, *Jarrett* (1977) did not observe any significant differences between plasma cyclic AMP content of depressed patients (19.9 ± 1.4 nM) and that of controls (19.1 ± 1.2 nM). When the control group was further subdivided into euthymic and schizophrenic patients, the euthymic patients had significantly lower plasma cyclic AMP than schizophrenic patients. Euthymic patients were not significantly different from depressed patients.

It is difficult to draw any conclusions about changes in plasma cyclic AMP in psychiatric illness from these two studies. The discrepancy between the two may be attributed to the method used, and it is not clear whether whole-blood cyclic AMP is related to plasma cyclic AMP concentrations.

In recent years, use of provocative tests has offered a useful strategy for the study of biological abnormalities in psychiatric illness. Apomorphine-, insulin- or L-dopa-induced growth hormone release has been considered as an index of hypothalamic or pituitary function in psychiatric patients (see *Martin,* 1973). Since apomorphine-induced GH release appears to be mediated through central DA receptors, this test provides information about control of DA activity (*Pandey et al.,* 1977). Other tests, e.g., precursor challenge, also have been employed (*Frazer, Pandey,* and *Mendels,* 1973).

Ebstein et al. (1976) recently reported that administration of epinephrine causes an increase in plasma cyclic AMP levels which is inhibited by administration of Li^+. They suggested that the observed increase is a result of the responses produced by β-adrenergic receptors. *Belmaker et al.* (1976) observed that epinephrine-stimulated cyclic AMP levels in plasma after haloperidol treatment, though lower, were not significantly different from the untreated subjects.

The use of adrenergic receptor response *in vivo* appears to be important for studies of adrenergic receptor sensitivity in affective illness. Even though the responses produced are primarily peripheral, such studies may provide information if a generalized adrenergic receptor defect is involved. Even though *Ebstein et al.* (1976) did not observe a significant relationship between epinephrine-induced increase in plasma cyclic AMP levels and clinical diagnosis, this method needs to be evaluated further because the number of subjects used was quite small. Other adrenergic agonists, e.g., isoproterenol, could be used for the study of such responses.

CONCLUSIONS

There has been an increasing awareness during the past several years that cyclic nucleotides may be involved in the pathogenesis of affective illness and schizophrenia. The cyclic nucleotides and the catecholamines are closely related; cyclic AMP has been shown to mediate the actions of catecholamines, and adenylate cyclase appears to be an integral part of the adrenergic receptors. Therefore, the cyclic nucleotide system offers a useful indirect tool for studies of the catecholamine dysfunction in these diseased states.

Two basic approaches have been used in studies of the nucleotide system in psychiatric disorders. First, direct human studies have examined whether an abnormality of the cyclic nucleotide system exists in these states and whether treatment with psychotropic drugs alters the cyclic nucleotide system. The second approach has been to examine the basic mechanisms by which psychotropic drugs produce such changes and to determine whether or not the therapeutic effects are mediated through the cyclic nucleotides.

In human studies, peripheral fluids such as urine and plasma have been used for the determination of cyclic AMP, and platelets and leukocytes for studies of adenylate cyclase. Initial studies with urinary cyclic AMP were encouraging because depressed patients were found to excrete lower amounts of cyclic AMP, and because the excretion of cyclic AMP changed during the shift from depression to mania. Changes in clinical status were also found to be associated with changes in urinary cyclic AMP. Psychotropic drugs such as lithium were shown to alter the excretion of cyclic AMP. It appears fairly well established that urinary cyclic AMP is altered in affective illness, although some studies have failed to confirm this observation. However, because of several limitations, these results should be interpreted with caution. The major source of urinary cyclic AMP probably is not the brain, and urinary cyclic AMP may reflect only peripheral metabolic changes caused by stress or motor activity. Studies with plasma cyclic AMP did not show abnormalities in patients with either affective illness or schizophrenia.

Some efforts have been made to study adenylate cyclase in human platelets and leukocytes. Platelet adenylate cyclase and its responsiveness to PGE_1 and NE were not significantly different from normal levels in patients with affective illness and schizophrenia, but a subgroup of patients diagnosed as acute schizophrenics exhibited increased response to PGE_1-stimulation of platelet adenylate cyclase. These results, however, are preliminary. Whereas platelet adenylate cyclase offers an index of α-adrenergic activity, leukocyte adenylate cyclase is stimulated by NE and isoproterenol, and this reponse has been shown to be β-adrenergic. Direct studies on binding of human platelets with the α-adrenergic antagonist dihydroergocryptine, and of leukocytes with the β-adrenergic antagonist dihydroalprenolol, also confirm the previous reports that platelets are associated with α- and leukocytes with β-adrenergic receptors. In this respect, the stimulation characteristics of leukocyte adenylate cyclase closely resemble those of brain adenylate cyclase.

These observations, coupled with the negative results obtained in studies of platelet adenylate cyclase, prompted us to study the leukocyte adenylate cyclase system in man. Initial results suggest that isoproterenol stimulation of leukocyte adenylate cyclase may be reduced in patients with unipolar depression. The investigations with platelet and leukocyte adenylate cyclase, although not providing any definitive results about an abnormality of adenylate cyclase, are at least suggestive. The leukocyte adenylate cyclase studies should be repeated in other laboratories. One of the alternate strategies of investigating abnormalities in leukocyte adenylate cyclase levels would be a direct quantitation of β-adrenergic receptors in leukocyte preparations from patients with affective illness.

Williams, Snyderman, and *Lefkowitz* (1976) have shown, by binding studies with [^{3}H]-DHA, that adenylate cyclase in human leukocytes is closely related to the β-adrenergic receptors. It is thus possible to quantitate and compare the number of β-adrenergic receptors in human leukocytes by means of ^{3}H-DHA binding. Such studies are currently in progress in our laboratories.

Cyclic AMP studies in the CSF of patients with psychiatric disorders have not been conclusive; it has not been possible to relate the earlier findings of decreased urinary cyclic AMP in depression to a decreased turnover of cyclic AMP in the brain. Animal studies have been helpful in elucidating the effects of psychotropic drugs on the adenylate cyclase-cyclic AMP system, but they have provided no evidence that the therapeutic effects of the psychotropic drugs are mediated by cyclic AMP.

Recent studies suggest that the cyclic nucleotide system may play an important role in brain function and in the regulation of mood states. However, clinical studies elucidating such a role for cyclic AMP are few, and often are difficult to interpret. Alternate strategies of cyclic nucleotide studies in humans must be developed to determine if changes in mood state are mediated through the cyclic AMP system, and whether or not psychotropic drugs produce their therapeutic effects by altering the nucleotide system.

REFERENCES

Abdulla, Y.H., and Hamadah, K. 1970. 3′,5′ Cyclic adenosine monophosphate in depression and mania. Lancet I: 378–381

Alston, W.C., Patel, K.R., and Kerr, J.W. 1974. Response of leukocyte adenyl cyclase to isoprenaline and effect of alpha blocking drugs in extrinsic bronchial asthma. Brit Med J 1:90–95

Aschroft, G.W., Eccleston, D., Murray, L.G., and Glen, A.I.M. 1972. Modified amine hypothesis for the aetology of affective illness. Lancet II:573–577

Banerjee, S.P., Kung, L.S., Riggi, S.J., and Chanda, S.K. 1977. Development of β-adrenergic receptor subsensitivity by antidepressants. Nature 268:455–456

Belmaker, R.M., Ebstein, R.P., Schoenfeld, H., and Rimon, R. 1976. The effect of haloperidol on epinephrine-stimulated adenylate cyclase in humans. Psychopharmacology 49:215–217

Biederman, J., Rimon, R., Ebstein, R., Belmaker, R.H., and Davidson, J.R. 1977. Cyclic AMP in the CSF of patients with schizophrenia. Brit J Psychiat 130:64–67

Bliss, E.L., Ailion, T., and Zwanziger, J. 1968. Metabolism of norepinephrine, serotonin and dopamine in rat brain with stress. J Pharmacol Exp Ther 164:122–134

Bourne, H.R., Lehrer, R.I., Cline, M.J., and Melmon, K.L. 1971. Cyclic 3′,5′-adenosine monophosphate in the human leukocyte. Synthesis, degradation, and effects on neutrophil candidacidal activity. J Clin Invest 50:920–929

Bourne, H.R., and Melmon, K.L. 1971. Adenyl cyclase in human leukocytes. Evidence for activation by separate β-adrenergic and prostaglandin receptors. J Pharmacol Exp Ther 178::1-7

Broadus, A.E., Kaminsky, N.E., Hardmann, J.G., Sutherland, E.W., and Liddle, G.W. 1970a. Kinetic parameters and renal clearances of plasma adenosine 3',-5'-monophosphate and guanosine 3',5'-monophosphate in man. J Clin Invest 49:2222-2236

Broadus, A.E., Kaminsky, N., Northcut, R.C., Hardmann, J.G., Sutherland, E.W., and Liddle, G.W. 1970b. Effects of glucagon on adenosine 3',-5'-monophosphate and guanosine 3',5',-monophosphate in human plasma and urine. J Clin Invest 49: 2237-2245

Brown, E.M., Aurbach, G.D., Hauser, D., and Troxler, F. 1976. β-adrenergic receptor interactions. Characterization of iodohydroxybenzylpindolol as a specific ligand. J Biol Chem 25:1232-1238

Brown, B.L., Salway, J.G., Albano, J.D.M., Hullin, R.P., and Ekins, R.P. 1972. Urinary excretion of cyclic AMP and manic-depressive psychosis. Brit J Psychiat 120: 405-408

Bylund, B.B., and Snyder, S.H. 1976. Beta adrenergic receptor binding in membrane preparations from mammalian brain. Mol Pharmacol 12:568-580

Carenzi, A., Gillin, C., Guidotti, A., Schwartz, M.A., Trabucchi, M., and Wyatt, R.J. 1975. Dopamine-sensitive adenylate cyclase in human caudate nucleus. Arch Gen Psychiat 32:1056-1059

Chasin, M., Rivkin, I., Mamrak, F., Samaniego, S.G., and Hess, S.M. 1971. Alpha and beta-adrenergic receptors as mediators of accumulation of cyclic AMP in specific areas of guinea pig brain. J Biol Chem 246:3037-3041

Cramer, H., Goodwin, F.K., Post, R.M., and Bunney, W.E., Jr. 1972. Effects of probenecid and exercise on cerebrospinal-fluid cyclic A.M.P. in affective illness. Lancet I:1346-1347

Creese, I., Burt, D.R., and Snyder, S.H. 1976. Dopamine receptor binding predicts clinical and pharmacological potencies of antischizophrenic drugs. Science 192:481-483

Deguchi, T., and Axelrod, J. 1973. Supersensitivity and subsensitivity of the β-adrenergic receptor in pineal gland regulated by catecholamine transmitter. Proc Nat Acad Sci USA 70:2411-2414

Deykin, D., and Parris, E. 1972. Cyclic AMP and Platelet Function. N Engl J Med 286:358-363

Ebstein, R., Belmaker, R., Grunhaus, L., and Rimon, R. 1976. Lithium inhibition of adrenaline-stimulated adenylate cyclase in humans. Nature 259:411-413

Eccleston, D., Loose, R., Prillar, I.A., and Sugdin, R. F. 1970. Exercise and urinary excretion of cyclic AMP. Lancet 11:25

Frazer, A., Pandey, G.N., and Mendels, J. 1973. Metabolism of tryptophan in depressive disease. Arch Gen Psychiat 29:528-535

Frazer, A., Pandey, G., Mendels, J., Neeley, S., Kane, M., and Hess, M.E. 1974. The effect of tri-iodothyronine in combination with imipramine on ^{3}H-cyclic AMP production in slices of rat cerebral cortex. Neuropharmacology 13:1131-1140

Geisler, A., Bech, P., Johannesen, M., and Rafaelsen, O.J. 1976. Cyclic AMP levels in cerebrospinal fluid in manic-melancholic patients. Neuropsychobiology 2:211-220

Gorman, R.R., Bunting, S., and Miller, O.V. 1977a. Modulation of human platelet adenylate cyclase by prostacyclin. Prostaglandins 13:377-388

Gorman, R.R., Bunting, S., and Miller, O.V. 1977b. Effect of prostacyclin on cyclic AMP concentrations in human platelets. Prostaglandins 13:389-397

Hamadah, K., Holmes, H., Barker, G.B., Hartman, G.C., and Parke, D.V.W. 1972. Effect of electric convulsion therapy on urinary excretion of 3',5' cyclic adenosine monophosphate. Brit Med J 674:439-441

Hendley, E.D., and Welch, B.L. 1975. Electroconvulsive shock: Sustained decrease in norepinephrine uptake in a reserpine model of depression. Life Sci 16:45-54

Hansen, O. 1972. Blood nucleoside and nucleotide studies in mental disease. Brit J Psychiat 121:341-350

Huang, M., and Daly, J.W. 1972. The accumulation of cyclic adenosine monophosphate in incubated slices of brain tissue. I. Structure-activity relationships of agonists and antagonists of biogenic amines, and of tricyclic tranquilizers and antidepressants. J Med Chem 15:458–463

Hullin, R.P., Salway, J.G., Allsopp, M.N.E., Barnes, G.D., Albano, J.D.M., and Brown, B.L. 1974. Urinary cyclic AMP in the switch process from depression to mania. Brit J Psychiat 125:457–458

Iversen, L.I. 1977. Catecholamine sensitive adenylate cyclases in nervous tissues. J Neurochem 29:5–12

Jarrett, D.B. 1977. Cyclic AMP in severe depression. In *Handbook of Studies on Depression,* ed. G.D. Burrows, pp. 343–366. Excerpta Medica. Amsterdam: Elsevier/North Holland Biomedical Press

Jenner, F.A., Sampson, G.A., Thompson, E.A., Somerville, A.R., Beard, N.A., and Smith, A.A. 1972. Manic-depressive psychosis and urinary excretion of cyclic AMP. Brit J Psychiat 121:236–237

Kafka, M.S., Tallman, J.F., and Smith, C.C. 1978. Alpha-adrenergic receptors on human platelets. Life Sci 21:1429–1438

Kakiuchi, S., and Rall, T.W. 1968. The influence of chemical agents on the accumulation of cyclic AMP in slices of rabbit cerebellum. Mol Pharmacol 4:367–378

Kakiuchi, S., Rall, T.W., and McIlwain, H. 1969. The effect of electrical stimulation upon the accumulation of adenosine 3′,5′-phosphate in isolated cerebral tissue. J Neurochem 16:485–491

Kodama, T., Matsukado, Y., Suzuki, T., Tanaka, S., and Shimizu, H. 1971. Stimulated formation of adenosine 3′,5′-monophosphate by desipramine in brain slices. Biochim Biophys Acta 252:165–170

Kuo, J.F., and De Renzo, E.C. 1969. A comparison of the effects of lipolytic and antilipolytic agents on adenosine 3′,5′-monophosphate levels in adipose cells as determined by prior labelling with adenine-8^{14}C. J Biol Chem 244:2252–2260

Lefkowitz, R.J. 1976. The β-adrenergic receptor. Life Sci 18:461–472

Logsdon, P.J., Carnright, D.V., Middleton, E., Jr., and Coffey, R.G. 1973. The effect of phentolamine on adenylate cyclase and on isoproterenol stimulation in leukocytes from asthmatic and nonasthmatic subjects. J Allergy Clin Immunol 52:148–157

Logsdon, P.J., Middleton, E., Jr., and Coffey, R.G. 1972. Stimulation of leukocyte adenyl cyclase by hydrocortisone and isoproterenol in asthmatic and nonasthmatic subjects. J Allergy Clin Immunol 50:45–56

Lust, W.D., Goldberg, N.D., and Passonneau, J.V. 1976. Cyclic nucleotides in murine brain. The temporal relationship of changes induced in adenosine 3′,5′-monophosphate and guanosine 3′,5′-monophosphate following maximal electroshock or decapitation. J Neurochem 26:5–10

Martin, J.B. 1973. Neural regulation of growth hormone secretion. N Engl J Med 288: 1384–1393

Moyes. I.C.A. 1972. E.C.T. and cAMP excretion. Brit Med J 3:829

Moyes, R.B., and Moyes, I.C.A. 1976a. Urinary adenosine 3′,5′-cyclic monophosphate: effects of electroconvulsive therapy. Brit J Psychiat 129:173–177

Moyes, I.C.A., and Moyes, R.B. 1976b. Further developments in the study of 3′,5′ cyclic adenosine monophosphate in relation to psychiatric illness. Postgrad Med J 52:110–114

Mue, S., Ise, T., Ono, Y., and Kisaburo, A. 1975. Leukocyte adenyl cyclase activity in human bronchial asthma. Tohoku T Exp Med 117:135–142

Murphy, D.L., Donnelly, C., and Moskowitz, J. 1973. Inhibition by lithium of prostaglandin E_1 and norepinephrine effects on cyclic adenosine monophosphate production in human platelets. Clin Pharmacol Ther 14:810–814

Murphy, D.L., Donnelly, C., and Moskowitz, J. 1974. Catecholamine receptor function in depressed patients. Am J Psychiat 131:1389–1391

Nathanson, J.A. 1977. Cyclic nucleotide and nervous system function. Physiol Rev 57:157–256

Naylor, G.J., Stansfield, D.A., Whyte, S.F., and Hutchinson, F. 1974. Urinary excretion of adenosine 3′,5-cyclic monophosphate in depressive illness. Brit J Psychiat 125: 275–279

Owen, P., and Moffat, A.C. 1973. Variation of cyclic-A.M.P. excretion with urine volume. Lancet II:1205
Palmer, G.C. 1972. Increased cyclic AMP response to norepinephrine in the rat brain following 6-hydroxydopamine. Neuropharmacology 11:145–149
Palmer, G.C. 1976. Influence of tricyclic antidepressants on the adenylate cyclase-phosphodiesterase system in the rat cortex. Neuropharmacology 15:1–7
Palmer, G.C., Robison, G.A., Manian, A.A., and Sulser, F. 1972. Modification by psychotropic drugs of the cyclic AMP response to norepinephrine in the rat brain *in vitro*. Psychopharmacologia 23:201–211
Palmer, G.C., Robison, G.A., and Sulser, F. 1971. Modification by psychotropic drugs of the cyclic adenosine monophosphate response to norepinephrine in rat brain. Biochem Pharmacol 20:236–239
Pandey, G.N., Casper, R., Chang, S., and Davis, J.M. 1978a. Platelet adenylate cyclase in affective illness and schizophrenia, in preparation.
Pandey, G.N., Garver, D.L., Dysken, M., and Davis, J.M. 1978b. β-adrenergic receptor function in affective illness. Abs Proc Amer Psych Assoc 131(b). New Research, Amer J Psychiat
Pandey, G.N., Garver, D.L., Tamminga, C., Ericksen, S., Ali, S.I., and Davis, J.M. 1977. Postsynaptic supersensitivity in schizophrenia. Am J Psychiat 134:518–522
Pandey, G.N., Heinze, W.J., Jones, F.D., and Davis, J.M. 1976. The effect of electroconvulsive shock on brain adenylate cyclase activity. Trans Amer Soc Neurochem 7:111
Paul, M.I., Cramer, H., and Bunney, W.E., Jr. 1971. Urinary adenosine 3′,5′-monophosphate in the switch process from depression to mania. Science 171:300–303
Paul, M.I., Ditzion, B.R., and Janowsky, D.S. 1970b. Affective illness and cyclic-A.M.P. excretion. Lancet I:88
Paul, M.I., Ditzion, B.R., Pauk, G.L., and Janowsky, D.S. 1970a. Urinary adenosine 3′,5′-monophosphate excretion in affective disorders. Amer J Psychiat 126:1493–1497
Post, R.M., Cramer, H., and Goodwin, F.K. 1977. Cyclic AMP in cerebrospinal fluid of manic and depressive patients. Psychol Med 7:599–605
Prange, A.J., Jr., Wilson, I.C., Knox, A., McClane, T.K., and Lipton, M.A. 1970. Enhancement of imipramine by thyroid stimulating hormone. Clinical and theoretical implications. Am J Psychiat 127:191–199
Robison, G.A., Arnold, A., and Hartmann, R.C. 1969a. Divergent effects of epinephrine and prostaglandin E_1 on the level of cyclic AMP in human blood platelets. Pharmacol Res Comm 1:325
Robison, G.A., Butcher, R.W., and Sutherland, E.W. 1969b. On the Relation of hormone receptors to adenyl cyclase. In *Fundamental Concepts in Drug-Receptor Interactions*, eds. J.F. Danielli, J.F. Morau, and D.J. Triggle, pp. 59–81. New York: Academic Press
Robison, G.A., Coppen, A.J., Whybrow, P.C., and Prange, A.J. 1970. Cyclic A.M.P. in affective disorders. Lancet II:1028–1029
Sarai, K., Frazer, A., Brunswick, D., and Mendels, J. 1978. Desmethylimipramine induced decrease in β-adrenergic receptor binding in rat cerebral cortex. Biochem Pharmacol, in press
Scott, R.E. 1970. Effects of prostaglandins, epinephrine and NaF on human leukocyte, platelet and liver adenyl cyclase. Blood 35:514–516
Seeman, P., Lee, T., and Chau-Wong, K. 1976. Antipsychotic drug doses and neuroleptic/dopamine receptors. Nature 261:717–718
Shimizu, H., Daly, J.W., and Creveling, C.R. 1969. A radioisotopic method for measuring the formation of adenosine 3′,5′-monophosphate in incubated slices of brain. J Neurochem 16:1609–1619
Siggins, G.R., Hoffer, B.J., and Bloom, F.E. 1969. Cyclic AMP: Possible mediator for norepinephrine effects on cerebellar Purkinje cells. Science 165:1018–1020
Sinanan, K., Keatinge, A.M.B., Beckett, P.G.S., and Love, W.C. 1975. Urinary cyclic AMP in 'endogenous' and 'neurotic' depression. Brit J Psychiat 126:49–55
Somerville, A.R. 1973. Biochem Soc Spec Publ 1:127–132
Spitzer, R., Endicott, J., and Robbins, L. 1975. Research diagnostic criteria (RDC)

for a selected group of functional disorders, 2nd ed. New York: New York State Psychiatric Institute, Biometrics Research
Sutherland, E.W., and Robison, G.A. 1966. Metabolic effects of catecholamines. Section A. The role of cyclic-3′,5′-AMP in response to catecholamines and other hormones. Pharmacol Rev 18:145-161
Sutherland, E.W., Robison, G.A., and Butcher, R.W. 1968. Some aspects of the biological role of adenosine 3′,5′-monophosphate (cyclic AMP). Circulation 37:279-306
Vetulani, J., Stawarz, R.J., Dingell, J.V., and Sulser, F. 1976a. A possible common mechanism of action of antidepressant treatments. Naunyn-Schmiedeberg's Arch Pharmacol 293:109-114
Vetulani, J., Stawarz, R.J., and Sulser, F. 1976b. Adaptive mechanisms of the noradrenergic cyclic AMP generating system in the limbic forebrain of the rat: Adaptation to persistent changes in the availability of norepinephrine (NE). J Neurochem 27:661-666
Vetulani, J., and Sulser, F. 1975. Action of various antidepressant treatments reduces reactivity of noradrenergic cyclic AMP-generating system in limbic forebrain. Nature 257:495-496
Wang, Y.C., Pandey, G.N., Mendels, J., and Frazer, A. 1974a. Effect of lithium on prostaglandin E_1-stimulated adenylate cyclase activity of human platelets. Biochem Pharmacol 23:845-855
Wang, Y.C., Pandey, G.N., Mendels, J., and Frazer, A. 1974b. Platelet adenylate cyclase responses in depression: Implications for a receptor defect. Psychopharmacologia 36:291-300
Williams, R.H., Barish, J., and Ensinck, J.W. 1972. Hormone effects upon cyclic nucleotide excretion in man. Proc Soc Exp Biol Med 139:447-454
Williams, L.T., Snyderman, R., and Lefkowitz, R.J. 1976. Identification of β-adrenergic receptors in human lymphocytes by (—)[^{3}H] alprenolol binding. J Clin Invest 57:149-155

Sensitivity of Noradrenergically Mediated Cyclic AMP Responses in the Brain

*H. Ryan Wagner**,
*Gene C. Palmer***, *and James N. Davis**
*Departments of Medicine and Pharmacology
Duke University School of Medicine
Veterans Administration Hospital
Durham, North Carolina 27705
**Department of Pharmacology
University of South Alabama
College of Medicine
Mobile, Alabama 36688

INTRODUCTION

Noradrenergic neurons in the central nervous system have the capacity to exhibit considerable plasticity in response to altered levels of neuronal activity. At presynaptic sites changes in neuronal impulse flow may alter rates of transmitter release, reuptake, enzymatic synthesis or degradation. Postsynaptic adaptations may include changes in the density of receptor sites, altered activity in the receptor-coupled enzyme, adenylate cyclase, or in 3′,5′-cyclic adenosine monophosphate (cAMP)-deactivating phosphodiesterases. Depending on the antecedent physiological conditions, such changes may be mediated by short-term regulatory or allosteric modifications which occur and fade within a matter of minutes or by long-term inductive processes which develop and extinguish over a period of days. These modifications may occur singularly or in combination, but their summated effects inevitably reflect homeostatic pressures toward the restoration of some preexisting functional level of neuronal tone.

The noradrenergically activated α- and β-receptor-coupled adenylate cyclases (*Schultz* and *Daly*, 1973a) in this system represent two of the few identified receptor-coupled biochemical responses in the brain and for this reason have been extensively studied since they were first described by *Kakiuchi* and *Rall* (1968). Recently, understanding of this receptor-effector complex has been further extended with the use of newly developed radioligand binding techniques to demonstrate adenylate

The authors have carried out research in this area supported by the NIH (MH06058, NS06233, NS13101) and the VA (1680). Ms. Liz Inscoe provided expert secretarial assistance in the preparation of this manuscript.

cyclase-coupled β-adrenergic membrane receptors in brain[1] (*Alexander, Davis*, and *Lefkowitz*, 1975). Based on such studies, it now appears that noradrenergically induced cAMP sensitivity may be regulated by changes in 1) the density or affinity of membrane receptors, 2) the coupling of the receptor and its catalytic moiety, 3) the activities of adenylate cyclase or phosphodiesterase, 4) protein activators and modulators of these enzymes, or 5) the availability of ATP or GTP to the membrane receptor complex. Against this background, we review data on postsynaptic modifications occurring in noradrenergic receptor-coupled cAMP generating systems in the brain during and following periods of altered neuronal activity (i.e., during states of supersensitivity and subsensitivity in noradrenergically elicited cAMP responses).

In this regard we propose that adaptive changes in noradrenergically mediated cAMP responses are of two basic types, acute and chronic. Acute changes in receptor response are triggered by rapid changes in the concentration of norepinephrine (NE) at postsynaptic receptor sites. Such changes develop and are reversed in a matter of minutes and most probably reflect allosteric or regulatory modification of sites in the receptor-effector chain. In contrast, chronic changes in receptor response develop following prolonged alterations in nerve impulse flow through adrenergic neurons. Development of and recovery from chronic effects occur over a course of days and most probably reflect inductive (or repressive) adjustments in the absolute number of some molecular component(s) in the receptor-effector response chain. Accordingly, this review is organized into three major sections, addressing chronic noradrenergic supersensitivity (acute noradrenergic supersensitivity has not been substantially demonstrated), acute noradrenergic subsensitivity and chronic noradrenergic subsensitivity. Within each section we discuss requisite conditions for development and extinction of the state, receptor characteristics, pharmacological specificity, anatomical and species specificity, and finally the underlying molecular mechanisms of each condition.

---

[1]By contrast to peripheral tissues, norepinephrine also elicits cAMP increases in the brain by interacting with α-adrenergic receptors (*Perkins* and *Moore*, 1973). Attempts to identify these α-adrenergic receptors by radioligand binding have been less than satisfying. Several radioligands bind to sites on brain membranes with some of the characteristics expected for an α-adrenergic receptor, but none of these brain binding sites have been clearly correlated with peripheral α-adrenergic receptors or receptor-mediated responses (*Davis et al.*, 1977; *Greenberg* and *Snyder*, 1978; *U'Pritchard, Greenberg*, and *Snyder*, 1977). Our own experience suggests that the binding of the ligand, [^{3}H]-WB 4101, may correlate with α-adrenergic components of noradrenergically mediated accumulations of cAMP (*Davis et al.*, 1978). However, for the purposes of this review we have included only those radioligand binding studies which measure brain β-adrenergic receptors.

CHRONIC SUPERSENSITIVITY IN NORADRENERGICALLY ELICITED cAMP RESPONSES

Noradrenergic supersensitivity in brain cAMP generating systems describes a condition of increased hormonal sensitivity induced by reduced levels of neuronal input at the postsynaptic junction. Exposure of supersensitive tissue to noradrenergic agonists results in increased production of cAMP as contrasted with the amount of cAMP elicited by agonists in control tissue not subjected to episodes of reduced neuronal activity. The state has been experimentally induced in animals by chemical sympathectomy with 6-hydroxydopamine (6-OHDA) (*Blumberg et al.*, 1976; *Dismukes et al.*, 1975; *Huang, Ho,* and *Daly*, 1973; *Kalisker, Rutledge*, and *Perkins*, 1973; *Nahorski*, 1977; *Palmer*, 1972; *Palmer* and *Scott*, 1974; *Skolnick* and *Daly*, 1976, 1977; *Sporn et al.*, 1976), through reserpine-induced depletion of brain monoamines (*Blumberg et al.*, 1976; *Dismukes* and *Daly*, 1974; *Kakiuchi, Rall*, and *McIlwain*, 1969; *Baudry, Martes,* and *Schwartz*, 1976; *Nahorski*, 1977; *Palmer, French,* and *Narod,* 1976a; *Palmer, Sulser,* and *Robison*, 1973; *Palmer, Wagner,* and *Putnam,* 1976b; *Wagner et al.*, 1978; *Williams* and *Pirch*, 1974), following withdrawal from a schedule of chronic ethanol ingestion (*French et al.*, 1974, 1975; *Israel, Kimura,* and *Kuriyama,* 1972), and after electrolytic lesions of brain catecholamine tracts (*Dismukes et al.*, 1975; *Eccelston,* 1973). Although these states each reflect conditions of enhanced sensitivity, they vary from one another in many of their characteristics.

Rates of Appearance and Disappearance

The rate of development of noradrenergic supersensitivity in brain cAMP systems varies according to the type of experimental manipulation used in its induction. In the case of 6-OHDA, *Kalisker et al.* (1973) found increased accumulations of cAMP in slice preparations of rat cerebral cortex in response to 1 μM norepinephrine (NE) 48 hr after the second of two intraventricular injections of 6-OHDA. Responses to 30 μM NE were not elevated at this time but were significantly increased 96 hr following treatment. The EC_{50} values for NE-elicited cAMP accumulations were decreased at both 48 and 96 hr. Addition of cocaine to the system reduced the EC_{50} for NE-induced cAMP accumulations in slices obtained from unlesioned control animals but did not affect the EC_{50} in 6-OHDA-lesioned animals. Maximum responses to NE were not changed by cocaine in either the control or experimental conditions, nor did 6-OHDA treatment alter the EC_{50} for isoproterenol (ISO)-induced accumulations of cAMP. Since cocaine is known to block NE reuptake into presynaptic neurons and since ISO is a poor substrate for reuptake, increased NE sensitivity of slices from lesioned animals after 48 hr appeared to reflect a loss of presynaptic reuptake mechanisms during the degeneration of 6-OHDA-lesioned nerve terminals.

In support of this view, the authors noted a high correlation (r = 0.91) between increased effectiveness of low (1 μM) concentrations of NE to stimulate cAMP production and the percentage inhibition of [^{3}H]-NE uptake in cortical slices obtained from lesioned animals. In contrast with the altered state of the system at the earlier time, the enhanced response to 30 μM NE after 96 hr included a twofold increase in maximum NE-elicited cAMP accumulations. The latter effect was not enhanced by cocaine and occurred at a time when degeneration of the presynaptic terminals as measured by the percentage inhibition of [^{3}H]-NE uptake was complete, suggesting that increased responsiveness after 96 hr occurred through postsynaptic alterations.

From these data, *Kalisker et al.* (1973) concluded that changes in noradrenergically coupled cAMP responses following 6-OHDA treatment reflect two independent processes: 1) a fast-developing increase in NE sensitivity occurring as presynaptic reuptake capacity is lost during the degeneration of nerve terminals, and 2) a longer developing (4-day) increase in maximal NE-elicited cAMP capacity of the system related to the induction of postsynaptic noradrenergic supersensitivity.

Recovery from 6-OHDA-induced postsynaptic supersensitivity is an even slower process and, in fact, may not occur. In this regard, *Skolnick* and *Daly* (1977) found elevations in adrenergic cAMP sensitivity in slices of rat cerebral cortex up to 25 days after treatment of animals with 6-OHDA (longer time intervals were not studied), and *Dismukes* and *Daly* (1975) noted adrenergic cAMP supersensitivity in slices made from either cerebral cortex or midbrain in 4- and 5-month-old rats lesioned at birth with 6-OHDA. Using a similar paradigm, *Palmer* and *Scott* (1974) reported adrenergic supersensitivity in slices of cerebral cortex 35 days post partum in rats treated at birth with 6-OHDA. Increased cAMP accumulations were also observed in rats lesioned on post partum days 5 and 6 but not in animals lesioned on post partum day 14. The latter observation suggests caution in interpreting data on the time course of noradrenergic cAMP supersensitivity obtained from immature animals as it may be confounded with developmental variables.

Rates of development of reserpine-induced noradrenergic supersensitivity in brain cAMP generating systems generally parallel 6-OHDA-induced rates. In this regard, *Dismukes* and *Daly* (1974) observed significant increases in NE-elicited accumulations of cAMP in slice preparations of rat cerebral cortex 24 hr following the second of two intraperitoneal (i.p.) 5.0 mg/kg injections of reserpine; responsiveness 5 hr after the first injection was not significantly increased. Once developed, supersensitivity was still demonstrable on the ninth day following the end of treatment but not on the sixteenth day. Using a slightly different treatment schedule,

Baudry, Martes, and *Schwartz* (1976) administered 5.0 mg/kg (i.p.) of reserpine to rats on the first day of treatment and 1.0 mg/kg on each subsequent day. Under this schedule significant increases in NE-elicited cAMP responsiveness occurred on the second day of treatment and reached a maximum by the fourth day. In the guinea pig, *Kakiuchi et al.* (1969) found significant increases in the cAMP accumulations induced by electrical stimulation of a slice preparation of cerebral cortex following reserpine treatments (i.p.) of 0.5 mg/kg and 2.0 mg/kg at 44- and 20- hr intervals before decapitation, while noradrenergic supersensitivity in cAMP responsiveness to ISO was reported (*Nahorski,* 1977) in a slice preparation from the cerebral hemispheres of neonatal chicks 1 day following the second of two subcutaneous (s.c.) injections of reserpine (2.5 mg/kg/day). Finally, *Palmer et al.* (1976b) did not find noradrenergic cAMP supersensitivity until the fourth day of treatment in slice preparations of cerebral cortex from rats injected (i.p.) daily with 2.5 mg/kg reserpine. Although varying with species and dose schedules, these data suggest that reserpine-induced supersensitivity in brain cAMP systems requires a period of 48 to 96 hr after the beginning of treatment for development.

In addition to 6-OHDA and reserpine, noradrenergic cAMP supersensitivity also has been induced by electrolytically lesioning the medial forebrain bundle (*Dismukes et al.,* 1975) and locus ceruleus (*Eccleston,* 1973). In the former case, increased noradrenergic cAMP sensitivity was found 2 days after lesioning. Maximum responses occurred on day 9 and were still significantly elevated 3 wk later (the longest time interval tested). Similarly, *Eccleston* (1973) reported noradrenergic cAMP supersensitivity in samples of cerebral cortex ipsilateral to a locus ceruleus lesion 4 wk after lesioning (shorter intervals were not described). Possible mediation of supersensitivity at presynaptic sites (see above) was not directly tested in either study. However, in the former study maximum cAMP responses to both ISO and NE were enhanced 2 wk post lesion, suggesting a postsynaptic locus since ISO is a poor substrate for presynaptic reuptake mechanisms. Finally, development of noradrenergic cAMP supersensitivity following withdrawal from chronic ethanol ingestion has not been examined parametrically with respect to time but is known to occur within 3 days following the onset of withdrawal (*French* and *Palmer,* 1973).

In summary of these rate data, supersensitivity of noradrenergically mediated cAMP responses in the brain develop, depending on the species and mode of induction, within 2 to 4 days following interruption in neuronal activity. Enhanced sensitivity developing at earlier time points in 6-OHDA-treated animals and perhaps in electrolytically lesioned animals reflects the loss of presynaptic reuptake mechanisms and is not related to postsynaptic receptor-mediated processes. Once developed, the response is enduring, persisting for at least 9 days after cessation of reserpine treatments and for as long as 5 mo following 6-OHDA lesions.

Changes in Maximum Responses and Agonist Affinities

The slow rate of development of noradrenergic cAMP supersensitivity in brain suggests that the phenomenon occurs by an increase in the synthesis of some molecular component in the receptor-effector chain. Consistent with this position, increases in the maximum response of the brain cAMP generating system to saturating amounts of adrenergic agonists were reported in slice preparations of rat cerebral cortex following 6-OHDA treatment (*Huang et al.*, 1973; *Kalisker et al.*, 1973; *Sporn et al.*, 1976), in cortical slices from reserpine-treated rats (*Baudry et al.*, 1976; *Dismukes* and *Daly*, 1974; *Palmer et al.*, 1973,1976b) and neonatal chicks (*Nahorski*, 1977), and in rats during ethanol withdrawal (*French* and *Palmer*, 1973).

With regard to 6-OHDA-induced supersensitivity, *Kalisker et al.* (1973), *Nahorski* (1977) and *Sporn et al.* (1976) observed no change in the apparent affinity of cAMP-generating systems for noradrenergic agonists apart from changes related to the loss of presynaptic nerve endings following 6-OHDA treatments (see above). At the receptor level, *Sporn et al.* (1976) reported no change in the K_i for ISO displacement of the β-adrenergic ligand, [^{125}I] hydroxybenzylpindolol (HYP), following 6-OHDA treatment, despite increased membrane receptor density. Although the latter observations are consistent with an increased maximum response, the percentage increase in receptor number was inadequate to account for the total increase in cAMP responsiveness. In the F-344 strain of rats after 6-OHDA treatment, the same increase in membrane receptor density was noted by *Skolnick et al.* (1978) as in Sprague-Dawley rats, but in contrast to the Sprague-Dawley rats no increase in cyclic AMP responsiveness was found in F-344 brain slices. These findings may reflect methodological limitations involved in comparing membrane receptor binding with tissue slice cyclic AMP accumulations. Alternatively, they may be explained in part by the fact that both α- and β-receptors appear to mediate cyclic AMP accumulations although only β-adrenergic membrane receptors were measured in the latter experiments (see below).

Reserpine-induced noradrenergic cAMP supersensitivity appears to involve similar processes. Alterations in agonist affinity (EC_{50}) were not observed in slice preparations of cerebral cortex obtained from reserpine pretreated rats (*Dismukes* and *Daly*, 1974) or neonatal chicks (*Nahorski*, 1977) despite increases in maximal noradrenergic cAMP response capacities in each system. In apparent conflict with these findings *Palmer*,

French, and *Narod* (1976) found significant increases in the EC_{50} of slice preparations of cerebral cortex obtained from rats pretreated with reserpine for 4 days. Maximum responses to NE were not significantly elevated. The same group (*French et al.,* 1975) also reported shifts in agonist affinity during the supersensitive state produced by ethanol withdrawal. It is important to note that rats in the latter two studies were anesthetized prior to reserpine injections and prior to decapitation. As the effects of the anesthetic on noradrenergic cAMP supersensitivity are unknown, the latter data must be interpreted with caution. Thus, the evidence seems to indicate that reserpine-induced noradrenergic cAMP supersensitivity occurs through processes similar to those involved in the 6-OHDA-induced state, i.e., through an increase in the maximum response capacity of the system with no change in agonist affinity.

α-Adrenergic versus β-Adrenergic Receptors

Both α- and β-adrenergically mediated cAMP responses in supersensitivity are well investigated. *Huang et al.* (1973) reported that the α-adrenergic antagonist, phentolamine, inhibited NE-induced accumulations of cAMP in slice preparations of cerebral cortex obtained from 6-OHDA-treated Sprague-Dawley rats by 40-50%. In this same system, the β-adrenergic antagonist, propranolol, inhibited NE-induced responses by 80–85%, while cAMP responses elicited by ISO were blocked completely. Percent inhibition of NE-elicited responses by the two antagonists remained the same in control and 6-OHDA lesioned animals. In agreement with the latter observation, *Kalisker et al.* (1973) also reported finding that 6-OHDA-induced noradrenergic cAMP supersensitivity was mediated by proportional increases in both α- and β-receptor-coupled components. Similar changes may occur during ethanol-induced noradrenergic cAMP supersensitivity (*French* and *Palmer,* 1973), although high basal levels and facilitation of NE-induced cAMP accumulations by the α-adrenergic antagonist, phenoxybenzamine, make the latter data difficult to interpret (see above for possible methodological artifacts related to anesthetic use in the latter study).

In contrast with the 6-OHDA-induced state in Sprague-Dawley rats, enhanced noradrenergic cAMP responsiveness following reserpine treatment is mediated exclusively by increases in β-adrenergic responses. Addition of propranolol to slice preparations of rat cerebral cortex eliminated reserpine-induced increments in NE-elicited cAMP responsiveness (*Dismukes* and *Daly,* 1974; *Palmer et al.* 1976b). Conversely, phentolamine was without effect on the reserpine-enhanced portion of the NE response. Although unresolved, differences between reserpine treatment and 6-OHDA treatment in Sprague-Dawley rats do not reflect the loss of presynaptic elements in 6-OHDA-treated animals, as noradrenergic cAMP supersensitivity induced by electrolytic lesions of the median forebrain bundle also appear to be mediated predominantly by increases in β-adrenergic components (*Dismukes et al.*, 1975).

Molecular Mechanisms

Despite considerable study, the molecular mechanism(s) underlying supersensitivity phenomena remain vague. In this regard basal levels of cAMP in Sprague-Dawley rats were unaffected following 6-OHDA treatment (*Dismukes* and *Daly*, 1975; *Huang et al.*, 1973; *Kalisker et al.*, 1973; *Nahorski*, 1977; *Palmer*, 1972; *Skolnick* and *Daly*, 1977), reserpine (*Baudry et al.*, 1976; *Dismukes* and *Daly*, 1974; *Nahorski*, 1977; *Palmer et al.*, 1976a & b), electrolytic lesions of the medial forebrain bundle (*Dismukes et al.*, 1975) or during withdrawal from chronic ethanol treatment (*French* and *Palmer*, 1973). NaF-stimulated activity of the catalytic component of adenylate cyclase was not altered by exposure to 6-OHDA (*Kalisker et al.*, 1973), reserpine (*Nahorski*, 1977; *Palmer et al.* 1976b) or by chronic ethanol ingestion (*Kuriyama* and *Israel*, 1973). Similarly, decreased phosphodiesterase activity did not affect supersensitivity induced by 6-OHDA (*Huang et al.*, 1973) or reserpine (*Dismukes* and Daly, 1974; *Nahorski*, 1977) as estimated from the ability of phosphodiesterase inhibitors to alter noradrenergic responses. However, *Kalisker et al.* (1973) observed small (25%) but significant decreases in both the low and high activity forms of phosphodiesterase as assayed directly in preparations of cerebral cortex from 6-OHDA-treated rats. In contrast, *Nahorski* (1977) found no difference in direct assays of phosphodiesterase activity in samples of cerebral cortex obtained from reserpine-treated neonatal chicks. Based on these data, the role of phosphodiesterases in noradrenergic supersensitivity, although equivocal, does not appear to be central.

Of the three studies investigating involvement of membrane receptors in the supersensitivity condition, *Sporn et al.* (1976) reported a 30% increase in the maximum density of β-adrenergic receptor sites as measured with the radioligand, HYP, in the cerebral cortex of 6-OHDA-treated rats. At the same time, ISO-induced accumulations of cAMP were elevated 80% in lesioned animals. In addition to the failure to observe correspondence between percent changes in receptor density and biological responsiveness (i.e., ISO-induced cAMP accumulations), the K_i for ISO displacement of HYP was 10-fold larger than the EC_{50} for ISO-induced accumulations of cAMP. *Skolnick et al.* (1978) also observed a 25% increase in the density of β-adrenergic receptor binding without a change in affinity and without a correlation to the changes in cyclic AMP responses in two different strains of rats. In contrast with the findings of *Sporn et al.* (1976) and *Skolnick et al.* (1978), *Nahorski* (1977) saw no change in the binding of the β-adrenergic ligand, [^{3}H]-propranolol, to membranes made from the cerebral cortex of neonatal chicks treated *in vivo* with either reserpine or 6-OHDA. Despite a lack of effect at the receptor level, ISO-induced cAMP responses in slices of cerebral cortex prepared from the same animals were significantly elevated.

These discrepancies could reflect 1) the involvement of α-adrenergic receptors in the supersensitivity process, 2) the existence of nonlinear relationships between receptor occupation and enzyme activation, or 3) technical artifacts (see below) related to differences between the experimental conditions used in measurements of receptor binding and cAMP-generating capacity in brain systems. It should be noted that the equivocal nature of the data does not allow any conclusions about the molecular basis of supersensitivity to be drawn at this time.

Pharmacological Specificity

Supersensitivity in cAMP responsiveness developing after interruption of noradrenergic activity is relatively specific. Increased cAMP accumulations after 6-OHDA treatments have been reported in response to NE, ISO and ISO-adenosine combinations but not to dopamine (DA), serotonin (5-HT), veratradine, adenosine or to combinations of adenosine-5-HT or histamine-5-HT (*Huang et al.*, 1973; *Kalisker et al.*, 1973). In slices from reserpine-treated animals, increased cAMP accumulations were not found in response to veratradine or 5-HT or to DA or histamine in the presence of phosphodiesterase inhibitors. At the same time responses to NE or DA were enhanced. *Palmer, Wagner,* and *Putnam* (1976b) reported increased sensitivity in broken cell preparations of rat cerebral cortex adenylate cyclase to NE and DA following reserpine treatment. DA supersensitivity was not seen in neuronal or glial-enriched fractions obtained from reserpine-treated animals. cAMP accumulations in adult rats lesioned neonatally with 6-OHDA showed enhanced hormonal sensitivity to NE, ISO, NE-adenosine and prostaglandin E_1 but not to adenosine (*Dismukes* and *Daly*, 1975). Supersensitivity of cAMP responses following electrolytic lesions of the medial forebrain bundle occurred in response to NE, ISO and histamine plus the phosphodiesterase inhibitor, isobutylmethylxathine (IBMX), while sensitivity to adenosine, veratradine, prostaglandin E_1, 5-HT and IBMX was not increased (*Dismukes et al.*, 1975).

From these data, interruption of adrenergic transmission appears to produce a relatively specific increase in the sensitivity of cAMP responsiveness to noradrenergic agents. Sensitivity to adenosine by itself or in combination with various agents also has been noted, although the role of adenosine in adrenergically coupled cAMP systems remains unclear

(*Schultz* and *Daly*, 1973b). This selectivity is even more impressive given the relative lack of selectivity (for noradrenergic systems) of many of the techniques used in the induction of the supersensitive state.

Anatomical and Species Specificity

Noradrenergic cAMP supersensitivity was convincingly demonstrated in the rat cerebral cortex following treatment of animals with 6-OHDA (*Huang et al.*, 1973; *Kalisker et al.*, 1973; *Palmer*, 1972; *Palmer* and *Scott*, 1974; *Skolnick* and *Daly*, 1977; *Sporn et al.*, 1976) or reserpine (*Baudry et al.*, 1976; *Dismukes* and *Daly*, 1974; *Palmer et al.*, 1973, 1976a & b; *Wagner et al.*, 1978), after electrolytic lesions of the medial forebrain bundle (*Dismukes et al.*, 1975) and during withdrawal from chronic ethanol ingestion (*French* and *Palmer*, 1973; *French et al.*, 1975). Noradrenergic cAMP supersensitivity was also reported following reserpine treatment in the cerebral cortex of the rabbit (*Kakiuchi* and *Rall*, 1968) and neonatal chick (*Nahorski*, 1977) and in the chick cerebral cortex following 6-OHDA treatment (*Nahorski*, 1977). Looking between two rat strains, *Skolnick* and *Daly* (1977) reported finding 6-OHDA-induced cAMP supersensitivity in slice preparations of cerebral cortex from Sprague-Dawley rats but not in the cortical samples from F-344 rats, a second rat strain characterized by abnormally high responsiveness of noradrenergically mediated cAMP systems. Interestingly, neither strain demonstrated noradrenergic supersensitivity in slices prepared from mesencephalic tissue, despite profound depletion of endogenous hypothalamic NE.

Despite the consistency with which supersensitivity has been demonstrated in the cerebral cortex, the phenomenon does not appear to occur uniformly throughout the brain. In this regard, *Dismukes* and *Daly* (1975) reported enhanced NE sensitivity in cortical and midbrain slices from adult rats which had been injected neonatally with 6-OHDA but not in slice samples from the pons-medulla or cerebellum. *Palmer* (1972) found increased noradrenergic cAMP sensitivity following 6-OHDA administration to adult rats in slice preparations of rat cerebral cortex, brain stem and hypothalamus but not in slices prepared from cerebellar tissue. *Palmer et al.* (1973) also reported enhanced noradrenergic sensitivity after reserpine treatment in slice preparations from the cerebral cortex, hypothalamus and hippocampus but not in slices from the brainstem, midbrain, cerebellum and striatum. Electrolytic lesions of the medial forebrain bundle produced enhanced noradrenergic cAMP responsiveness in slice preparations from rat cerebral cortex, cerebellum and hippocampus but not from the midbrain (*Dismukes et al.*, 1975). Finally, *Palmer*, *Wagner*, and *Putnam* (1976b) reported enhanced sensitivity to NE in neuronal fractions obtained from the cerebral cortex of reserpine-treated rats but not in glial-enriched fractions obtained from the same tissue source.

Based on these data, noradrenergic supersensitivity of brain cAMP responses occurs at neuronal sites in the cerebral cortex of numerous species. Supersensitivity has been less consistently observed in slice preparations from the rat midbrain, pons-medulla and hypothalamus. Finally, although noradrenergic cAMP supersensitivity has been demonstrated in a number of species, the failure of *Skolnick* and *Daly* (1977) to observe the phenomenon in the F-344 rat strain cautions against broad generalizations about the universality of chronic supersensitivity.

Summary

Noradrenergic supersensitivity of brain cAMP systems occurs in a number of species and within discrete brain areas. It develops 48 to 72 hr following interruption of noradrenergic transmission to postsynaptic elements and is relatively long-lasting once developed. Although speculative, rates of development are consistent with inductive rather than allosteric processes. The condition is relatively specific for noradrenergic agonists and may be mediated by both α- and β-adrenergic receptor-coupled responses, depending on techniques of induction. The underlying molecular mechanisms are poorly understood, although changes in phosphodiesterase activity and postsynaptic receptor density have been implicated to some degree in the mediation of this phenomenon.

ACUTE SUBSENSITIVITY IN NORADRENERGICALLY ELICITED cAMP RESPONSES

Noradrenergic subsensitivity of brain cAMP systems describes a condition of reduced responsiveness in postsynaptic noradrenergically mediated adenylate cyclase responses resulting from prior exposure of the system to noradrenergic agonists. Subsensitivity has been induced by both direct exposure of brain slices to noradrenergic agonists and following treatment of animals with pharmacological agents which facilitate noradrenergic transmission, (e.g., antidepressant drugs and amphetamines).

Unlike noradrenergic supersensitivity, which develops over a number of days, postsynaptic noradrenergic subsensitivity comprises two separate phenomena, one short-term and one long-term. We consider first the characteristics of acute noradrenergic cAMP subsensitivity.

Rates of Appearance and Disappearance

Parametric studies have not been conducted on the development of acute subsensitivity. Despite this, it is clear from available data that the phenomenon occurs rapidly. *Schultz* and *Daly* (1973c) reported decreased sensitivity in slice preparations from guinea pig cerebral cortex to combinations of histamine-NE following prior exposure of the slices to combinations of histamine-NE or adenosine-NE. Although not tested at earlier intervals, desensitization was complete within 10 min after initial hormone exposure. In a similar study using rats, *Schultz* and *Daly* (1973a) reported subsensitive responses to NE after a 5-min preexposure. Using an *in vivo* paradigm with neonatal chicks, *Nahorski* (1977) reported decreased ISO-induced accumulations of cAMP in slice preparations of cerebral cortex in chicks exposed to ISO 3 hr previously.

Development of acute noradrenergic subsensitivity has been studied parametrically at the receptor level. Using rat cerebral cortex, *Wagner, Bartolome,* and *Davis* (1977) and *Wagner* and *Davis* (1977) observed that ISO-induced accumulations of cAMP in rat cortex slices were reduced by a 30-min prior exposure of the slices to 10 μM ISO. These investigators noted decreased binding of the β-adrenergic radioligand, [^{3}H]-dihydroalprenolol (DHA), to membrane preparations obtained from the slices preincubated with 10 μM ISO. Decreased DHA binding was apparent within 10 min after exposure of slices to ISO, with maximum effects obtained after 20 min. Recovery of DHA binding sites occurred within 10 to 20 min following removal of ISO from the bath and could be markedly facilitated by exposure of desensitized slices to agents known to cause cell depolarization (*Wagner* and *Davis*, 1977).

Acute noradrenergic subsensitivity of cAMP responses is developed within 5 to 20 min after exposure of slice preparations to agonist agents (*Wagner et al.*, 1977). Recovery of hormonal sensitivity (as estimated from the time course of receptor recovery observed in ISO-treated slices) also occurs rapidly (10 to 20 min) and indicates that both development of acute noradrenergic cAMP subsensitivity and its subsequent reversal are rapid processes occurring within minutes following exposure of brain slices to noradrenergic agonists. Rapid changes in membrane receptor density appear to be at least partly involved in short-term subsensitivity.

Changes in Maximum Responses and Agonist Affinities

Little work has been directed at characterizing states of decreased hormonal sensitivity in brain cAMP-generating systems. In the rabbit cerebellum (*Kakiuchi* and *Rall,* 1968), rat cerebral cortex (*Schultz* and *Daly*, 1973a; *Wagner* and *Davis*, in preparation) and guinea pig cerebral cortex (*Schultz* and *Daly*, 1973c), reduced accumulations of cAMP were obtained to concentrations of NE or ISO in excess of those required for maximal stimulation of the cAMP system in control animals. Although these findings suggest that maximum cAMP responses are decreased in the subsensitive state, none of these works reported concentration-response curves. In the only case where this was done, *Nahorski* (1977) reported decreases in maximal ISO-induced accumulations of cAMP in slice preparations of cerebral cortex obtained from neonatal chicks previously treated with ISO, while the EC_{50} remained unaltered. Determination of [^{3}H]-propranolol binding in these animals showed a decrease in maximum receptor density with no change in the affinity of ISO for the receptor site. *Wagner et al.* (1977) also reported decreased density of DHA binding sites in ISO-exposed slices of rat cerebral cortex with no change in K_D. Subsequent exposure of ISO-treated slices to depolarizing agents resulted in increased receptor density, again without affinity change (*Wagner* and *Davis*, 1977).

In summary of these limited data, it appears that acute noradrenergic cAMP subsensitivity occurs through a decrease in the maximal responsiveness of the system, while the EC_{50} remains unaltered.

Pharmacological Specificity

Kakiuchi and *Rall* (1968) reported decreased responsiveness of slice preparations obtained from rabbit cerebellum to NE following preexposure of those slices to NE. The response to histamine at the same time was unaffected. Less specific effects were reported by *Schultz* and *Daly* (1973c) who found that exposure of slice preparations of guinea pig cerebral cortex to histamine-NE combinations produced reduced responsiveness to histamine and to histamine-NE combinations but not to adenosine or NE. Exposure of rat cerebral cortex to NE led to decreased responses to NE upon rechallenge of the slices with NE but not to adenosine-NE combinations.

Molecular Mechanisms

Schultz and *Daly* (1973c), working with slice preparations of guinea pig cerebral cortex, and *Nahorski* (1977), working with slice preparations of neonatal chick cerebral cortex, reported that treatment of noradrenergically subsensitive slices with phosphodiesterase inhibitors eliminated the acute subsensitive condition. In the guinea pig cortex phosphodiesterase-induced restoration of cAMP responsiveness was partial, but in the chick brain inhibition of phosphodiesterase activity produced complete restoration of ISO-induced cAMP responsiveness. Despite these data, chick brain phosphodiesterase activity was unchanged from control levels in ISO-treated chicks as assessed by direct assay (*Nahorski*, 1977). In concomitant investigations of β-adrenergic receptor binding, *Nahorski* (1977) observed a 30% decrease in receptor density (as measured with [^{3}H]-propranolol) at a time when ISO-induced cAMP accumulations were decreased by approximately 70%. Basal cAMP levels in slice preparations were not altered by exposure to ISO. Similarly, ISO exposure did not alter guanyl nucleotide- or NaF-stimulated activity of adenylate cyclase. *Wagner* and *Davis* (in preparation) made similar observations in slice preparations of rat cerebral cortex. In the latter work exposure of slices to ISO for 30 min reduced ISO-induced cAMP accumulations by 65% at a time when DHA binding was decreased by only 25%. Basal cAMP levels were unaltered by exposure to ISO.

Summary

Exposure of noradrenergically coupled cAMP systems to β-adrenergic agonists leads to a rapid decrease in the ability of the system to produce cAMP when subsequently rechallenged with agonists. The phenomenon develops quickly (5-20 min) and is rapidly reversed following removal of the agonist. Mediation of acute subsensitivity at the molecular level appears to involve decreases in the activity of cAMP-deactivating phosphodiesterases and in the density of β-adrenergic membrane receptor sites.

CHRONIC SUBSENSITIVITY IN NORADRENERGICALLY ELICITED cAMP RESPONSES

Chronic exposure of brain β-adrenergic receptor-coupled cAMP generating systems to pharmacological agents which facilitate noradrenergic transmission leads to decreased responsiveness of the system to subsequent stimulation by noradrenergic agonists. Chronic exposure to amphetamines (*Baudry et al.*, 1976; *Martes, Baudry,* and *Schwartz*, 1975), antidepressant drugs (*Frazer et al.*, 1974; *Vetulani et al.*, 1976a; *Vetulani, Stawarz,* and *Sulser*, 1976b; *Vetulani* and *Sulser*, 1975), monoamine oxidase inhibitors (*Vetulani et al.*, 1976b) or electroconvulsive shock (*Vetulani et al.*, 1976a; *Vetulani* and *Sulser*, 1975) produced conditions of chronic noradrenergic cAMP subsensitivity. In fact, these states may also include acute allosteric modifications (see below) which potentially confound assessment of slower developing rates of inductive changes. Despite this, rates of recovery from subsensitivity differ profoundly between acute and chronic states and provide the distinguishing characteristics between these conditions.

Rates of Appearance and Disappearance

Development of chronic noradrenergic cAMP subsensitivity after treatment of animals with antidepressant agents (commonly tricyclic drugs) is a relatively slow process. *Vetulani et al.* (1976b) observed noradrenergic subsensitivity in cAMP systems in slice preparations of cerebral cortex from rats treated daily for 4 to 8 wk with the tricyclic drug, desipramine, or the antidepressant drug, iprindole. Animals treated for 1 to 2 wk did not show subsensitive responses. In contrast, *Schultz* (1976) found decreased noradrenergic cAMP sensitivity in rats on the sixth day following oral ingestion of the neuroleptic drug, chlorpromazine, or the tricyclic drug, imipramine (both at 20 mg/kg/day). Similarly, *Frazer et al.* (1974) reported decreased noradrenergic sensitivity in slices of cerebral cortex from rats treated for only 5 days with imipramine (20 mg/kg/day, i.p.). The failure of *Vetulani et al.* (1976b) to obtain subsensitivity at similar time points remains unresolved, although it may reflect anatomical differences related to use of limbic forebrain tissue in the latter study. In this regard, chronic noradrenergic cAMP subsensitivity was also found in slices of limbic forebrain obtained from rats treated with the monoamine oxidase inhibitors (MAOI), pargyline or nialamide, for 21 days but not in rats injected on the same schedule for 14 days. Recovery of sensitivity to control levels in the former animals occurred 9 days after cessation of drug treatment (*Vetulani et al.*, 1976b).

In contrast with the appearance of subsensitivity in MAOI or amphetamine-treated rats, development of chronic noradrenergic subsensitivity of brain cAMP systems following amphetamine treatment is a rapid process. *Baudry et al.* (1976) and *Martes et al.* (1975) reported significant decreases in NE-induced cAMP accumulations in slices of cerebral cortex from mice which had ingested amphetamine in their drinking water (100 μg/ml) only 5 hr before sacrifice. Noradrenergic responsiveness decreased with continued treatment to a minimum of 70% of control levels by the sixth day where it thereafter remained for 15 days (the longest interval tested) (*Baudry et al.*, 1976). Recovery of noradrenergic sensitivity after 6 days of drug treatment was a slow process, with full recovery still not evident 11 days after withdrawal in some animals.

Chronic noradrenergic cAMP subsensitivity was also observed (*Vetulani et al.*, 1976a; *Vetulani* and *Sulser*, 1975) in slice preparations of limbic forebrain obtained from rats subjected to electroconvulsive shock treatments (ECS). Subsensitivity appeared 1 day after the first ECS session and remained evident on each subsequent day during a sequence of ECS administration. Noradrenergic sensitivity still had not returned to control levels 8 days after the last of 8 daily ECS treatments (*Vetulani et al.*, 1976b). Finally, noradrenergic cAMP subsensitivity was also reported after 14 wk of ethanol ingestion (*French et al.*, 1974). Course of onset for the effect was not investigated, nor was it clear that such subsensitivity differed from the short-term of subsensitivity described following acute ethanol ingestion (*French et al.*, 1974; see above) as information on recovery rates was not reported.

In several cases failure to distinguish between acute and chronic noradrenergic cAMP subsensitivity has precluded assessment of the rate of development of the effect. Thus, the rapid onset of subsensitivity observed during amphetamine treatment, after acute ethanol ingestion or after ECS administration probably includes short-term (allosteric) modifications of the noradrenergic cAMP generating system which give way during schedules of chronic treatment to long-term (inductive) effects. Only changes of the latter type mediate the slow recovery of noradrenergic cAMP sensitivity seen after cessation of treatments. In contrast with the above, noradrenergic cAMP subsensitivity elicited by chronic administration of antidepressant agents is clearly a long-term phenomenon requiring at least 1 (and possibly as many as 4) wk for development.

Changes in Maximum Responses and Agonist Affinities

There is limited evidence to suggest that chronic noradrenergic cAMP subsensitivity occurs through a decrease in the maximum noradrenergically elicited cAMP responses with no change in agonist affinity. *Frazer et al.* (1974) and *Schultz* (1976) reported decreases in the maximum accumulations of cAMP obtained in response to high (100 μM) concentrations of NE in slices from imipramine-treated rats. *Banerjee et al.* (1977) reported decreased β-adrenergic membrane receptor density in desipramine-treated rats as measured with the radioligand, [^{3}H]-DHA; receptor affinity remained unchanged in the latter study. *Baudry et al.* (1976) and *Martes et al.* (1975) also found decreases in maximal NE-elicited accumulations of cAMP in amphetamine-treated mice, with EC_{50} values unchanged between experimental and control animals. In the only exception to these data, *French et al.* (1975) reported an increase in the EC_{50} for NE-elicited cAMP accumulations in slice preparations from cerebral cortex of rats during a sequence of chronic ethanol ingestion. Maximum NE-elicited cAMP accumulations were not significantly different between control and experimental animals (however, see above for possible confounding of the latter work with anesthetic effects).

With the exception of the latter report, the evidence suggests that chronic noradrenergic cAMP subsensitivity is mediated by a decrease in the maximal noradrenergic response capacity of the system reflecting a decrease in membrane receptor density; receptor affinities remain unaltered. Despite apparent similarities with acute processes, the two phenomena most probably are mediated by different mechanisms (see below).

Molecular Mechanisms

Basal cAMP levels were unaltered following chronic treatment with antidepressant drugs (*Frazer et al.*, 1974; *Vetulani et al.*, 1976a; *Vetulani* and *Sulser*, 1975), electroconvulsive shock (*Vetulani et al.*, 1976a;

Vetulani and *Sulser*, 1975), amphetamines (*Baudry et al.*, 1976; *Martes et al.*, 1975) or ethanol ingestion (*French et al.*, 1975). A significant increase in basal cAMP levels was noted after 14 days of treatment with the MAOI, nialamide, but had returned to control levels after 21 days of treatment. Basal levels following chronic treatment with another MAOI, pargyline, did not differ from control levels at any time (*Vetulani et al.*, 1976b). However, *Palmer* (1973) and *Palmer et al.*, (1973) found that acute injections of pargyline *in vivo*, as well as *in vitro* additions, resulted in enhanced basal levels of cAMP in slice preparations of rat brain. *Baudry et al.* (1976) found an elevation in NE elicited cAMP accumulations in amphetamine-treated mice in the presence of the phosphodiesterase inhibitor, IBMX. However, the percentage elevation in responsiveness was the same between control and experimental animals, with noradrenergic cAMP sensitivity still reduced by approximately 30% in samples of rat cerebral cortex from amphetamine-treated rats.

In the only receptor binding study in the area (see above), *Banerjee et al.* (1977) observed decreased density in [^{3}H]-DHA binding sites in animals treated with desipramine for 6 wk. Receptor affinity was not affected. Based on these limited data, decreased receptor density appears to mediate in part chronic noradrenergic cAMP subsensitivity. Phosphodiesterase inhibition has had no affect on chronic subsensitivity in the only instance where the effect was investigated. Beyond these observations there is little understanding of the molecular bases of the phenomenon.

Pharmacological Specificity

The pharmacological selectivity of chronic noradrenergic cAMP subsensitivity has been investigated only by one group (*Baudry et al.* 1976; *Martes et al.* 1975). In these studies, hormonally elicited cAMP accumulations in response to DA, 5-HT or adenosine in mice chronically treated with amphetamines did not differ significantly from levels obtained in control animals. Based on these limited data, there is an apparent restriction of the phenomenon to noradrenergic agonists. (The authors also point out that these data argue against mediation of subsensitivity by nonspecific activation of phosphodiesterases.)

Anatomical and Species Specificity

Chronic noradrenergic subsensitivity was reported in the cerebral cortex of mice (*Baudry et al.*, 1976; *Martes et al.*, 1975) and rats (*Frazer et al.*, 1974; *French et al.*, 1974, 1975; *Schultz*, 1976) and in the limbic forebrain (septal nuclei, olfactory tubercles, olfactory nuclei, accumbens nuclei and anterior amygdaloid nuclei) of rats (*Vetulani et al.*, 1976a & b; *Vetulani* and *Sulser*, 1975). No other data are available on species or anatomical specificity of chronic state noradrenergic subsensitivity.

Summary

Rates of onset of long-term noradrenergic subsensitivity of brain cAMP systems induced by amphetamines, ECS or alcohol ingestion are difficult to assess because of confounding with acute effects. In contrast, treatment with antidepressant drugs requires at least one week for onset and in one instance was not observed prior to the fourth week during a schedule of daily desipramine injections. In all cases, recovery from chronic noradrenergic cAMP subsensitivity is a slow process, requiring one to two weeks for restoration of control levels. The phenomenon appears specific to noradrenergic compounds, although the relative contribution of various receptor types has not been studied. At a molecular level, altered receptor density has been observed, although the decrease in receptor number is inadequate to account for the entire decrease in noradrenergic cAMP sensitivity on a one-for-one basis.

CONCLUSIONS

It is clearly established that noradrenergic cAMP systems in the brain are responsive to changed levels of neuronal activity. Decreased neuronal activity induces conditions of enhanced sensitivity in noradrenergically elicited cAMP responses, while increased levels of neuronal input conversely lead to states of decreased hormonal sensitivity.

In each condition, altered responsiveness appears to occur through changes in the maximum response capacity of the system, while agonist affinities remain unchanged. In cases of chronic noradrenergic cAMP supersensitivity and subsensitivity, development of the effect occurs over a period of days and, once developed, is relatively permanent even after the restoration of normal levels of neuronal activity. In both cases, time courses are consistent with inductive processes, as in the increased synthesis (or degradation, as appropriate) of the molecular components of the system. Acute subsensitivity, in contrast, appears to develop and reverse too rapidly for mediation by inductive processes; instead, short-term changes may involve allosteric modifications of the cyclic AMP generating system.

Noradrenergic cAMP subsensitivity (of both types) and supersensitivity remain poorly understood. There are limited and equivocal data implicating alterations in both phosphodiesterase activity and β-adrenergic receptor density in these phenomena. In almost all instances response changes in noradrenergically elicited cAMP accumulations are of an order of 20–30%. In this regard, it is possible that receptor occupation and agonist activation of the biological response (cAMP production) are related in something other than a one-to-one fashion. Moreover, it seems likely that adaptive phenomena are mediated at several levels in the chain of events between receptor binding and biological effect, of which receptor density is but one component. Present discrepancies between receptor binding and hormonally elicited cAMP accumulations may reflect procedural artifacts arising from current inability to measure these two parameters under identical conditions (binding is assayed in membrane preparations while brain cAMP accumulations are typically measured in broken cell or slice preparations). Discrepancies between membrane binding and slice cyclic AMP probably also reflect the complex mechanism controlling cyclic AMP levels in brain, including the presence of at least two receptors capable of mediating norepinephrine responses.

Resolution of these issues will come from continued study. With a greater knowledge of the mechanism of subsensitivity and supersensitivity of brain cAMP generating systems, the molecular basis of such diverse processes as memory, stress responses, drug tolerance, drug withdrawal and mental illness may be better understood.

REFERENCES

Alexander, R.W., Davis, J.N., and Lefkowitz, R.J. 1975. Direct identification and characterization of β-adrenergic receptors in the rat brain. Nature 258:437-440

Banerjee, S.P., Kung, L.S., Riggi, S.J., and Chanda, S.K. 1977. Development of β-adrenergic receptor subsensitivity by antidepressants. Nature 268:455 - 456

Baudry, M., Martes, M.-P., and Schwartz, J.-C. 1976. Modulation in the sensitivity of noradrenergic receptors in the CNS studied by the responsiveness of the cyclic AMP system. Brain Res 116:111-124

Blumberg, J.B., Vetulani, J., Stawarz, R.J., and Sulser, F. 1976. The noradrenergic cyclic AMP generating system in the limbic forebrain. Pharmacological characterization *in vitro* and possible role of limbic noradrenergic mechanisms in the mode of action of antipsychotics. Europ J Pharmacol 37:357-366

Davis, J.N., Strittmatter, W., Hoyler, E., and Lefkowitz, R.J. 1977. [^{3}H] dihydroergocryptine binding in rat brain. Brain Res 132:327-336

Davis, J.N., Arnett, C.B., Hoyler, E., Stalvey, L.P., Daly, J.W., and Skolnick, P. 1978. Brain α-adrenergic receptors. Comparison of [^{3}H] WB 4101 binding with norepinephrine-stimulated cyclic AMP in rat cerebral cortex. Brain Res, in press

Dismukes, R.K., and Daly, J.W. 1975. Altered responsiveness of adenosine 3′,5′-monophosphate-generating systems in brain slices from adult rats after neonatal treatment with 6-hydroxydopamine. Exp Neurol 49:150-160

Dismukes, K., and Daly, J.W. 1974. Norepinephrine-sensitive systems generating adenosine 3′,5′-monophosphate. Increased responses in cerebral cortical slices from reserpine-treated rats. Mol Pharmacol 10:933-940

Dismukes, R.K., Ghosh, P., Creveling, C.R., and Daly, J.W. 1975. Altered responsiveness of adenosine 3′,5′-monophosphate-generating systems in rat cortical slices after lesions of the medial forebrain bundle. Exp Neurol 49:725-735

Eccleston, D. 1973. Cyclic AMP and a possible animal model of receptor supersensitivity. In *Frontiers in Catecholamine Research,* eds. E. Usdin and S. H. Snyder, pp. 1083-1084. New York:Pergamon Press

Frazer, A., Pandey, G., Mendels, J., Neeley, S., Kane, M., and Hess, M.E. 1974. The effect of tri-iodothyronine in combination with imipramine on [^{3}H]-cyclic AMP production in slices of rat cerebral cortex. Neuropharmacology 13:1131-1140

French, S.W., and Palmer, D.S. 1973. Adrenergic supersensitivity during ethanol withdrawal in the rat. Res Comm Chem Path Pharmacol 6:651-663

French, S.W., Palmer, D.S., Narod, M.E., Reid, P.E., and Ramey, C.W. 1975.

Noradrenergic sensitivity of the cerebral cortex after chronic ethanol ingestion and withdrawal. J Pharmacol Exp Ther 194:319-326

French, S.W., Reid, P.E., Palmer, D.S., Narod, M.E., and Ramey, C.W. 1974. Adrenergic subsensitivity of the rat brain during chronic ethanol ingestion. Res Comm Chem Path Pharmacol 9:575-578

Greenberg, D.A., and Snyder, S.H. 1978. Pharmacological properties of [^{3}H] dihydroergocryptine binding sites associated with α-noradrenergic receptors in rat brain membranes. Mol Pharmacol 14:38-49

Huang, M., Ho, A.K.S., Daly, J.W. 1973. Accumulation of adenosine 3′,5′-monophosphate in rat cerebral cortical slices. Mol Pharmacol 9:711-717

Israel, M.A., Kimura, H., and Kuriyama, K. 1972. Changes in activity and hormonal sensitivity of brain adenyl cyclase following chronic ethanol administration. Experientia 28:1322-1323

Kakiuchi, S., and Rall, T.W. 1968. The influence of chemical agents on the accumulation of adenosine 3′,5′-phosphate in slices of rabbit cerebellum. Mol Pharmacol 4:367-378

Kakiuchi, S., Rall, T.W., and McIlwain, H. 1969. The effect of electrical stimulation upon the accumulation of adenosine 3′,5′-phosphate in isolated cerebral tissue. J Neurochem 16:485-491

Kalisker, A., Rutledge, C.O., and Perkins, J.P. 1973. Effect of nerve degeneration by 6-hydroxydopamine on catecholamine-stimulated adenosine 3′,5′-monophosphate formation in rat cerebral cortex. Mol Pharmacol 9:619-629

Kuriyama, K., and Israel, M.A. 1973. Effect of ethanol administration on cyclic 3′,5′-adenosine monophosphate metabolism in brain. Biochem Pharmacol 22:2919-2922

Martes, M.-P., Baudry, M., and Schwartz, J.-C. 1975. Subsensitivity of noradrenaline-stimulated cyclic AMP accumulation in brain slices of d-amphetamine-treated mice. Nature 225:731-733

Nahorski, S.R. 1977. Altered responsiveness of cerebral Beta adrenoceptors assessed by adenosine cyclic 3′,5′-monophosphate formation and [^{3}H]-propranolol binding. Mol Pharmacol 13:679-689

Palmer, D.S., French, W.S., and Narod, M.E. 1976a. Noradrenergic subsensitivity and supersensitivity of the cerebral cortex after reserpine treatment. J Pharmacol Exp Ther 196:167-171

Palmer, G.C. 1972. Increased cyclic AMP response to norepinephrine in the rat brain following 6-hydroxydopamine. Neuropharmacol 11:145-149

Palmer, G.C. 1973. Influence of amphetamines, protriptyline and pargyline on the time course of the norepinephrine-induced accumulation of cyclic AMP in rat brain. Life Sci 12:345-355

Palmer, G.C., and Scott, H.R. 1974. The cyclic AMP response to noradrenaline in young adult rat brain following post-natal injections of 6-hydroxydopamine. Experientia 30:520-521

Palmer, G.C., Sulser, F., and Robison, G.A. 1973. Effects of neurohumoral and adrenergic agents on cyclic AMP levels in various areas of the rat brain *in vitro*. Neuropharmacol 12:327-337

Palmer, G.C., Wagner, H.R., and Putnam, R.W. 1976b. Neuronal localization of the enhanced adenylate cyclase responsiveness to catecholamines in the rat cerebral cortex following reserpine injections. Neuropharmacol 15:695-702

Perkins, J.P., and Moore, M.M. 1973. Characterization of the adrenergic receptors mediating a rise in cyclic 3′,5′-adenosine monophosphate in rat cerebral cortex. J Pharmacol Exp Therap 185:371-378

Schultz, J. 1976. Psychoactive drug effects on a system which generates cyclic AMP in brain. Nature 261:417-418

Schultz, J., and Daly, J.W. 1973a. Accumulation of cyclic adenosine 3′,5′-monophosphate in cerebral cortical slices from rat and mouse: stimulatory effects of α- and β-adrenergic agents and adenosine. J Neurochem 21:1319-1326

Schultz, J., and Daly, J.W. 1973b. Cyclic adenosine 3′,5′-monophosphate in guinea pig cerebral cortical slices. J Biol Chem 248:843-852

Schultz, J., and Daly, J.W. 1973c. Cyclic adenosine 3′,5′-monophosphate in guinea pig cerebral cortical slices. J Biol Chem 248:860–866

Skolnick, P., and Daly, J.W. 1976. Interaction of clonidine with pre- and post-synaptic adrenergic receptors of rat brain: effects on cyclic AMP-generating systems. Europ J Pharmacol 39:11–21

Skolnick, P., and Daly, J.W. 1977. Strain differences in responsiveness of norepinephrine-sensitive adenosine 3′,5′-monophosphate-generating system in rat brain slices after intraventricular administration of 6-hydroxydopamine. Europ J Pharmacol 41:145–152

Skolnick, P., Stalvey, L.P., Daly, J.W., Hoyler, E., and Davis, J.N. 1978. Binding of α- and β-adrenergic ligands to cerebral cortical membranes. Effect of 6-hydroxydopamine treatment and relationship to the responsiveness of cyclic AMP-generating systems in two rat strains. Eur J Pharmacol 47:201–210

Sporn, J.R., Harden, T.K., Wolfe, B.B., and Molinoff, P.B. 1976. β-Adrenergic receptor involvement in 6-hydroxydopamine-induced supersensitivity in rat cerebral cortex. Science 194:624–626

U'Pritchard, D.C., Greenberg, D.A., and Snyder, S.H. 1977. Binding characteristics of a radiolabeled agonist and antagonist at central nervous system *alpha* noradrenergic receptors. Mol Pharmacol 13:454–473

Vetulani, J., Stawarz, R.J., Dingell, J.V., and Sulser, F. 1976a. A possible common mechanism of action of antidepressant treatments. Naunyn-Schmiedeberg's Arch Pharmacol 293:109–114

Vetulani, J., Stawarz, R.J., and Sulser, F. 1976b. Adaptive mechanisms of the noradrenergic cyclic AMP generating system in the limbic forebrain of the rat. Adaptation to persistent changes in the availability of norepinephrine (NE). J Neurochem 27:661–666

Vetulani, J., and Sulser, F. 1975. Action of various antidepressant treatments reduces reactivity of noradrenergic cyclic AMP-generating system in limbic forebrain. Nature 257:495–496

Wagner, H.R., Bartolome, M., and Davis, J.N. 1977. Beta-adrenergic receptor desensitization in brain slices. Trans Amer Soc Neurochem 8:193

Wagner, H.R., and Davis, J.N. 1977. Voltage dependent modulation of β-adrenergic receptor membrane binding in rat brain slices. Soc Neurosci Abs 3:463

Wagner, H.R., Palmer, G.C., Hall, T.L., and Putnam, R.W. 1978. Decreased spontaneous motor activity parallels reserpine-induced supesensitivity of catecholamine responsive rat cerebral cortex adenylate cyclase. Behav Biol 23:225–229

Williams, B.J., and Pirch, J.H. 1974. Correlation between brain adenylate cyclase activity and spontaneous motor activity in rats after chronic reserpine treatment. Brain Res 68:227–234

Neuropharmacological Control of Pineal Gland Cyclic Nucleotide Systems

*Samuel J. Strada and Michael W. Martin**
Department of Pharmacology
The University of Texas Medical School at Houston
Houston, Texas 77025

INTRODUCTION

The pineal gland functions as a physiological transducer which converts neural input into endocrine output. In higher organisms, information about environmental lighting is converted into neural signals by specialized photoreceptors in the eyes. After processing in the central nervous system, these signals either drive or modulate a variety of neuroendocrine regulatory mechanisms. Information about environmental lighting is relayed to the pineal gland, at least in the rat and presumably in other mammalian species, via a multisynaptic pathway from the retina. Postganglionic noradrenergic fibers originating in the superior cervical ganglia constitute the final segment of the pathway. This pathway controls circadian rhythms of certain biogenic amines and metabolic enzymes in the pineal gland. For example, the expression of a circadian rhythm in indole metabolism depends on the intact innervation of the pineal gland.

The rat pineal gland has been a valuable experimental tissue to study environmental and neural influences on postsynaptic biochemical events. It affords a particularly useful model for studying the relationships between cyclic 3′,5′-AMP and the function of a gland and offers an excellent opportunity for determining neuropharmacological effects on the adenylate cyclase-cyclic 3′,5′-AMP system. The pineal contains relatively high activities of adenylate cyclase and cyclic 3′,5′-AMP phosphodiesterase (*Weiss* and *Costa,* 1968a; *Weiss* and *Strada,* 1972), and relatively high concentrations of cyclic 3′,5′-AMP (*Ebadi, Weiss,* and *Costa,* 1970, 1971a & b). Thus, one might suspect cyclic 3′,5′-AMP to be intimately involved in the function of the gland. Moreover, the adenylate cyclase of the pineal gland is stimulated specifically by the phar-

***Supported by a NIH Predoctoral Fellowship (1T32 GM07405)**

macologically active catecholamines and, since adenylate cyclase is functionally linked to the β-adrenergic receptor complex (*Robison, Butcher,* and *Sutherland,* 1967, 1971), the pineal gland provides a useful model for studying the adrenergic receptor. Finally, since the pineal gland is innervated principally, if not exclusively, by sympathetic nerve fibers (*Kappers,* 1960), the effect of increased or decreased sympathetic input on the cyclic 3′,5′-AMP system can be determined.

A study of the factors influencing the cyclic 3′,5′-AMP system of the pineal gland gains physiological importance considering findings that this organ, through the release of antigonadotropic substances such as melatonin, may be responsible for the photic regulation of gonadal activity (*Wurtman, Axelrod,* and *Kelly,* 1968; *Mess,* 1968; *Reiter* and *Fraschini,* 1969) and that cyclic 3′,5′-AMP is involved in the formation of melatonin in the pineal gland (*Klein* and *Berg,* 1970; *Weiss* and *Crayton,* 1970a; *Berg* and *Klein,* 1971; *Wurtman, Shein,* and *Larin,* 1971; *Strada et al.,* 1972).

In this report we will discuss some of the conditions found to alter the activities of nucleotide cyclases and concentrations of cyclic nucleotides of the rat pineal gland. We will show that the hormone-sensitive adenylate cyclase-cyclic 3′,5′-AMP system of the pineal can be influenced both acutely and chronically by several neural and hormonal factors. Modification of sympathetic input to the pineal gland can be shown to induce changes in the degree to which norepinephrine activates the pineal adenylate cyclase system. Finally, we will discuss the involvement of the cyclic 3′,5′ nucleotide systems in the production of antigonadotropic principles of the rat pineal gland and attempt to relate the factors controlling the concentration of these cyclic nucleotides in the pineal gland to the physiological function of the gland.

EFFECT OF SYMPATHETIC NERVE ACTIVITY ON THE CYCLIC 3′,5′-AMP SYSTEM OF RAT PINEAL GLAND

The pineal gland is innervated by sympathetic nerve fibers whose cell bodies lie in the superior cervical ganglion (*Kappers,* 1960). There are, therefore, various ways to modify the neuronal input to this gland: the primary innervation of the pineal gland can be abolished by surgical ablation of the superior cervical ganglia (denervation); nerve activity can be reduced by severing the preganglionic sympathetic nerve fibers to these ganglia (decentralization); a chemical sympathectomy of the ganglia can be performed by administering 6-hydroxydopamine to newborn animals (*Angeletti,* 1971; *Jaim-Etcheverry* and *Zieher*, 1971); and finally, one can physiologically decrease sympathetic neuronal activity in the pineal by exposing rats to continuous light (*Taylor* and *Wilson,* 1970). Using these procedures we have shown that decreasing the sympathetic input to the pineal gland alters the norepinephrine-induced stimulation of adenylate cyclase activity in pineal gland homogenates and the ability of the catecholamine to elevate the concentration of cyclic 3′,5′-AMP in cultured pineal glands. The acute effects of sympathetic neural activity on the cyclic 3′,5′-AMP system of the rat pineal gland will be discussed first.

ACUTE ACTIVATION OF THE ADENYLATE CYCLASE-CYCLIC 3′,5′-AMP SYSTEM OF RAT PINEAL GLAND

To establish a causal relationship between the action of a hormone with changes in the cellular concentration of cyclic 3′,5′-AMP, the hormone must exert specificity in increasing the concentration of cyclic 3′,5′-AMP. It is now well known that although many tissues possess adenylate cyclase activity, there are differences with which the adenylate cyclase of various tissues responds to hormones (*Robison, Butcher,* and *Sutherland,* 1971). *Weiss* and *Costa* (1968b) first showed that adenylate cyclase of pineal gland homogenates is stimulated specifically by the pharmacologically active catecholamines. We recently examined the interaction between various effectors of pineal gland adenylate cyclase, including guanyl nucleotides.

The adenylate cyclase activity of the rat pineal gland is stimulated 3- to 6-fold above basal levels by β-adrenergic agonists and 70% by dopamine but not by histamine or prostaglandin E_1. The order of potency of β-adrenergic agonists to stimulate pineal adenylate cyclase is isoproterenol > epinephrine ≥ norepinephrine ≫ dopamine.

Sodium fluoride (NaF) produces the largest increase in pineal adenylate cyclase activity measured *in vitro* (Table 1), although the magnitude of the stimulation varies somewhat from preparation to preparation. If the activity in the presence of 10 mM NaF represents total catalytic capacity of the system, then 40–50% of the total catalytic capacity is activated by 1-norepinephrine at maximally effective concentrations of the agonist (Table 1, see below). GTP and its synthetic analogue, 5′-guanylyl imidodiphosphate Gpp(NH)p, both of which are known to facilitate catecholamine stimulation of adenylate cyclase in a variety of tissues (*Rodbell et al.,* 1975; *Brown et al.,* 1976), also increase the activation of pineal adenylate cyclase by norepinephrine. GTP has only a minimal effect alone or in combination with norepinephrine, but Gpp(NH)p stimulates the enzyme 2- to 3-fold and displays a synergistic effect on the activation by norepinephrine (Table 1, Fig. 1). Typically, 80% of the NaF-stimulated activity can be detected in the presence of norepinephrine and Gpp(NH)p (Fig. 1). The guanine nucleotide appears to facilitate receptor-cyclase coupling rather than modifying the affinity of the enzyme for

Table 1. Activation of Rat Pineal Gland Sonicate Adenylate Cyclase Activity

Agonist	Adenylate Cyclase Activity (pmoles cAMP/min/mg protein)	Basal Activity (%)
Basal	95 ± 7	100
NE (100 μM)	317 ± 5	330
GTP (100 μM)	128 ± 3	135
Gpp(NH)p (100 μM)	193 ± 10	200
NE + GTP	354 ± 33	370
NE + Gpp(NH)p	596 ± 21	630
NaF (10 mM)	734 ± 57	770

Pineal glands were obtained from adult male rats maintained in a 12 hr light-12 hr dark cycle. The pineals were rinsed with cold 50 mM Tris, 0.32 M sucrose, pH 7.4, pooled, and disrupted by sonication for 20 sec at a setting of 20 W with a Bronwill Biosonik IV sonicator fitted with a microprobe. The sonicates were assayed immediately for adenylate cyclase activity by the method of *Thompson, Williams,* and *Little* (1973). The assay reaction mixture consisted of α-^{32}P-ATP (1.2×10^{-6} cpm), 1.0 mM ATP, 5.0 mM MgCl, 1.0 mM cyclic AMP, 0.75 mg/ml bovine serum albumin, 15.6 mM phosphocreatine, 5 units creatine phosphokinase, 50 mM Tris-Cl (pH 7.4), 0.05 mM EGTA, and 0.2 mM 3-isobutyl-1-methylxanthine, and sonicate equivalent to 0.2 pineal in a final volume of 0.2 ml. Catecholamines were added to the assay in 0.1% ascorbic acid. The reaction was initiated by the addition of 0.05 ml of pineal sonicate and the samples were incubated for 10 min at 30°C. The reaction was terminated by addition of 0.05 ml 1 N acetic acid. Samples were diluted to 1.0 ml with 50 mM Tris-Cl (pH 7.4) containing (8-^{3}H)-cyclic AMP. ^{32}P-cyclic AMP was isolated by adsorption chromatography on dry Al_2O_3-MnO_2 columns; recoveries of labeled cyclic AMP were 70–80%. Values shown are the mean ± SEM from three different experiments with duplicate or triplicate determinations in each. NE indicates 1-norepinephrine.

norepinephrine since activity is increased at all concentrations of norepinephrine tested with no apparent effect on the concentration of norepinephrine required for half-maximal activation.

A direct and dose-dependent effect of norepinephrine to rapidly and markedly increase the concentration of cyclic 3′,5′-AMP is also seen in the intact gland (Table 2). About a 6-fold increase in the endogenous concentration of cyclic 3′,5′-AMP is evident within 5 min, and a maximum increase is seen 10–15 min after the administration of norepinephrine (Fig. 2). The characteristics of this stimulation, like that found in pineal gland homogenates, indicate that the norepinephrine-induced increase in the concentration of cyclic 3′,5′-AMP more closely resembles a β-adrenergic receptor action than an alpha effect. Thus, the β-adrenergic blocking agent, propranolol, inhibits completely the norepinephrine-induced elevation of cyclic 3′,5′-AMP, whereas the α-adrenergic blocker, phentolamine, is totally ineffective.

McAfee, Schorderet, and *Greengard* (1971) have shown that electrical stimulation of the fibers innervating the superior cervical ganglia increases the concentration of cyclic 3′,5′-AMP in these ganglia. Although

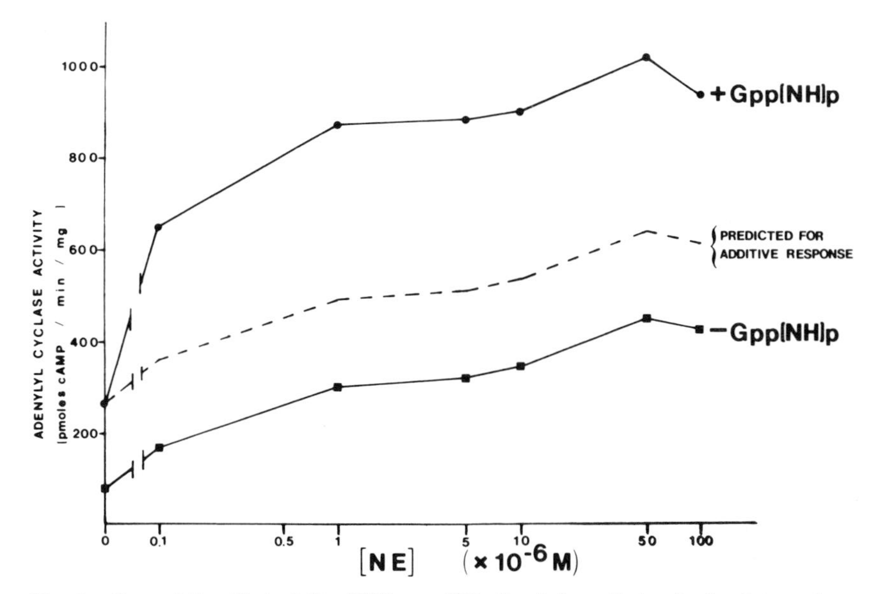

Fig. 1. Synergistic effect of Gpp(NH)p on NE stimulation of pineal adenylate cyclase. Pineal adenylate cyclase activities were determined in response to varying concentrations of 1-norepinephrine as described in Table 1, in the presence (•) and absence (■) of Gpp(NH)p (10 μM). Also shown (---) are the predicted additive effects of Gpp(NH)p and norepinephrine.

Table 2. Effect of Norepinephrine on the Concentration of Cyclic AMP of Rat Pineal Glands

Norepinephrine (μM)	Cyclic AMP (pmoles/gland)
0	11 ± 1 (24)
1	18 ± 5 (3)
5	49 ± 12 (6)[a]
10	74 ± 12 (11)[b]
50	100 ± 6 (8)[b]
100	144 ± 14 (8)[b]

Pineal glands were cultured in an *in vitro* system as described previously (*Strada et al.*, 1972). The glands were removed after decapitation and dissected free of adhering tissue. Within 10 min the pineals were placed on self-supporting stainless steel screens in small concave vessels containing 0.6 ml of medium. The basic culture medium used was modified BGJ (Grand Island Biological Co.) with the addition of penicillin (100 units/ml), streptomycin (10 μg/ml), bovine serum (1 mg/ml), ascorbic acid (0.1 mg/ml) and glutamine (0.2 mM). The glands were incubated for 10 min in a humidified atmosphere of 95% oxygen and 5% carbon dioxide at 37°C containing the indicated concentrations of 1-norepinephrine bitartrate. At the end of the incubation period, the pineal glands were homogenized in cold 0.4 N perchloric acid and assayed for cyclic 3′,5′-AMP content by the method of *Ebadi, Weiss,* and *Costa* (1971b). Each figure represents the mean ± SE. The values in parentheses indicate the number of experiments.

[a] $P < 0.01$ compared with no norepinephrine.

[b] $P < 0.001$ compared with no norepinephrine.

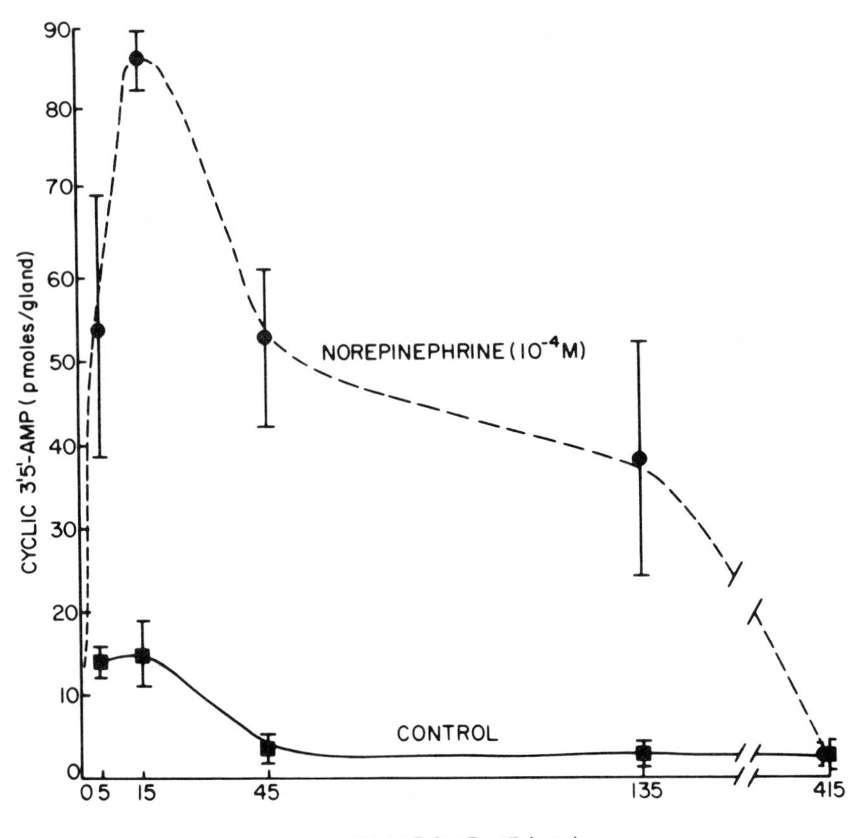

Fig. 2. Effect of norepinephrine on the concentration of cyclic 3′,5′-AMP in cultured pineal glands. Pineal glands were obtained from adult male rats maintained in alternating periods of 12 hr light and 12 hr dark. Rats were decapitated 5 hr after the beginning of the light cycle. The glands were incubated for various times in the absence or presence of norepinephrine (0.1 mM). At the end of each incubation period, the glands were assayed for cyclic 3′, 5′-AMP content. Each point represents the mean value of four experiments ± SE. From *Strada et al.*, 1972.

the adrenergic neurotransmitter, norepinephrine, increases the activity of adenylate cyclase when added to pineal gland homogenates and elevates the endogenous levels of cyclic 3′,5′-AMP in cultured pineal gland, it remains to be demonstrated that direct sympathetic nerve activity changes acutely the norepinephrine-sensitive adenylate cyclase-cyclic 3′,5′-AMP system of the pineal gland. However, following the acute systemic administration of norepinephrine either *in vivo* or *in vitro*, a reduction in the hormone-sensitive cyclase response can be observed. Moreover, there is evidence that altered sympathetic activity can chronically alter the cyclic

3′,5′-AMP system of the rat pineal. These subjects will be discussed in the following sections.

Acute Regulation of Pineal Gland Adenylate Cyclase

Recent investigations have demonstrated that after initial stimulation by β-adrenergic agonists, adenylate cyclase becomes refractory or desensitized to subsequent challenge by catecholamines (*Remold-O'Donnel*, 1974; *Mickey, Tate*, and *Lefkowitz*, 1975; *Lefkowitz* and *Williams*, 1978; *Morishima et al.*, 1978). Thus, after prolonged exposure of the tissue to catecholamine, either *in vivo* or *in vitro*, there is a progressive loss of responsiveness of the β-adrenergic receptor-coupled adenylate cyclase to further stimulation. The loss of catecholamine-sensitive enzyme activity is highly specific; no comparable losses in basal, NaF-stimulated or prostaglandin E (PGE_1)-stimulated activities usually are observed in preparations rendered refractory to catecholamines.

Lefkowitz and colleagues have used a radiolabeled, competitive, β-adrenergic receptor antagonist, ^{3}H-alprenolol, to identify and characterize β-adrenergic binding sites in membrane preparations from a variety of tissues (*Mukherjee et al.*, 1975, 1976; *Lefkowitz* and *Williams*, 1978). The properties of these binding sites satisfy established criteria for a β-adrenergic receptor site (*Cuatrecasas*, 1974). These newly developed techniques have been used to show that catecholamine-induced desensitization of adenylate cyclase in frog erythrocytes, cardiac membranes and rat pineal gland is accompanied by a parallel decline in the number of specific binding sites (*Lefkowitz* and *Williams*, 1978; *Kebabian et al.*, 1975; *Zatz et al.*, 1976). These authors suggest that the loss of functional membrane receptor sites leads to a reduced adenylate cyclase response to the catecholamine. They have postulated that the interaction of agonist with its specific receptor leads to a reversible conformational change in the receptor which prevents subsequent agonist binding and activation of adenylate cyclase.

We have examined the desensitization of rat pineal gland adenylate cyclase induced by exposure to isoproterenol both *in vivo* and in an *in vitro* organ culture system. Some of our findings are discussed below.

Desensitization of Pineal Gland
Adenylate Cyclase to Catecholamines In Vivo

The effect of β-adrenergic agonist exposure on the responsiveness of rat pineal gland adenylate cyclase activity was tested by administering a single subcutaneous injection of d,l-isoproterenol (5 mg/kg) 2 hr before removing the glands. The responsiveness of the enzyme system of pineal gland sonicates to norepinephrine was determined *in vitro*. Isoproterenol pretreatment *in vivo* had no effect on the basal activity *in vitro* nor on activity stimulated by NaF (Fig. 3). The adenylate cyclase response to norepinephrine alone ($P < 0.01$) and that of norepinephrine and Gpp(NH)p ($P < 0.001$) were reduced by 48% and 40%, respectively. The reduced response to norepinephrine caused by *in vivo* isoproterenol pretreatment could not be overcome even at high doses of norepinephrine (Fig. 4) or isoproterenol. The dose of norepinephrine for half-maximal activation of adenylate cyclase activity was unchanged. Even though Gpp(NH)p was able to significantly enhance the stimulation produced by norepinephrine, it could not overcome the effect of prior *in vivo* exposure to isoproterenol.

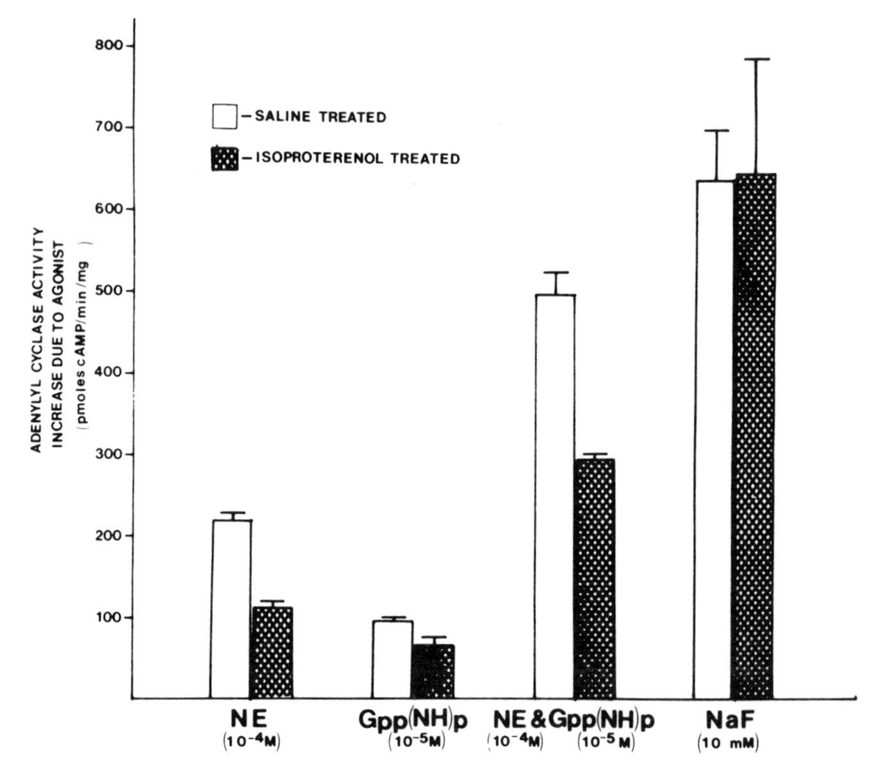

Fig. 3. Desensitization of adenylate cyclase induced by prior *in vivo* exposure to isoproterenol. Rats were administered either 5 mg/kg of d,1-isoproterenol in a 0.9% saline solution or vehicle 2 hr before sacrifice. Shown is the response of adenylate cyclase to the various agonists (increase over basal activity) measured in sonicates of pineal glands from control (□) and isoproterenol-treated (■) animals. Values are the mean ± SEM of two experiments, each with duplicate determinations. *$P < 0.01$, NE (10^{-4}), isoproterenol-treated compared to control; **$P < 0.001$, NE and Gpp(NH)p, isoproterenol-treated compared to control; ***Not significant, NaF, isoproterenol-treated compared to control.

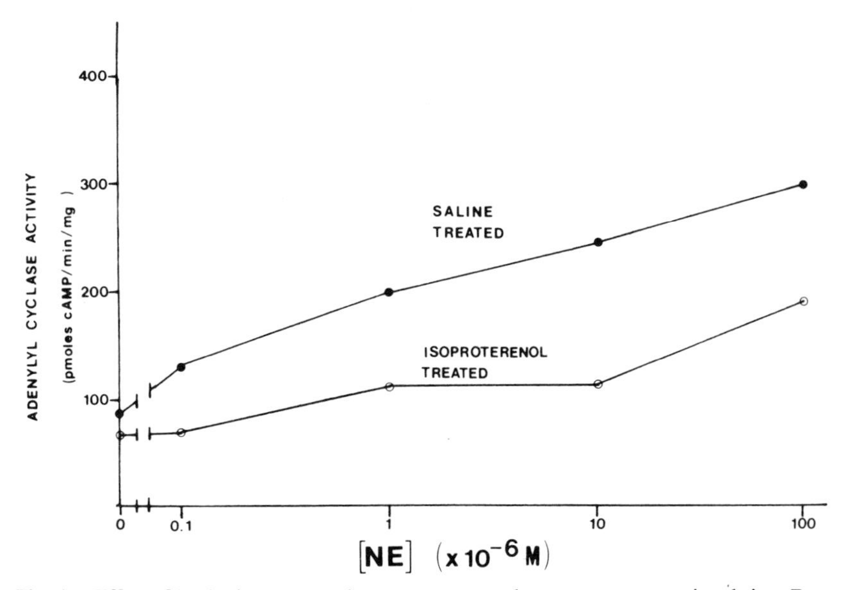

Fig. 4. Effect of *in vivo* isoproterenol pretreatment on the response to norepinephrine. Rats were injected with either isoproterenol (5 mg/kg, s.c.) or saline; 2 hr later the pineals were removed, washed, sonicated and adenylate cyclase activity measured. Values are the mean of triplicate determinations for a typical experiment.

Desensitization of Pineal Gland Adenylate Cyclase Activity in Vitro

Pineal glands were also exposed to isoproterenol (0.1 mM) *in vitro* for 2 hr. Following the pretreatment period the pineals were washed thoroughly with medium and Tris-sucrose buffer to remove residual isoproterenol, sonicated and adenylate cyclase activity was measured. The 2-hr exposure to isoproterenol rendered the pineal adenylate cyclase almost totally refractory to stimulation by norepinephrine (Fig. 5). Glands exposed to isoproterenol produced only 15% of the control response to norepinephrine. Again, Gpp(NH)p was synergistic with norepinephrine but unable to overcome the desensitization effect. The adenylate cyclase response in the presence of both norepinephrine and Gpp(NH)p was reduced to 50% of that of control. Basal activities were the same in control and desensitized glands. The *in vitro* adenylate cyclase responses to NaF and to Gpp(NH)p in the pineals desensitized *in vitro* were reduced slightly from those of control glands. The reductions observed in these activities were much smaller than those observed for the norepinephrine-stimulated responses.

Desensitization to catecholamines developed rapidly in the rat pineal gland and the degree of desensitization was dependent upon the concentration of agonist to which the gland was exposed (Fig. 6). Exposure of the

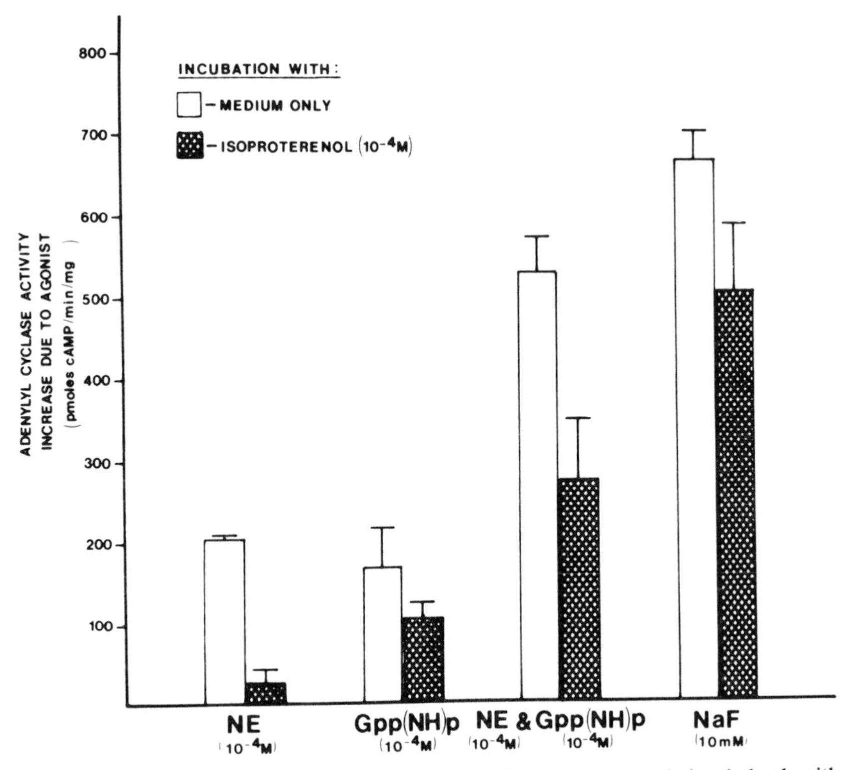

Fig. 5. Densensitization of adenylate cyclase induced by incubation of pineal glands with isoproterenol in organ culture. Pineal glands were incubated in organ culture as described in Table 2 for 1 hr in medium alone. Medium containing d,l-isoproterenol (0.1 mM) and 0.1% ascorbic acid, or medium plus ascorbic acid vehicle only, was then added to the cultures and incubated for an additional 2 hr at 37°C. Pineals were removed, washed twice with fresh medium and once with 50 mM Tris-Cl/0.32 M sucrose (pH 7.4), sonicated and activity determined as described in Table 1. Shown is the increase over basal activity in pineals cultured in medium only (□) or in medium with 0.1 mM isoproterenol (■). The results are the mean ± SEM of two experiments, with duplicate determinations.

glands to isoproterenol (0.1 mM) in culture caused 50% loss in the adenylate cyclase responsiveness within 30 min and 90% within 2 hr. Glands exposed to a 10-fold lower concentration of isoproterenol displayed 85% of the control response after 30 min of exposure but were desensitized to 30% of the control response by 2 hr (Fig. 6). Isoproterenol was minimally effective at 1 μM and maximally effective at 100 μM under these culture conditions.

The desensitization process was reversible if exposure to the agonist was discontinued. In Figure 7 pineal glands were treated for 2 hr with isoproterenol (0.01 mM) *in vitro*. The pineals were then washed thoroughly with fresh medium and incubated an additional 8 hr in

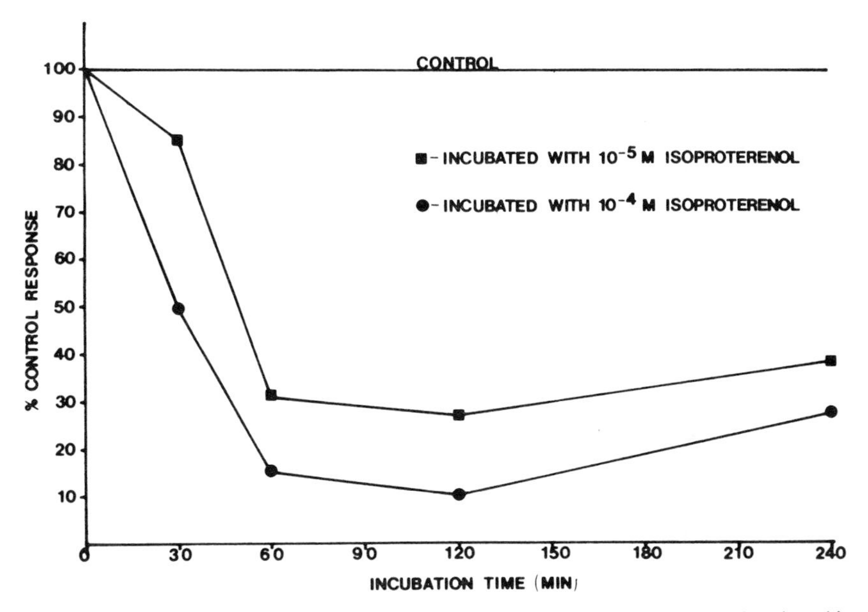

Fig. 6. Time course of the development of desensitization. Pineal glands were incubated in organ culture as described in Table 2. After 1 hr preincubation in medium, the glands were incubated in medium with or without isoproterenol (0.1 mM) for the times indicated. Pineals were then processed as in Table 1. Values are the percent of the control response to isoproterenol (0.1 mM) expressed as the mean of triplicate determinations.

medium with no isoproterenol. The adenylate cyclase response to norepinephrine alone or combined with Gpp(NH)p was restored to 85% of control levels. Similar experiments have shown a variable degree of resensitization (30–90%) at 8 or 12 hr after agonist removal. Therefore, the recovery from a fully desensitized state occurs at a much slower rate than its development, requiring 8 or more hr to achieve a complete resensitization.

Chronic Alternation of the Adenylate Cyclase-Cyclic 3′,5′-AMP System of Rat Pineal Gland

Effect of Decentralization or Ablation of the Superior Cervical Ganglia

In the rat, bilateral removal of the superior cervical ganglia or decentralization of these ganglia interrupts the principal pathway through which the effects of environmental lighting reach the pineal (*Kappers,* 1960; *Moore et al.,* 1967; *Moore* and *Rapport,* 1971). Taking advantage of this anatomical situation, we studied the role played by light and the sympathetic nervous system in inducing changes in adenylate cyclase activity.

The adenylate cyclase of pineal gland homogenates of denervated pineal

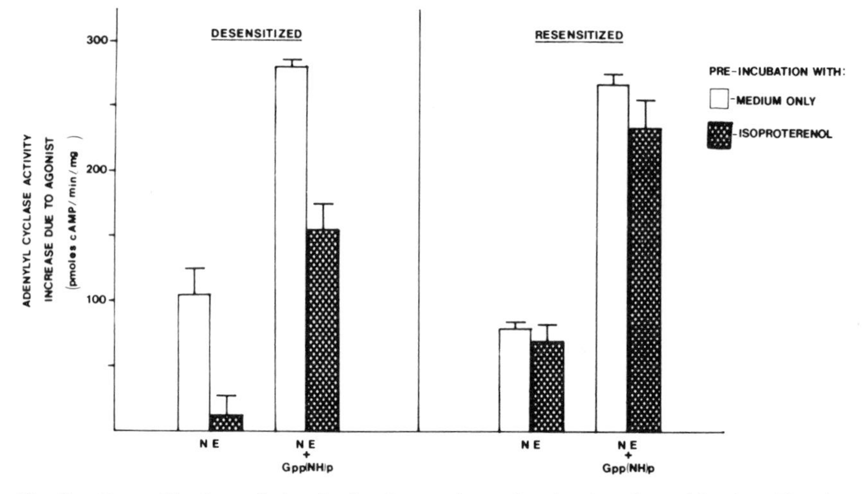

Fig. 7. Resensitization of pineal adenylate cyclase after *in vitro* desensitization. Pineals were cultured for 1 hr in medium alone. Glands were then incubated without (□) or with (■) 0.1 mM isoproterenol for 2 hr. Following this 2-hr desensitization period, half the control and treated glands were removed from culture, washed, sonicated and assayed. The remaining glands were washed 5 times with fresh medium and then incubated an additional 8 hr in fresh medium without isoproterenol and then processed similarly. Adenylate cyclase activity was determined in the presence of 100 μM norepinephrine or 100 μM norepinephrine and 100 μM Gpp(NH)p. Values are mean ± SEM of triplicate determinations.

glands is more responsive to norepinephrine than that of control animals (*Weiss* and *Costa*, 1967). Moreover, maximal as well as submaximal concentrations of norepinephrine are potentiated by denervation, suggesting that denervation results in an increased quantity of the norepinephrine-sensitive enzyme. These effects were shown to be of a chronic nature since an increased responsiveness to norepinephrine was observed only after 4 wk had elapsed between ganglionectomy and the measurement of enzymatic activity (*Weiss*, 1969). If this effect on the adenylate cyclase-cyclic 3′,5′-AMP system is related to decreased nerve activity afferent to the pineal gland rather than to the absence of the intact ganglia, decentralization of the preganglionic cervical sympathetic nerves would be expected to induce similar changes on the pineal adenylate cyclase enzyme. Table 3 shows that chronic decentralization causes an increased activation of the enzyme by both norepinephrine and sodium fluoride. Neither denervation nor decentralization significantly altered the basal adenylate cyclase activity (the activity measured in the absence of stimulant).

From these results, the addition of norepinephrine to denervated pineal glands would be expected to cause a greater increase in the endogenous concentration of cyclic 3′,5′-AMP. To test this hypothesis, the superior cervical ganglia of adult male rats were removed bilaterally. Ten weeks after surgical denervation, the pineal glands of sham and of ganglionec-

Table 3. Effects of Chronic Decentralization of the Superior Cervical Ganglia on Norepinephrine and Sodium Fluoride-induced Enhancement of Adenylate Cyclase Activity of Rat Pineal Gland

	Cyclic 3′,5′-AMP Formed (pmoles/mg of protein/min) Increase over control value due to	
Treatment	Norepinephrine	NaF
Sham operated	73 ± 9	176 ± 24
Decentralized	150 ± 20[a]	404 ± 58[b]

The superior cervical ganglia of 30-day-old rats were decentralized bilaterally by cutting the preganglionic nerves 7 wk prior to use. Adenylate cyclase activity of pineal homogenates was determined by the procedure of *Krishna, Weiss,* and *Brodie* (1968) in the absence of any drug or in the presence of either norepinephrine (10^{-4} M) or NaF (10^{-2} M). Each value represents the mean of four experiments ± SEM. From *Weiss* and *Kidman,* 1969.
[a] $P < 0.05$ compared with the sham-operated control rats.
[b] $P < 0.01$ compared with the sham-operated control rats.

tomized animals were cultured *in vitro* in the absence or presence of 50 μM norepinephrine. The administration of the catecholamine to control glands produced a 5-fold increase in the endogenous concentration of cyclic 3′,5′-AMP, whereas it elicited a more than 14-fold increase when added to glands whose sympathetic innervation had been removed by ganglionectomy (Fig. 8).

Effect of Chemical Sympathectomy with 6-Hydroxydopamine

Chemical sympathectomy with 6-hydroxydopamine has been widely used for studying adrenergic mechanisms (*Thoenen* and *Tranzer,* 1968; *Furness et al.,* 1970; *Uretsky,* and *Iversen,* 1970; *deChamplain,* 1971), and a typical denervation supersensitivity to catecholamines after 6-hydroxydopamine has been shown in several organs (*Haeusler, Haefley,* and *Thoenen,* 1969; *Nadeu, deChamplain,* and *Tremlay,* 1971; *Wagner* and *Trendelenburg,* 1971). *Angeletti* (1971) and *Jaim-Etcheverry* and *Zieher* (1971) have shown that the administration of 6-hydroxydopamine to rats shortly after birth produces an irreversible chemical sympathectomy. This procedure provides an effective means of chronically removing sympathetic input to the pineal gland and has allowed us to investigate further the phenomenon of supersensitivity to norepinephrine of the adenylate cyclase-cyclic 3′,5′-AMP system of the pineal.

6-Hydroxydopamine (50 μg/g body weight) was administered subcutaneously daily for 7 days to newborn rats. Nine weeks later we removed the pineal glands and measured the concentration of cyclic 3′,5′-AMP in response to various concentrations of norepinephrine. As can be seen

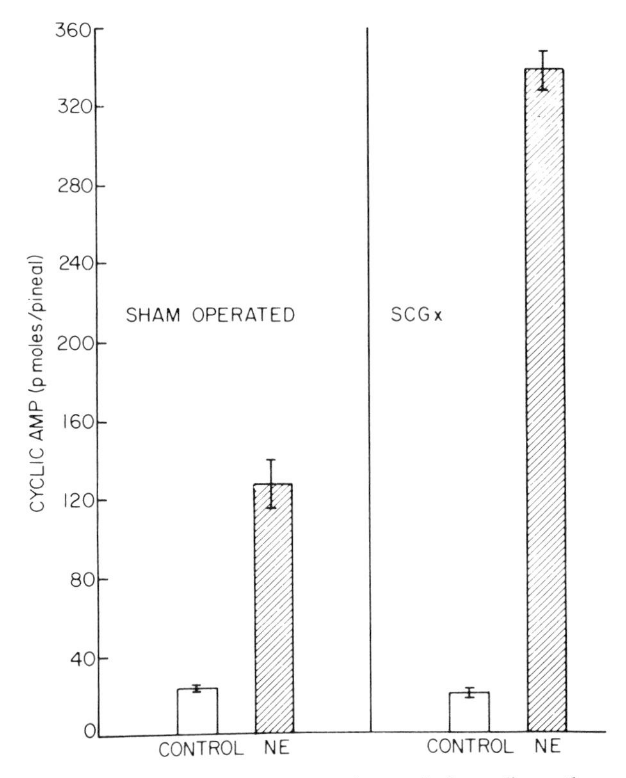

Fig. 8. Effect of bilateral removal of the superior cervical ganglia on the norepinephrine-induced elevation of cyclic 3′,5′-AMP in cultured rat pineal glands. The superior cervical ganglia of adult male rats were removed bilaterally by surgical ablation. The rats were maintained in alternating periods of 12 hr light and 12 hr dark. Ten weeks after removal of the ganglia and 5 hr after the start of the light cycle, the pineal glands were cultured in incubation medium in the absence or presence of norepinephrine (50 μM). The glands were removed 10 min after the addition of norepinephrine and assayed for cyclic 3′,5′-AMP content. Each value represents the mean $\pm$ SE of five experiments. $P < 0.001$ norepinephrine (sham) vs norepinephrine (ganglionectomized). From *Strada* and *Weiss*, 1974.

(Table 4), norepinephrine increased the concentration of cyclic 3′,5′-AMP in pineals of both 6-hydroxydopamine- and sham-treated rats. The basal concentrations of cyclic 3′,5′-AMP were not appreciably affected by 6-hydroxydopamine treatment, but the responses to submaximal as well as to maximal concentrations of norepinephrine were greater in pineal glands of 6-hydroxydopamine-treated animals than in litter mate controls.

Effect of Continuous Light on the Adenylate Cyclase-Cyclic 3′,5′-AMP System of Rat Pineal Gland

Long-term exposure of rats to continuous light decreases sympathetic in-

Table 4. Effect of 6-Hydroxydopamine Treatment of Newborn Animals on the Norepinephrine-induced Elevation of Cyclic AMP in Cultured Rat Pineal Glands

Norepinephrine (μM)	Sham	6-Hydroxydopamine (cyclic AMP pmoles/gland)
0	15 ± 2 (8)	9 ± 1 (8)
1	27 ± 3 (6)	71 ± 14 (6)[a]
5	49 ± 12 (6)	263 ± 30 (6)[b]
50	75 ± 12 (12)	289 ± 36 (12)[b]

Daily doses of 6-hydroxydopamine (50 μg/g) were injected subcutaneously to newborn animals for 7 days. The rats were kept in alternating periods of 12 hr light and 12 hr dark. Nine weeks after the last injection and 5 hr after the start of the light cycle, the rats were decapitated and the pineal glands were removed and incubated for 10 min with varying concentrations of norepinephrine. The figures represent the mean values ± SE. The numbers in parentheses indicate the number of experiments. From *Strada* and *Weiss*, 1974.
[a] $P < 0.01$ compared with control animals.
[b] $P < 0.001$ compared with control animals.

put to the pineal (*Taylor* and *Wilson*, 1970) and induces changes in the pineal adenylate cyclase system similar to those observed after chronic denervation of the pineal gland. Neither denervation nor light exposure significantly alters the basal enzymatic activity, but both conditions cause an increased activation of pineal adenylate cyclase in response to norepinephrine or sodium fluoride (*Weiss*, 1969). We extended these studies by showing that norepinephrine produces a greater increase in the endogenous concentration of cyclic 3′,5′-AMP in cultured pineal glands of rats that had been previously exposed to continuous light.

Adult male Sprague-Dawley rats were kept in continuous light for 10 days or in a 12 hr light–12 hr dark cycle. Five hours after the start of the last cycle, the animals were decapitated and the pineal glands cultured *in vitro*, in the absence or presence of 50 μM norepinephrine. The basal concentrations of cyclic 3′,5′-AMP of pineal glands of animals exposed to continuous light did not differ appreciably from those of rats kept in the normal light-dark cycle, but the response to norepinephrine was greater in those animals kept in continuous light. Whereas norepinephrine added to pineal cultures of animals kept in normal light-dark cycles increased the concentration of cyclic 3′,5′-AMP 4-fold, it elevated the levels of cyclic 3′,5′-AMP about 9-fold in those kept in continuous light (Fig. 9).

Is the Presence of the Sympathetic Nervous System Required for the Development of a Norepinephrine-Sensitive Adenylate Cyclase?

The fact that acute and chronic changes in sympathetic tone modulate the sensitivity of adenylate cyclase to norepinephrine stimulation poses the interesting question of whether sympathetic nerves might also influence the

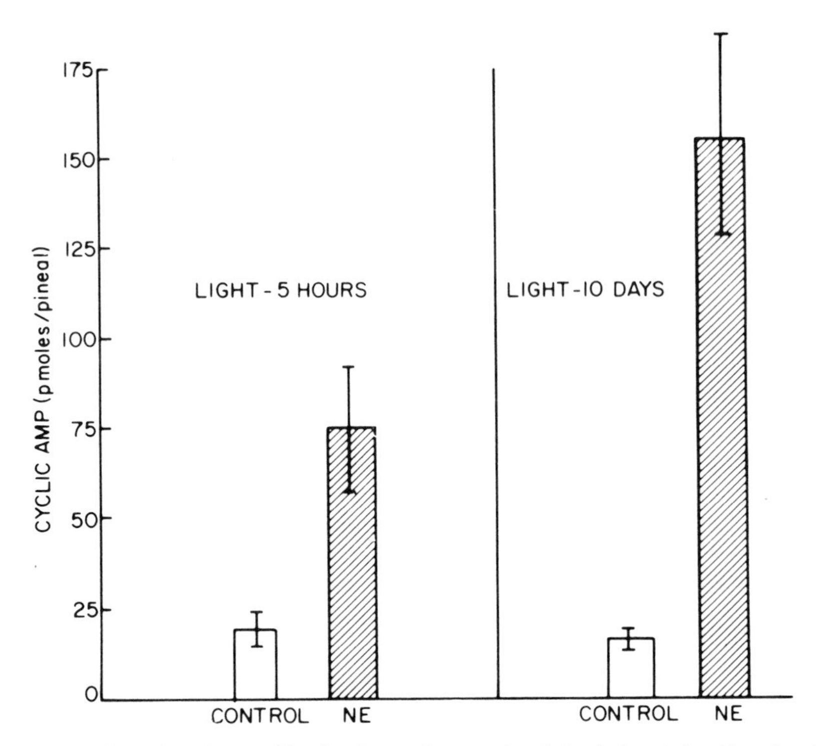

Fig. 9. Effect of continuous illumination on the norepinephrine-induced elevation of cyclic 3′,5′-AMP in cultured rat pineal glands. Male rats either were kept in continuous light for 10 days or were maintained in alternating periods of 12 hr light and 12 hr dark. Five hours after the start of the light cycle, the rats were decapitated and the pineal glands removed. The pineal glands were cultured in incubation medium in the absence or presence of norepinephrine (50 μM). The glands were removed 10 min after the addition of norepinephrine and assayed for cyclic 3′,5′-AMP content. Each value represents the mean ± SEM of four experiments. $P < 0.05$ norepinephrine (normal light-dark) vs norepinephrine (constant light). From *Weiss* and *Strada,* 1972.

development of a norepinephrine-sensitive adenylate cyclase system. To answer this question, we took advantage of the fact that pineal adenylate cyclase is not responsive to the action of norepinephrine 1 day after birth but develops its responsiveness soon thereafter (*Weiss,* 1971). Earlier studies had shown that neither ganglionectomy of newborn rats nor administering the antiserum to nerve growth factor (*Levi-Montalcini* and *Booker,* 1960) to rats shortly after birth prevented the development of the norepinephrine-sensitive adenylate cyclase system in the pineal (*Weiss,* 1970, 1971). Another approach to investigating the need for the sympathetic nerves in developing the responsiveness of adenylate cyclase to norepinephrine would be to remove the sympathetic input by chemical means. Our experiments showing that norepinephrine increased the concentration of cyclic 3′,5′-AMP in pineal glands of rats that had been treated at birth with 6-hydroxydopamine provided additional support for the concept that a norepinephrine-sensitive adenylate cyclase can develop in the absence of the sympathetic nervous system. In fact, the evidence suggests that prolonged sympathetic nervous activity causes a diminution of the catecholamine-sensitive enzyme system.

Summary and Conclusions

The rat pineal gland provides an excellent model system for studying the control of cellular responses through a β-adrenergic receptor-adenylate cyclase system. Cyclic AMP mediates the induction of an enzyme, serotonin N-acetyltransferase, in the rat pineal gland (*Strada et al.*, 1972) and a characteristic feature of this response is a diurnal variation in its sensitivity to adrenergic stimulation (*Deguchi* and *Axelrod,* 1973a & b; *Romero* and *Axelrod,* 1975). Thus the pineal gland can be either supersensitive or subsensitive to stimulation by norepinephrine; the relative responsiveness displayed by the gland is determined by the degree and duration of prior stimulation of the receptor by norepinephrine (*Strada* and *Weiss,* 1974; *Axelrod,* 1974). The mechanisms responsible for controlling these changes in responsiveness apparently involve both the systems which modulate cyclic AMP levels and the system which utilizes cyclic AMP to induce specific protein synthesis (*Kebabian et al.*, 1975; *Axelrod,* 1974; *Romero, Zatz,* and *Axelrod,* 1975).

Our results suggest that the mechanisms responsible for regulating the responsiveness of the pineal gland to noradrenergic stimulation involve coupling of the β-adrenergic receptor-cyclase system. Exposure of the pineal gland to a β-adrenergic agonist either *in vivo* or *in vitro* results in a rapid loss in the ability of the gland to respond to adrenergic stimulation. Thus, norepinephrine not only stimulates adenylate cyclase but also acts to regulate the sensitivity of the receptor-cyclase system to further stimulation. The greater the interaction of agonist with receptor, either by exposure for a longer time or by exposure to a higher concentration of agonist, the less sensitive the system becomes to further stimulation by the agonist. The fact that basal and NaF-stimulated enzyme activities are unaffected by the prior exposure to catecholamine demonstrates that the loss of cyclase activity is not due to a loss of catalytic activity. Furthermore, the dose-dependent nature of the development of desensitization suggests that it is the interaction of agonist with receptor which is the necessary first step for induction of the process. Since no apparent change occurs in the concentration of norepinephrine required for half-maximal activation, it is probable that desensitization is not due to a change in the affinity of the receptor for norepinephrine. It is not possible to answer this directly since there are no good direct methods to measure agonist binding or adrenergic receptor content. One consequence of altered receptor properties during desensitization could be a functional uncoupling of enzyme activation from adrenergic receptors.

Studies showing that catecholamine exposure leads to a reduction in the amount of antagonist specific binding (*Kebabian et al.*, 1975; *Lefkowitz* and *Williams*, 1978) without appreciable affects on the affinity of binding are consistent with this interpretation. We are unable to obtain significant stimulation of the pineal adenyate cyclase activity with PGE_1, so that the effects of catecholamine pretreatment on the activity induced by another agonist-receptor system could not be tested. Our data support the hypothesis that control of cyclic AMP synthesis is agonist-specific, in terms of both the initial stimulation of activity and the subsequent attenuation of the initial response.

Several conditions which reduce or abolish the sympathetic input to the pineal gland, i.e., chronic denervation by superior cervical ganglionectomy, decentralization of the superior cervical ganglia, immunosympathectomy with the antiserum to the nerve growth factor, chemical sympathectomy with 6-hydroxydopamine, or long-term exposure to light, all induce similar changes on the pineal adenylate cyclase-cyclic 3′,5′-AMP system. It is tempting, therefore, to speculate and to generalize that the sympathetic input to adrenergic structures can modify the response of adenylate cyclase (coupled to the proposed adrenergic receptor) to norepinephrine (the adrenergic neurotransmitter). In terms of this model, sympathetic activity causes the release from the nerve terminals of two substances, each of which influences adenylate cyclase activity. The actions of these substances, however, are quite different, both temporally and mechanistically. One substance (norepinephrine) causes an immediate but short-lasting stimulation of adenylate cyclase activity, probably by inducing a conformational change in the enzyme system which leads to an "uncoupling" of the enzyme and the receptor. The other substance (as of yet purely hypothetical) acts more slowly, causing a reduction in the quantity of hormone-sensitive adenylate cyclase, possibly by interfering with the *de novo* synthesis of the enzyme. The end result is that the organism possesses the means of protecting itself from excessive sympathetic tone by reducing the synthesis or functional coupling of an adrenergic receptor, and can induce the production of more or better coupled adrenergic receptor in the wake of abnormally low sympathetic activity.

RELEVANCE OF THE ADENYLATE CYCLASE-CYCLIC 3′,5′-AMP SYSTEM TO THE FUNCTION OF THE PINEAL GLAND

An understanding of the physiological interrelationships between the pineal gland and its target organ(s), primarily the gonads, and the involvement of the adenylate cyclase-cyclic 3′,5′-AMP system in this interaction should shed light on the role the pineal gland plays in gonadal function. Earlier findings showing that gonadal steroids influence pineal adenylate cyclase activity support the concept that the cyclic 3′,5′-AMP system is the common component linking the two organs (*Weiss* and *Crayton,* 1970a & b). The degree to which norepinephrine activates the adenylate cyclase of pineal glands depends on the stage of ovarian function, with estrogen playing the predominant role in controlling enzyme activity. The results suggest that estradiol released from the ovary inhibits the norepinephrine-sensitive adenylate cyclase system of rat pineal glands, causing a decreased production of cyclic 3′,5′-AMP and a decreased synthesis of methoxyindoles such as melatonin in the pineal gland. Since melatonin (*Wurtman, Axelrod,* and *Chu*, 1963) and 5-methoxytryptophol (*McIsaac, Farrell,* and *Taborsky,* 1964) have antigonadotropic activity, this series of reactions linking the pineal and ovary constitute a positive feedback control mechanism, thereby enabling the organism to rapidly regulate the functional activity of the gonads.

The involvement of the neurotransmitter, norepinephrine, as well as the adenylate cyclase-cyclic 3′,5′-AMP system in the production of melatonin has been supported by a number of findings. Norepinephrine added to cultured pineal glands increases the rate of synthesis of radioactive serotonin and melatonin from labeled tryptophan, an effect which is mimicked by dibutyryl cyclic 3′,5′-AMP (*Shein* and *Wurtman*, 1969; *Klein et al.,* 1970b; *Wurtman, Shein,* and *Larin,* 1971; *Berg* and *Klein,* 1971). Moreover, both norepinephrine and dibutyryl cyclic 3′,5′-AMP increase the activity of N-acetyltransferase, the enzyme that catalyzes the conversion of serotonin to N-acetylserotonin and which under certain circumstances becomes the rate-limiting reaction in the conversion of tryptophan to melatonin (*Klein et al.,* 1970a).

Our earlier findings that norepinephrine increases the concentration of cyclic 3′,5′-AMP under the same conditions in which melatonin production is enhanced have provided an important link in the sequence of events beginning with neuronal activity and ending with the release of melatonin by the pineal gland. Within 10 min after the addition of 0.1 mM norepinephrine, the concentration of cyclic 3′,5′-AMP in the whole gland approached 10^{-4} moles/kg of tissue. Even assuming equal distribution of the cyclic nucleotide in the pineal gland, this concentration (based on results with dibutyryl cyclic 3′,5′-AMP; *Berg* and *Klein,* 1971) should be sufficient to markedly increase the rate of conversion of tryptophan to melatonin. Thus, our results are compatible with a theory of a norepinephrine adenylate cyclase-cyclic 3′,5′-AMP-dependent mechanism being responsible for changes in the activity of N-acetyltransferase and the increased production of melatonin by the pineal gland (*Weiss* and *Crayton,* 1970a; *Klein* and *Berg,* 1970). This series of reactions leading to an increased formation of the methoxyindole in the pineal may be outlined as follows: The sensory input (darkness) increases the activity of the sympathetic neurons, causing the release of the neurotransmitter from sympathetic nerve terminals in the pineal gland. The released norepinephrine activates the hormone-sensitive adenylate cyclase system of the pineal gland, increasing the intracellular concentration of cyclic 3′,5′-AMP. Cyclic 3′,5′-AMP in turn stimulates the activity of N-acetyltransferase, causing a greater rate of formation of N-acetylserotonin and an increased production of melatonin.

In support of this model, it has been shown that environmental lighting decreases electrical activity (*Taylor* and *Wilson*, 1970) and reduces the concentration of norepinephrine in the pineal gland (*Wurtman et al.*, 1967). Light also decreases the activities of hydroxyindole-O-methyltransferase (*Axelrod, Wurtman,* and *Snyder,* 1965) and N-acetyltransferase (*Klein* and *Weller,* 1970) and reduces the concentration of melatonin in the pineal gland (*Tomatis* and *Orias,* 1967; *Lynch,* 1971). Additional support for the concept that melatonin production is mediated by changes in sympathetic nervous activity comes from the studies of *Volkman* and *Heller* (1971) showing that stimulation of preganglionic sympathetic fibers to the superior cervical ganglia increases the activity of pineal N-acetyltransferase. A similar study showing that sympathetic nerve stimulation increases the intracellular concentration of cyclic 3′,5′-AMP in the pineal gland would provide definitive proof for an involvement of the adenylate cyclase-cAMP system in the formation of antigonadotropic principles in the pineal gland. Such studies should also help in relating sensory input to the pineal with the physiological function of the target organ. It should also be stressed that indoles are not the only biologically active substances found in the pineal gland. Polypeptides with antigonadotropic and other biological properties have been isolated in the pineal gland (*Reiter* and *Vaughan,* 1977). The role(s) the cyclic nucleotide systems might play in the actions of these substances remains conjectural.

STUDIES OF THE CYCLIC GMP SYSTEM OF RAT PINEAL GLAND

Recent evidence suggests that the pineal gland may have value in studying the biological role of cyclic GMP. For example, we found that the tissue contains appreciable activities of both guanylate cyclase and cyclic GMP phosphodiesterase (*Strada, Kirkegaard,* and *Thompson,* 1976). Studies in ganglionectomized rats indicate that pineal gland guanylate cyclase is located postjunctionally in pineal cells. Recent works (*O'Dea* and *Zatz,* 1976; *Zatz* and *O'Dea,* 1977) suggest that cyclic GMP itself must also be synthesized presynaptically. Such studies have shown that the accumulation of cyclic GMP in cultured pineal glands has characteristics of both α-and β-adrenergic mediated events. Indirect results suggest that cyclic GMP may modulate the release of norepinephrine presynaptically. Since cyclic GMP systems in many tissues appear to be intricately linked to calcium stores (*Goldberg, O'Dea,* and *Haddox,* 1973), the ability of these small molecules to act in concert to modulate the release of neurotransmitters and polypeptide hormones (perhaps antigonadotropic peptides as well) is an interesting possibility. Since in most target tissues much less is known of the regulation of cyclic GMP than of cyclic AMP, it seems possible that the pineal gland may afford a good model system to explore in detail hormonal and neurohumoral control of the cyclic GMP system.

REFERENCES

Angeletti, P.U. 1971. Chemical sympathectomy in newborn animals. Neuroendocrinology 10:55-59

Axelrod, J. 1974. The pineal gland. A neurochemical transducer. Science 184:1341-1438

Axelrod, J., Wurtman, R.J., and Snyder, S.H. 1965. Control of hydroxyindole-O-methyltransferase activity in the rat pineal gland by environmental lighting. J Biol Chem 240:949-954

Berg, G.R., and Klein, D.C. 1971. Pineal gland in organ culture. II. Role of adenosine 3',5'-monophosphate in the regulation of radiolabeled melatonin production. Endocrinology 89:453-464

Brown, E.M., Fedak, S.A., Woodard, C.J., Aurbach, G.D., and Rodbard, D. 1976. Beta-adrenergic receptor interactions. Direct comparison of receptor interaction and biological activity. J Biol Chem 251:1239-1246

Cuatrecasas, P. 1974. Insulin receptors, cell membranes, and hormone action. Biochem Pharmacol 23:2353-2361

deChamplain, J. 1971. Degeneration and regrowth of adrenergic nerve fibers in the rat peripheral tissues after 6-hydroxydopamine. Can J Physiol Pharmacol 49:345-355

Deguchi, T., and Axelrod, J. 1973a. Superinduction of serotonin N-acetyltransferase and supersensitivity of adenyl cyclase to catecholamines in denervated pineal glands. Mol Pharmacol 9:612-618

Deguchi, T., and Axelrod, J. 1973b. Supersensitivity and subsensitivity of the β-adrenergic receptor in pineal gland regulated by catecholamine transmitter. Proc Natl Acad Sci USA 70:2411-2414

Ebadi, M.S., Weiss, B., and Costa, E. 1970. Adenosine 3',5'-monophosphate in rat pineal gland. Increase induced by light. Science 170:188-190

Ebadi, M.S., Weiss, B., and Costa, E. 1971a. Distribution of cyclic adenosine monophosphate in rat brain. Arch Neurol 24:353-357

Ebadi, M.S., Weiss, B., and Costa, E. 1971b. Microassay of adenosine 3',5'-monophosphate (cyclic AMP) in brain and other tissues by the luciferin-luciferase system. J Neurochem 18:183-192

Furness, J.B., Campbell, G.R., Gillard, S.M., Malmfors, T., Cobb, J.L.S., and Burnstock, G. 1970. Cellular studies of sympathetic denervation produced by 6-hydroxydopamine in the vas deferens. J Pharmacol Exp Therap 174:111-122

Goldberg, N.D., O'Dea, R.F., and Haddox, M.K. 1973. Cyclic GMP. Adv Cyclic Nucl Res 3:155-223

Haeusler, G., Haefley, W., and Thoenen, H. 1969. Chemical sympathectomy of the cat with 6-hydroxydopamine. J Pharmacol Exp Therap 170:50-61

Jaim-Etcheverry, G., and Zieher, L.M. 1971. Permanent depletion of peripheral norepinephrine in rats treated at birth with 6-hydroxydopamine. Eur J Pharmacol 13:272-276
Kappers, J.A. 1960. The development, topographical relations and innervation of the epiphysis cerebri in the albino rat. Zeit Zellafors Makros Anat 52:163-215
Kebabian, J.W., Zatz, M., Romero, J.A., and Axelrod, J. 1975. Rapid changes in rat pineal beta-adrenergic receptor. Alterations in 1-(^{3}H)-alprenolol binding and adenylate cyclase. Proc Natl Acad Sci USA 72:3735-3739
Klein, D.C., and Berg, G.R. 1970. Pineal gland. Stimulation of melatonin production by norepinephrine involves cyclic AMP mediated stimulation of N-acetyltransferase. Adv Biochem Psychopharmacol 3:241-263
Klein, D.C., Berg, G.R., and Weller, J. 1970a. Melatonin synthesis. Adenosine 3′,5′-monophosphate and norepinephrine stimulate N-acetyltransferase. Science 168:978-980
Klein, D.C., Berg, G.R., Weller, J., and Glinsmann, W. 1970b. Pineal gland. Dibutyryl cyclic adenosine monophosphate stimulation of labeled melatonin production. Science 167:1738-1740
Klein, D.C., and Weller, J. 1970. Input and output signals in a model neural system. The regulation of melatonin production in the pineal gland. In Vitro 6:197-204
Krishna, G., Weiss, B., and Brodie, B.B. 1968. A simple sensitive method for the assay of adenyl cyclase. J Pharmacol Exp Therap 163:379-385
Lefkowitz, R.J., and Williams, L.T. 1978. Molecular mechanisms of activation and desensitization of adenylate cyclase coupled beta-adrenergic receptors. Adv Cyclic Nucl Res 9:1-17
Levi-Montalcini, R., and Booker, B. 1960. Destruction of the sympathetic ganglia in mammals by an antiserum to a nerve growth protein. Proc Natl Acad Sci USA 46:384-391
Lynch, H.J. 1971. Diurnal oscillations in pineal melatonin content. Life Sci 10:791-795
McAfee, A.A., Schorderet, M., and Greengard, P. 1971. Adenosine 3′,5′-monophosphate in nervous tissue. Increase associated with synaptic transmission. Science 171:1156-1158
McIsaac, W.M., Farrell, G., and Taborsky, R.G. 1964. 5-methoxytryptophol. Effect on estrus and ovarian weight. Science 145:63-64
Mess, B. 1968. Endocrine and neurochemical aspects of pineal function. Int Rev Neurobiol 11:171-198
Mickey, J.V., Tate, R., and Lefkowitz, R.J. 1975. Subsensitivity of adenylate cyclase and decreased beta-adrenergic receptor binding after chronic exposure to (—)-isoproterenol *in vitro.* J Biol Chem 250:5727-5729
Moore, R.Y., Heller, A., Wurtman, R.J., and Axelrod, J. 1967. Visual pathway mediating pineal response to environmental light. Science 155:220-223
Moore, R.Y., and Rapport, R.L. 1971. Pineal and gonadal function in the rat following cervical sympathectomy. Neuroendocrinology 7:361-374
Morishima, I., Thompson, W.J., Robison, G.A., and Strada, S.J. 1978. Catecholamine desensitization in cultured fibroblasts. Mol Pharmacol, in press
Mukherjee, C., Caron, M.C., Coverstone, M., and Lefkowitz, R.J. 1975. Identification of adenylate cyclase-coupled beta-adrenergic receptors in frog erythrocytes with (—)-(^{3}H)alprenolol. J Biol Chem 250:4869-4876
Mukherjee, C., Caron, M.C., Mullikin, D., and Lefkowitz, R.J. 1976. Structure relationships of adenylate cyclase-coupled beta adrenergic receptors. Determination by direct binding studies. Mol Pharmacol 12:16-31
Nadeau, R.A., deChamplain, J., and Tremlay, G.M. 1971. Supersensitivity of the isolated rat heart after chemical sympathectomy with 6-hydroxydopamine. Can J Physiol Pharmacol 49:36-44
O'Dea, R.J., and Zatz, M. 1976. Catecholamine stimulated cyclic GMP accumulation in the rat pineal: Apparent presynaptic site of action. Proc Natl Acad Sci USA 73:3398-3402
Reiter, R.J., and Fraschini, F. 1969. Endocrine aspects of the mammalian pineal gland. A review. Neuroendocrinology 5:219-255
Reiter, R.J., and Vaughan, M.K. 1977. Pineal antigonadotrophic substances. Polypeptides and indoles. Life Sci 21:159-172

Remold-O'Donnel, E. 1974. Stimulation and desensitization of macrophage adenylate cyclase by prostaglandins and catecholamines. J Biol Chem 249:3615–3621

Robison, G.A., Butcher, R.W., and Sutherland, E.W. 1967. Adenyl cyclase as an adrenergic receptor. Ann N Y Acad Sci 139:703–723

Robison, G.A., Butcher, R.W., and Sutherland, E.W. 1971. *Cyclic AMP*. New York: Academic Press

Rodbell, M., Lin, M.C., Salomon, Y., Londos, C., Harwood, J.R., Martin, B.R., Rendell, M., and Berman, M. 1975. Role of adenine and guanine nucleotides in the activity and response of adenylate cyclase systems to hormones. Evidence for multisite transition states. Adv Cyclic Nucl Res 5:3–29

Romero, J.A., and Axelrod, J. 1975. Regulation of sensitivity to beta-adrenergic stimulation in induction of pineal N-acetyltransferase. Proc Natl Acad Sci USA 72:1661–1665

Romero, J.A., Zatz, M., and Axelrod, J. 1975. Beta-adrenergic stimulation of pineal N-acetyltransferase. Adenosine 3′,5′-cyclic monophosphate stimulates both RNA and protein synthesis. Proc Natl Acad Sci USA 72:2107–2111

Shein, H.M., and Wurtman, R.J. 1969. Cyclic adenosine monophosphate stimulation of melatonin and serotonin synthesis in cultured rat pineals. Science 166:519–520

Strada, S.J., Kirkegaard, L., and Thompson, W.J. 1976. Studies on rat pineal gland guanylate cyclase. Neuropharmacology 15:261–266

Strada, S.J., Klein, D.C., Weller, J., and Weiss, B. 1972. Effect of norepinephrine on the concentration of adenosine 3′,5′-monophosphate of rat pineal gland in organ culture. Endocrinology 90:1470–1475

Strada, S.J., and Weiss, B. 1974. Increased response to catecholamines of the cyclic AMP systems of rat pineal gland induced by decreased sympathetic activity. Arch Biochem Biophys 160:197–204

Taylor, A.N., and Wilson, R.W. 1970. Electrophysiological evidence for the action of light on the pineal gland in the rat. Experientia 26:267–269

Thoenen, H., and Tranzer, J.P. 1968. Chemical sympathectomy by selective destruction of adrenergic nerve endings with 6-hydroxydopamine. Naunyn-Schmiedebergs Arch Pharmacol 261:271–288

Thompson, W.J., Williams, R.H., and Little, S.A. 1973. Studies on the assay and activity of guanyl- and adenyl cyclase of rat liver. Arch Biochem Biophys 159:206–213

Tomatis, M.E., and Orias, R. 1967. Changes in melatonin concentration in pineal gland in rats exposed to continuous light or darkness. Acta Physiol Lat Amer 17:227–233

Uretsky, N.J., and Iversen, L.L. 1970. Effects of 6-hydroxydopamine on catecholamine containing neurons in the rat brain. J Neurochem 17:269–278

Volkman, P.H., and Heller, A. 1971. Pineal N-acetyltransferase activity. Effect of sympathetic stimulation. Science 173:839–840

Wagner, K., and Trendelenburg, U. 1971. Development of degeneration contraction and supersensitivity in the cat nictitating membrane after 6-hydroxydopamine. Naunyn-Schmiedeberg s Arch Pharmacol 270:215–236

Weiss, B. 1969. Effects of environmental lighting and chronic denervation on the response of adenyl cyclase of rat pineal gland to norepinephrine and sodium fluoride. J Pharmacol Exp Therap 168:146–152

Weiss, B. 1970. Factors affecting adenyl cyclase activity and its sensitivity to biogenic amines. In *Biogenic Amines as Physiological Regulators*, ed. J.J. Blum, pp. 35–73. Englewood Cliffs, New Jersey: Prentice Hall

Weiss, B. 1971. Ontogenetic development of adenyl cyclase and phosphodiesterase in rat brain. J Neurochem 18:469–478

Weiss, B., and Costa, E. 1967. Adenyl cyclase activity in rat pineal gland. Effects of chronic denervation and norepinephrine. Science 156:1750–1752

Weiss, B., and Costa, E. 1968a. Regional and subcellular distribution of adenyl cyclase and 3′,5′-cyclic nucleotide phosphodiesterase in brain and pineal gland. Biochem Pharmacol 17:2107–2116

Weiss, B., and Costa, E. 1968b. Selective stimulation of adenyl cyclase of rat pineal gland by pharmacologically active catecholamines. J Pharmacol Exp Therap 161:310–319

Weiss, B., and Crayton, J.W. 1970a. Neural and hormonal regulation of pineal adenyl cyclase activity. Adv Biochem Psychopharmacol 3:217–239
Weiss, B., and Crayton, J.W. 1970b. Gonadal hormones as regulators of pineal adenyl cyclase activity. Endocrinology 87:527–533
Weiss, B., and Kidman, A.D. 1969. Neurobiological significance of cyclic 3′,5′-adenosine monophosphate. Adv Biochem Psychopharmacol 1:131–164
Weiss, B., and Strada, S.J. 1972. Neuroendocrine control of the cyclic AMP systems of brain and pineal gland. Adv Cyclic Nucl Res 1:357–374
Wurtman, R.J., Axelrod, J., and Chu, E.W. 1963. Melatonin, a pineal substance. Effect on the rat ovary. Science 141:277–278
Wurtman, R.J., Axelrod, J., Kelly, D.E. 1968. *The Pineal.* New York: Academic Press
Wurtman, R.J., Axelrod, J., Sedvall, G., and Moore, R.Y. 1967. Photic and neural control of the 24-hour norepinephrine rhythm in the rat pineal gland. J Pharmacol Exp Therap 157:487–492
Wurtman, R.J., Shein, H.M., and Larin, F. 1971. Mediation by beta-adrenergic receptors of effect of norepinephrine on pineal synthesis of C^{14}-serotonin and C^{14}-melatonin. J Neurochem 18:1683–1687
Zatz, M., Kebabian, J.W., Romero, J.A., Lefkowitz, R.J., and Axelrod, J. 1976. Pineal beta adrenergic receptor. Correlation of binding of ^{3}H-1-alprenolol with stimulation of adenylate cyclase. J Pharmacol Exp Therap 196:714–722
Zatz, M., and O'Dea, R.F. 1977. Efflux of cyclic nucleotides from rat pineal. Release of guanosine 3′,5′-monophosphate from sympathetic nerve endings. Science 197:174–176

Genetic Determination of Cyclic AMP Level in Brain: Some Behavioral Implications

Albert Sattin

Department of Psychiatry
Institute of Psychiatric Research
Indianapolis, Indiana 46223

INTRODUCTION

The rationale for the work described and reviewed here arose from two sources. The first was the study of cyclic AMP accumulations in chopped tissue from mammalian brain. The second was a series of behavioral genetic studies.

The chopped tissue (or brain slice) studies arose from the earliest observations of Sutherland, Rall and associates (*Klainer et al.*, 1962) on the synthesis of cyclic AMP by brain tissue. The initial failure to observe consistent hormone effects in broken cell preparations led to the investigation of chopped tissue, which displayed large responses to catecholamines and numerous other substances of neurohumoral significance (*Kakiuchi* and *Rall,* 1968a & b). The rapid accumulations of cyclic AMP induced by various agents in chopped tissue generally reflect activation of cyclic AMP synthesis rather than inhibition of destruction (*Rall* and *Sattin,* 1970; *Schultz* and *Daly,* 1973a). The rapid and large cyclic AMP accumulations have been attributed to the activation of specific receptors (*Kakiuchi* and *Rall,* 1968a & b; *Sattin* and *Rall,* 1970; *Schultz* and *Daly,* 1973b; *Palmer, Sulser,* and *Robison,* 1973; *Perkins* and *Moore,* 1973; *Sattin, Rall* and *Zanella,* 1975). More recently, receptors have themselves been the object of study by competitive binding of the activator (*Maguire, Goldmann,* and *Gilman,* 1974; *Alexander, Davis,* and *Lefkowitz,* 1975; *Williams* and *Lefkowitz,* 1976). Furthermore, broken cell preparations of adenylate cyclase in brain have now been shown to display reproducible hormone effects (*Kebabian, Petzold,* and *Greengard,* 1972; *von Hungen* and *Roberts,* 1973; *Premont, Perez,* and *Bockaert,* 1977). However, the potentiative interactions be-

---

This work was supported by grant no. BMS 75-02955 from the National Science Foundation. The experimental work was performed in the Departments of Pharmacology and Psychiatry, School of Medicine, Case Western Reserve University. The author is grateful to Dr. James Norton for the statistical analysis. Drs. Richard Sprott and Michael Conneally provided helpful discussion. Monica Olszewski, Susan Strong, Evelyn Bennett and Kenneth Foltarz are thanked for their excellent technical assistance.

tween receptors that were originally described in chopped tissue have never been observed at the broken cell level and it is assumed that, whatever the underlying mechanisms, relatively intact membranes are required to show potentiation (*Sattin et al.,* 1975).

The accumulation of cyclic AMP in chopped neural tissue varies as a function of the neurohumoral agent applied, the anatomical region, the species, and the strain (*Rall* and *Sattin,* 1970; *Forn* and *Krishna,* 1971). Variation as a function of strain was suspected because of variable responses during work with random-bred strains. Side-by-side comparisons between three highly inbred strains of mice revealed striking differences in the responses to NE, adenosine and the combination of these agents (*Sattin,* 1975). This was of great interest because genetic crosses involving one of these strains, the C57BL/6, have provided data suggestive of specific genes coding for specific measurable behaviors in mice (*Oliverio, Eleftheriou,* and *Bailey,* 1973a & b; *Sprott,* 1974). The behaviors relevant to the neuropharmacological data are "initial exploratory activity" (*Oliverio et al.,* 1973a), "active avoidance learning" (*Oliverio et al.,* 1973b) and "passive avoidance learning" (*Sprott,* 1974).

The evidence for the first two behaviors was obtained using genetic crosses with mice of the BALB/c strain, while the third was obtained with the DBA/2 strain. Analysis with a battery of recombinant-inbred strains derived from the initial cross of C57BL/6 with BALB/c revealed that the transmission of active avoidance performance may be assigned to a gene on chromosome 9. The low learning rate is transmitted as a dominant trait by C57BL/6. Using the same genetic methodology, the transmission of initial exploratory activity may be assigned to a gene on chromosome 4. In this behavior, high activity is transmitted as a dominant by C57BL/6.

The third behavior, passive avoidance performance, was studied with standard genetic crosses between C57BL/6 and DBA/2. The data demonstrated dominant transmission of poor performance by C57BL/6. By coincidence, crosses between these two parental strains can be directly observed for transmission of two sets of coat-color alleles, brown-black, and dilute-dense, which are known to be located on chromosomes 4 and 9, respectively (*Stavnes* and *Sprott,* 1975). Passive avoidance performance was not associated with either set of alleles. Since the coat-color alleles are on the same two chromosomes (4 and 9) as the first two behaviors described above, the third behavior may have a separate genetic origin.

The C57BL/6 and DBA/2 strains were chosen for the cyclic AMP studies because of the ease of detecting possible linkage of a genetically transmitted cyclic AMP trait with either set of coat-color alleles. If documented, such linkage would suggest that the cyclic AMP trait is involved in the generation of the behavior to which it is linked. The original strain comparison suggested that a diminished response to NE was transmitted by C57BL/6J as a simple mendelian dominant (Table 1). It was further noted that the response to adenosine might also have been transmitted by C57BL/6J as a mendelian dominant (*Sattin,* 1975). A reported reduction in cyclic nucleotide phosphodiesterase in the supernatant fraction from cerebral cortex of the DBAs might have explained the increased responsiveness of DBA cortex to both NE and adenosine, but the initially observed strain difference in this enzyme activity was not repeatable (*Sattin,* unpublished observation). Therefore, variation in receptor-activated synthesis factors remains the most likely cause of strain variation.

Stalvey, Daly, and *Dismukes* (1976) pursued these hypotheses using additional Jackson strains of mice in a study designed to correlate NE- and adenosine-induced cyclic AMP accumulation with the behaviors described above. Using their own assessment of initial exploratory activity, they found an inverse correlation of spontaneous activity with adenosine-induced increase in cyclic AMP in chopped telencephalon tissue. The adenosine-induced increase in cyclic AMP was also negatively correlated with spontaneous open-field behavioral activity observed in another laboratory (*Southwick* and *Clark,* 1968) for six of their seven strains. However, the seventh strain, SEC/1ReJ, does correlate inversely with wheel-running activity (*Oliverio, Castellano,* and *Messeri,* 1972) (Fig. 1) using my original comparison of the adenosine effect in three strains (*Sattin,* 1975). The deviance of the SEC strain observed by *Stalvey et al.* (1976) might have been due to their using whole telencephalon rather than cerebral cortex, or to their routine use of Krebs-Ringer bicarbonate buffer containing 2.4 mM Ca^{++}. We have used a modified buffer containing 0.8 mM Ca^{++} (*Sattin* and *Rall,* 1970; *Sattin,* 1975; *Charnock,* 1963). Whichever the reason, it may be seen that our con-

Table 1. Effects of Norepinephrine (NE) on Chopped Cerebral Cortex from Mice of Two Strains and Their F_1 Hybrid

	Cyclic AMP (pmoles/mg protein)	
Strain	Control	NE
C57BL/6J	16.3 ± 0.5 (6)	26.2 ± 1.8 (12)
DBA/2J	13.2 ± 0.5 (6)	54.9 ± 4.0 (11)*
F_1 (CXD)	—	31.1 ± 3.4 (6)

After 50 min of preincubation, samples of chopped cortex were transferred to fresh medium and incubated for 10 min with no additions (control) or NE, 0.1 mM. Results from two experiments with females and two with males did not differ and the data were pooled. Data are means ± SEM. The number of incubation samples is in parentheses. Control data for F_1 were not obtained. *Higher than NE value for C57B1/6J; $P < 0.001$ by two-tailed *t* test. (Modified from *Sattin,* 1975.)

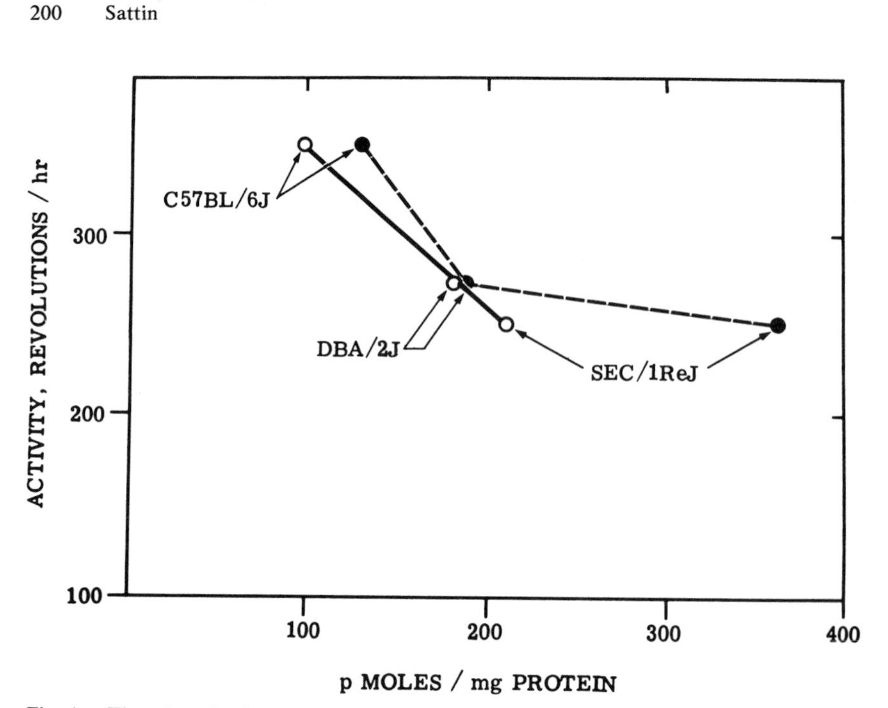

Fig. 1. The adenosine-induced maximal accumulation of cyclic AMP in chopped cerebral cortex (*Sattin*, 1975) and in chopped forebrain (*Stalvey et al.*, 1976) of three strains of mice plotted against wheel-running activity (*Oliverio et al.*, 1972). The solid line connects the data of *Sattin* (1975) and the dashed line connects that of *Stalvey et al.* (1976). See text for discussion of differences in incubation conditions.

ditions rectify the apparent anomaly that *Stalvey et al.* (1976) observed with the SEC strain. Thus, strain comparison studies support the hypothesis that initial exploratory activity, to which a gene on chromosome 4 has been assigned, is inversely related in mouse to the magnitude of the adenosine-induced cyclic AMP accumulation in telencephalon or in cerebral cortex.

In the same strain comparison study, *Stalvey et al.* (1976) failed to relate the NE-induced accumulation of cyclic AMP in telencephalon to active avoidance performance but did observe a negative correlation between this behavior and the NE response in a group of random-bred mice. This appeared to contradict the positive correlation that I initially observed using three strains. In the same report, however, *Stalvey et al.* (1976) actually showed that the NE effects in the random-bred mice were confounded with the effects of adenosine which were uncontrolled. When tissue from their inbred strains was incubated in the presence of theophylline (which blocks adenosine receptors), basal levels of cyclic AMP were reduced from 30% to 75%, depending on the strain. In the presence of theophylline, their basal values for the three strains used in my earlier comparison study were identical to the values which I obtained in the absence of theophylline. This demonstrates that, for these strains of mouse (C57BL/6J, DBA/2J and SEC/1ReJ), 1) basal levels of cyclic AMP in incubated whole telencephalon are the same as those in cerebral cortex, 2) in the conditions of *Stalvey et al.* (1976), unless theophylline is present there is no control over a substantial adenosine-activated component to the cyclic AMP accumulation that is attributed to NE or to another active agent, and 3) under our conditions of incubation this "background" effect of adenosine (see *Sattin, Rall,* and *Zanella,* 1975) is minimally present.

In their inbred strain comparison of the NE effect, *Stalvey et al.* (1976) did include theophylline, yet their NE-induced accumulations of cyclic AMP were 4- to 8-fold greater than those I reported. Again, either the anatomical or the Ca^{++} difference probably accounts for this. In the following section, the results of a mendelian genetic segregation study will provide indirect evidence for a positive correlation of NE-induced cyclic AMP accumulation and active avoidance performance in mice.

GENETIC SEGREGATION STUDY OF NE-INDUCED ACCUMULATION OF CYCLIC AMP IN MOUSE CEREBRAL CORTEX

Jackson-derived mice of the C57BL/6 and DBA/2 strains were raised under optimal conditions of temperature, humidity and 12-hr light cycle. The standard Wayne lab-blox feed was alternated weekly with 96W "Old Guilford Diet" (Emory Morse Co., Guilford, CT) for the DBA/2s. The mice were sacrificed at 2 mo. The cerebral cortex was dissected free of striatum and the temporal lobe plus hippocampus was discarded. Since magnification was not used, variable amounts of adjacent cortex were also discarded. The remaining cortex tissue was chopped on a McIlwain tissue chopper set at 0.312 mm. The methods for preincubation, incubation, fixation of tissue and assay of cyclic AMP and protein were as previously described (*Sattin*, 1975). Exposure to a maximal dose of NE, 0.1 mM, was used in an attempt to distinguish the individual mouse as a "high" or "low" accumulator of cyclic AMP.

The results for the two parental strains and three genetic crosses are depicted in Figure 2. These data were determined in individual animals, not in pooled tissue as in the original study (*Sattin*, 1975). However, the pattern of results for the two parental strains and the F_1 cross was the same. The F_1 mean was indistinguishable from that of the C57 parent, while the mean of the DBA parent was significantly higher than both (Table 2). This again suggests that low NE-induced accumulation of cyclic AMP is transmitted by C57BL/6 as a mendelian dominant in

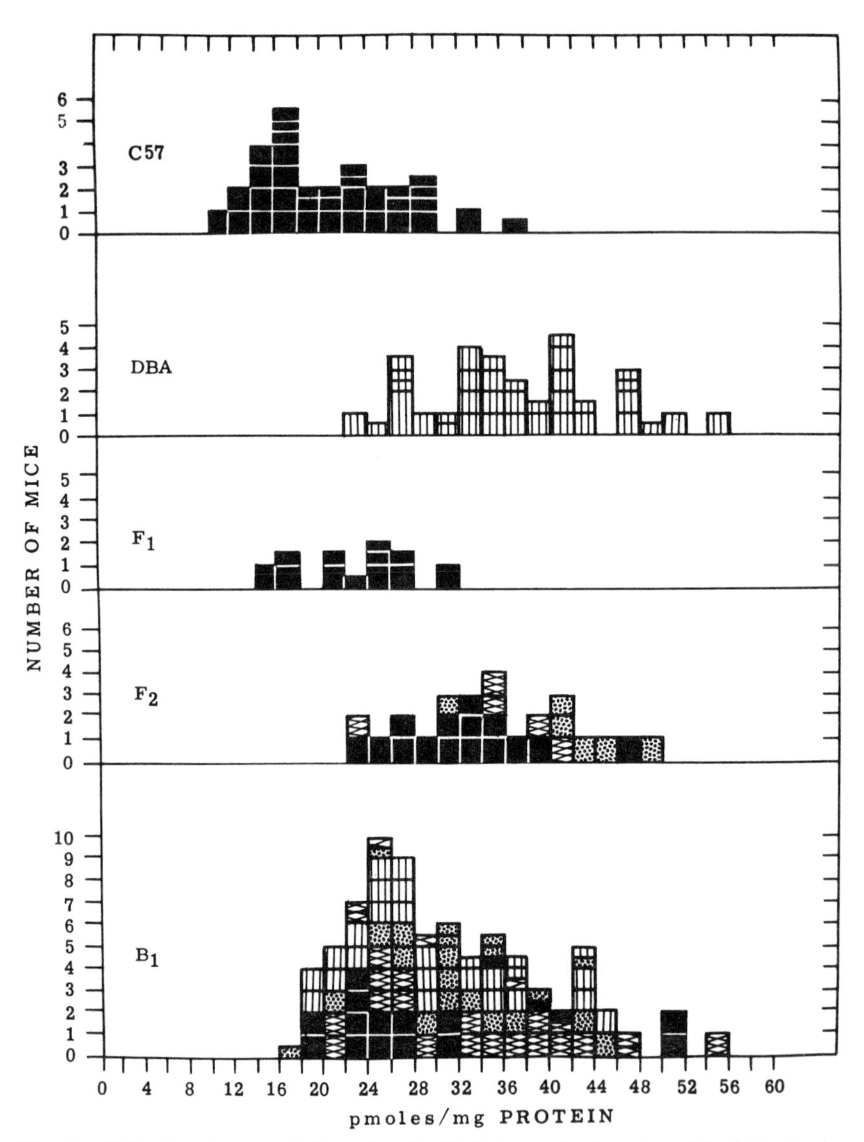

Fig. 2. NE-induced accumulation of cyclic AMP in cerebral cortex of individual mice of parental strains and genetic crosses. Procedures as described in text. B_1 represents the backcross of the F_1 to the recessive (DBA) parent. Each square represents an individual mouse. Full squares represent means of individual data points derived from the left and right cerebral cortex. (Each hemicortex is the mean of duplicate cyclic AMP determinations.) Five C57 and five DBA squares represent the entire cortex incubated as one sample. The half-squares represent mice in which only the right hemicortex was incubated. Colors are black in C57 and F_1 and dilute brown in DBA. In F_2 and B_1 (segregating populations) two additional colors are represented: brown (cross-hatched) and dilute black (stippled).

Table 2. Statistical Analysis of Population Means of Figure 2

Strain	C57	DBA	F_1	F_2	B_1
		pmoles/mg protein			
Mean	20.8	36.8	22.7	35.3	30.4
		t test			
C57 vs	—	$t = 10.3$*	$t = 0.84$	$t' = 7.8$*	$t' = 6.7$*
DBA vs	—	—	$t = 6.6$*	$t' = 0.82$	$t' = 4.7$*
F_1 vs	—	—	—	$t' = 5.3$*	$t' = 3.7$*
F_2 vs	—	—	—	—	$t = 2.9$†

For mice represented by both left and right hemicortices, the distribution of the absolute differences between the hemicortices was treated as the right half of a normal distribution. By setting a 99% tolerance limit (*Owen*, 1962), three outlying mice (one F_2 and two B_1s) were eliminated from further analysis. One-way analysis of variance among animals was used in each generation to obtain the means and variances because this has the effect of giving mice with two observations twice the weight of those with one observation while basing the degrees of freedom on the number of animals. Owing to a significant difference in the variances between the segregating and nonsegregating generations, the t' test was applied to comparison of means between these two populations (*Li*, 1964). *$P < 0.001$; †$P < 0.01$.

this pair of strains. The attempted genetic proof using data from the F_2 and B_1 populations was not entirely consistent with mendelian prediction. The mean of the F_2 was higher than expected and not significantly different from the DBA parent. On the other hand, the mean of the B_1 population lay between the parents and was significantly different from both (Table 2). Furthermore, the variance within the two segregating populations was statistically similar and both were significantly greater than the variances within the three nonsegregating populations, which were also statistically similar. Thus, the variance data and the B_1 mean are consistent with mendelian prediction but the F_2 mean is not.

Looking at the coat-color subgroups, only one significant difference (at the 0.05 level) was found: dilute vs nondilute in the F_2 population. In this population high NE-induced cyclic AMP accumulation segregated with the dilute allele. If we assume that an uncontrolled factor has shifted the mean of this population to the right, the coat-color data may be accepted as evidence favoring linkage of an NE-cyclic AMP trait to the dilute locus on chromosome 9. Active avoidance learning also has been assigned to this chromosome (*Oliverio et al.*, 1973b). Therefore, the coat-color segregation data in this population have indirectly demonstrated an association between low NE-induced accumulation of cyclic AMP and low active avoidance learning, both of which appear to be transmitted as dominant traits by C57BL/6 in its cross with DBA/2. Using a multiple strain-comparison approach, *Stalvey et al.*

(1976) failed to detect this association, but their NE data was contaminated with uncontrolled additive or potentiative effects of adenosine (see above). Even though theophylline was not used in the genetic data, my incubation conditions minimized the possibility that an adenosine background interfered with the NE effects.

If initial exploratory activity is associated with the black locus on chromosome 4, then according to the association of this behavior with adenosine-induced cyclic AMP accumulation (*Stalvey et al.*, 1976), the brown segregants should have yielded higher cyclic AMP values if adenosine contributed to the observed NE effects. This was not seen in either segregating population.

Anatomical dissection variation was probably the largest source of variance in all groups. This experimental variance may have obscured the expected bimodal distribution in the B_1 population where the means of the four coat-color subgroups were not significantly different from each other.

THE PHARMACOLOGY OF NE-INDUCED ACCUMULATION OF CYCLIC AMP IN MOUSE CEREBRAL CORTEX

What is the pharmacological nature of the catecholamine receptors responsible for the differences in accumulation of cyclic AMP in these strains? Results of agonist studies with pooled slices from C57BL/6 mice suggested that both α- and β-receptors were present (Fig. 3). Using EPI as a combined α-plus-β agonist, the α- and β-blockers phentolamine and propranolol were used in a side-by-side comparison of the two parental strains. It may be seen that the difference in the cyclic AMP accumulation by the two strains can be accounted for by the component that is blocked by phentolamine, although propranolol also eliminates the strain difference while even further reducing the cyclic AMP accumulation (Fig. 4) (*Sattin,* 1978). This pattern of response to α- and β-blocking agents is identical to that found in two previous reports for chopped brain tissue from rat (*Palmer, Sulser,* and *Robison,* 1973; *Perkins* and *Moore,* 1973).

We have interpreted such data as evidence for an interaction between a dependent α- and an independent β-receptor, i.e., activation of the α-receptor is dependent upon coactivation of a β-receptor. Another dependent α-receptor that has been characterized in guinea pig cortex requires coactivation of adenosine receptors (*Sattin, Rall,* and *Zanella,* 1975). In this pharmacologic context, which may be peculiar to brain, it appears that the quantitative difference between the C57BL/6 and DBA/2 lies in the magnitude of the response to α-receptor activation.

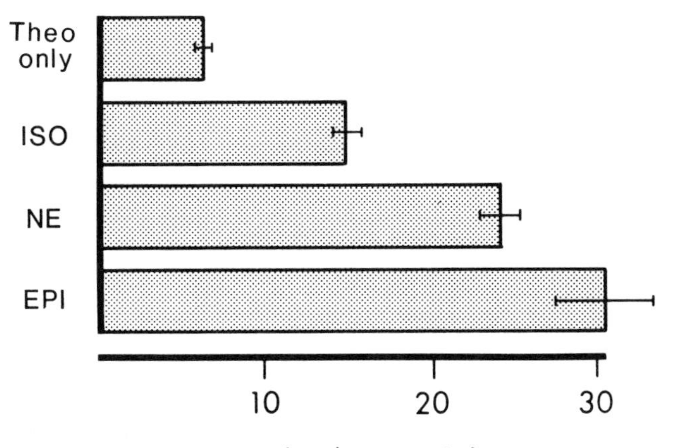

Fig. 3. Chopped cerebral cortex from C57BL/6 mice was preincubated as described in text, then exposed to 0.1 mM isoproterenol (ISO), norepinephrine (NE) or epinephrine (EPI) for 10 min before fixation and assay of cyclic AMP. All tissue samples were also exposed to theophylline (Theo), 0.5 mM, during the incubation period. Bars represent means ± SD of 4–6 samples. All differences were significant ($P < 0.005$).

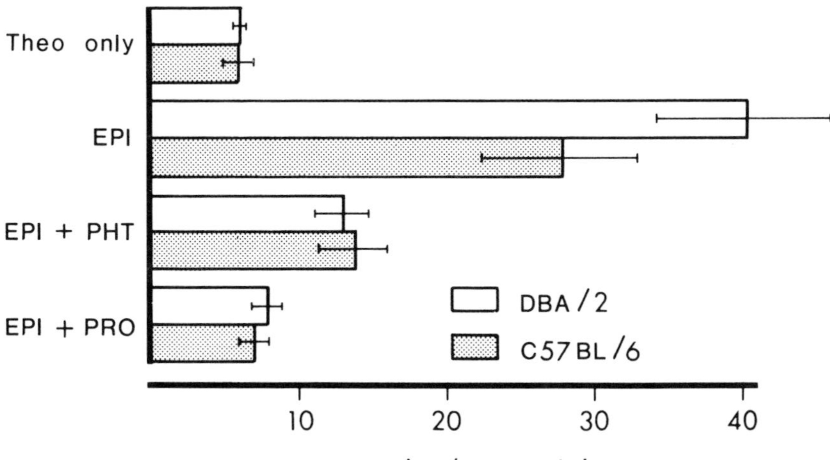

Fig. 4. Cerebral cortex tissue from mice of the two parental strains used in the segregation study (Fig. 2 and Table 2) was incubated as in Figure 3. Theophylline (Theo), 0.5 mM, was included in all samples. The "Theo only" values for both strains do not differ from corresponding "no Theo" controls (data not shown). Epinephrine (EPI), phentolamine (PHT) and propranolol (PRO) were added at 0.1 mM. Bars represent means ± SD of 6–11 samples. The two strains differed significantly in the EPI-induced accumulation of cyclic AMP ($P < 0.005$).

If this is true, the genetic data suggest that the C57BL/6 strain transmits as a dominant trait a smaller response to activation of the α-component of the β-plus-α catecholamine receptors in mouse cerebral cortex.

It should be noted that activation of adenosine-dependent receptors in the experiments of Figures 3 and 4 was excluded by the use of 0.5 mM theophylline in all incubations (*Sattin* and *Rall*, 1975; *Sattin*, 1978). Parallel pharmacological and genetic involvement of adenosine-dependent α-receptors (*Sattin*, *Rall*, and *Zanella*, 1975) is not ruled out. However, the inverse correlation of initial exploratory activity with adenosine-induced accumulation of cyclic AMP leads to the prediction that another segregation study of the same strain pair using adenosine instead of NE would display linkage with the black-brown locus on chromosome 4, to which the gene for this behavior has been assigned.

CONCLUDING REMARKS

In rat, cerebral cortical pyramidal tract cells have been characterized as neuronal elements that are "bidirectionally" controlled by AcCh and NE. The excitatory effect of the former may be mediated by cyclic GMP and the inhibitory effect of the latter by cyclic AMP (*Stone*, *Taylor*, and *Bloom*, 1975). These same neurons are profoundly inhibited by iontophoretic application of adenosine (*Phillis*, *Kostopoulos*, and *Limacher*, 1975). There is some evidence that NE and adenosine can potentiate each other's effect in this physiological system paralleling the neuropharmacological findings with cyclic AMP accumulation in chopped tissue (*Stone* and *Taylor*, 1978; *Sattin* and *Rall*, 1970). However, the physiology of adenosine poses a major problem. Although "purinergic neurons" have not been identified in CNS, we have suggested that ATP may be released together with NE or with other amine transmitters in CNS in analogy with the chromaffin granules of the adrenal medulla. Alternatively, adenosine may function as a metabolic negative feedback regulator of neuronal activity by virtue of the increased steady-state level resulting from increased utilization of ATP (*Sattin* and *Rall*, 1970). It might also function by a combination of these mechanisms.

The inverse relationship described in this review between adenosine receptor sensitivity and initial exploratory activity poses a physiological problem. If it can be assumed that this relationship has physiological meaning, the source of the adenosine is not yet known. If it comes from the neighborhood of large pyramidal cells of the type that has been studied by iontophoresis in rat, then these neurons would be expected to diminish their rate of firing in response to increased adenosine. The more sensitive these cells to adenosine, the more diminished their firing rates and the more diminished the motor activity at a given steady-state level of adenosine. It has been known for some time that adenosine and adenine nucleotides can abolish all motor activity when injected intraventricularly (*Feldberg* and *Sherwood,* 1954). Thus, this newly discovered neuropharmaco-behavioral relationship is at least consistent with our present limited physiological understanding of the effect of adenosine. The segregation data presented above add further support to the idea that active avoidance performance has a neuropharmacological substrate that is different from initial exploratory activity. The data suggest that the former behavior is indirectly linked to a gene controlling the magnitude of the NE-induced accumulation of cyclic AMP in cerebral cortex, probably via α-receptors.

It is interesting to note that α-receptor function in CNS has been physiologically related to rewarded behavior (*Wise, Berger,* and *Stein,* 1973). The active avoidance paradigm involves a reward that results from the initiation of an ambulatory movement. This contrasts with passive avoidance performance where the reward results from suppression of movement. The association of NE-induced cyclic AMP accumulation with active avoidance performance might be factitious. Additional genetic studies and more careful strain comparison studies are needed. If the functional relationships implied here receive further support, then these investigations should serve to clarify the role of NE and adenosine in the behavioral functioning of the CNS.

An additional variety of active avoidance performance, active avoidance "jump-up" performance, appears to be controlled by another gene. In contrast to the active avoidance ambulatory performance previously described, the DBA/2J transmits poor jump-up performance as a mendelian dominant in a cross with C57BL/6J (*Stavnes* and *Sprott,* 1975).

The spontaneous activity of three inbred strains of rat was inversely correlated with reported accumulation of cyclic AMP following incubation of chopped cerebral cortex with NE (*Skolnick* and *Daly,* 1974). The adenosine-induced accumulation of cyclic AMP in the cortex tissue was not related to this behavior. Striatum-plus-midbrain tissue from the same three inbred rat strains yielded a direct rather than indirect correlation of the spontaneous activity with the NE-induced cyclic AMP accumulation. This direct correlation was further corroborated by intraventricular injection studies followed by behavioral observations of the same strains (*Segal, Geyer,* and *Weiner,* 1975). This route of administration is not likely to directly influence cerebral cortex, but it is likely to affect the striatum and midbrain where catecholamines may stimulate locomotory behavior (*van Rossum, Broekkamp,* and *Pijnenburg,*

1977; *Iversen* and *Koob,* 1977), in contrast to the motor neurons of the cerebral cortex where their effect is probably inhibitory (*Stone et al.,* 1975).

These rat data appear to differ from the mouse data where adenosine- but not NE-induced accumulation of cyclic AMP in cortex is related (also inversely) to spontaneous activity. However, the rat cortex data may be interpreted retrospectively as a function of a combined NE plus adenosine response since those tissue incubations were done under the same conditions as those of *Stalvey et al.* (1976). Furthermore, the activity assessment in rat differs from that used with mice. In rat, the activity was scored only after the animals had been allowed a 2-day period of accommodation to the open-field situation where they were subsequently scored (*Segal, Kuczenski,* and *Mandell,* 1972).

SUMMARY

The following points may be made regarding the relationship of cyclic AMP responses to neuropharmacological agents and behavior:

1. A genetic segregation study in mice has suggested that a variety of active avoidance performance which was previously attributed to a gene on chromosome 9 is directly related to the magnitude of the NE-induced accumulation of cyclic AMP in the cerebral cortex.
2. The α-adrenergic receptor component of the NE-induced accumulation of cyclic AMP may be responsible for the differences observed in the cortex from the two parental strains used in this study. Therefore, this active avoidance performance may be mediated in part by activation of these cerebral cortical α-receptors.
3. The direct correlation of active avoidance performance with NE-receptor activation contrasts with an inverse correlation of adenosine receptor-mediated increase in cyclic AMP and initial exploratory activity in mice. It is suggested that a genetic locus on chromosome 4 controls the magnitude of the response to adenosine in cerebral cortex. The inhibition of exploratory behavior may be mediated in part by a physiological action of adenosine on the pyramidal tract cells and functionally related cells in cerebral cortex.

REFERENCES

Alexander, R.W., Davis, J.N., and Lefkowitz, R.J. 1975. Direct identification and characterization of beta-adrenergic receptors in the rat brain. Nature 258:437–440

Charnock, J.S. 1963. The accumulation of calcium by brain cortex slices. J Neurochem 10:219–223

Feldberg, W. and Sherwood, S.L. 1954. Injections of drugs into the lateral venticles of the cat. J Physiol 123:148-167
Forn, J. and Krishna, G. 1971. Effect of norepinephrine, histamine and other drugs on cyclic 3′,5′-AMP formation in brain slices of various animal species. Pharmacology 5:193-204
Iversen, S. and Koob, G.F. 1977. Behavioral implications of dopaminergic neurons in the mesolimbic system. In *Advances in Biochemical Psychopharmacology,* Vol. 16, eds. E. Costa and G.L. Gessa, New York: Raven Press
Kakiuchi, S. and Rall, T.W. 1968a. The influence of chemical agents on the accumulation of adenosine 3′,5′-phosphate in slices of rabbit cerebellum. Mol Pharmacol 4:367-378
Kakiuchi, S. and Rall, T.W. 1968b. Studies on adenosine 3′,5′-phosphate in rabbit cerebral cortex. Mol Pharmacol 4:379-388
Kebabian, J.W., Petzold, G.L. and Greengard, P. 1972. Dopamine-sensitive adenylate cyclase in caudate nucleus of rat brain and its similarity to the "dopamine receptor." Proc Nat Acad Sci USA, 69:391-401
Klainer, L.M., Chi, Y.-M., Friedberg, S.L., Rall, T.W. and Sutherland, E.W., Jr. 1962. Adenyl cyclase. IV. The effects of neurohormones on the formation of adenosine 3′,5′-phosphate by preparations from brain and other tissues. J Biol Chem 237: 1239-1243
Li, J.C.R. 1964. *Statistical Inference,* Vol. 1, Sec. 10.4. Ann Arbor, Mich.: Edwards Bros.
Maguire, M.E., Goldmann, P.H. and Gilman, A.G. 1974. The reaction of [^{3}H] norepinephrine with particulate fractions of cells responsive to catecholamines. Mol Pharmacol 10:563-581
Oliverio, A., Castellano, C. and Messeri, P. 1972. Genetic analysis of avoidance, maze, and wheel-running behaviors in the mouse. J Comp Physiol Psychol 79:459-473
Oliverio, A., Eleftheriou, B.E. and Bailey, D.W. 1973a. Exploratory activity. Genetic analysis of its modification by scopolamine and amphetamine. Physiol and Behav 10:893-899
Oliverio, A., Eleftheriou, B.E., and Bailey, D.W. 1973b. A gene influencing active avoidance performance in mice. Physiol Behav 11:498-501
Owen, D. B. 1962. *Handbook of Statistical Tables,* Sec. 5.4. Reading, Mass: Addison-Wesley
Palmer, G.C., Sulser, F. and Robison, G.A. 1973. Effects of neurohumoral and adrenergic agents on cyclic AMP levels in various areas of the rat brain *in vitro.* Neuropharmacology 12:327-337
Perkins, J.P. and Moore, M.M. 1973. Characterization of the adrenergic receptors mediating a rise in cyclic 3′, 5′-adenosine monophosphate rat cerebral cortex. J Pharmacol Exp Ther 185:371-378
Phillis, J.W., Kostopoulos, G.K. and Limacher, J.H. 1975. A potent depressant action of adenine derivatives on cerebral cortical neurones. Eur J Pharmacol 30:125-129
Premont, J., Perez, M. and Bockaert, J. 1977. Adenosine-sensitive adenylate cyclase in rat striatal homogenates and its relationship to dopamine—and Ca^{2+}—sensitive adenylate cyclases. Mol Pharmacol 13:1845-1850
Rall, T.W. and Sattin, A. 1970. Factors influencing the accumulation of cyclic AMP in brain tissue. In *Role of Cyclic AMP in Cell Function.* Advances in Psychopharmacology, Vol. 3, eds. E. Costa and P. Greengard, pp. 113-123. New York: Raven Press
Sattin, A., Rall, T.W. and Zanella, J. 1975. Regulation of cyclic adenosine 3′,5′-monophosphate levels in guinea-pig cerebral cortex by interaction of *alpha* adrenergic and adenosine receptor activity. J Pharmacol Exp Ther. 192:22-32
Sattin, A. and Rall, T.W. 1970. The effect of adenosine and adenine nucleotides on the cyclic adenosine 3′,5′-phosphate content of guinea pig cerebral cortex slices. Mol Pharmacol 6:13-23
Sattin, A. 1975. Cyclic AMP accumulation in cerebral cortex tissue from inbred strains of mice. Life Sci 16:193-204
Sattin, A. 1978. Genetic transmission of alpha-adrenergic responses in mouse cerebral cortex. In *Advances in Cyclic Nucleotide Research,* Vol. 9. New York: Raven Press

Schultz, J. and Daly, J.W. 1973a. Adenosine 3′,5′-monophosphate in guinea pig cerebral cortical slices. Effect of Alpha and Beta-adrenergic agents, histamine, serotonin and adenosine. J Neurochem 21:573–579

Schultz, J. and Daly, J.W. 1973b. Cyclic adenosine 3′,5′-monophosphate in guinea pig cerebral cortical slices. II. The role of phosphodiesterase activity in the regulation of levels of cyclic adenosine 3′,5′-monophosphate. J Biol Chem 248:853–859

Segal, D.S., Geyer, M.A. and Weiner, B.E. 1975. Strain differences during intraventricular infusion of norepinephrine. Possible role of receptor sensitivity. Science 189:301–303

Segal, D.S., Kuczenski, R.T. and Mandell, A.J. 1972. Strain differences in behavior and brain tyrosine hydroxylase activity. Behav Biol 7:75–81

Skolnick, P. and Daly, J.W. 1974. Norepinephrine—sensitive adenylate cyclases in rat brain: Relation to behavior and tyrosine hydroxylase. Science 184:175–177

Sprott, R.L. 1974. Passive-avoidance performance in mice: Evidence for single-locus inheritance. Behav Biol 11:231–237

Southwick, C.H. and Clark, L.H. 1968. Interstrain difference in aggressive and exploratory activity of inbred mice. Comm Behav Biol 1A:49–59

Stalvey, L., Daly, J.W. and Dismukes, R.K. 1976. Behavioral activity and accumulation of cyclic AMP in brain slices of strains of mice. Life Sci 19:1845–1850

Stavnes, K.L. and Sprott, R.L. 1975. Genetic analysis of active avoidance performance in mice. Psychol Rep 36:515–521

Stone, T.W. and Taylor, D.A. 1978. An electrophysiological demonstration of a synergistic interaction between norepinephrine and adenosine in the cerebral cortex. Brain Res 147:396–400

Stone, T.W., Taylor, D.A., and Bloom, F.E. 1975. Cyclic AMP and cyclic GMP may mediate opposite neuronal responses in rat cerebral cortex. Science 3:845–847

van Rossum, J.M., Broekkamp, C.L.E., and Pijnenburg, A.J.J. 1977. Behavioral correlates of dopamine function in the nucleus accumbens. In *Advances in Biochemal Psychopharmacology,* Vol. 16, eds. E. Costa and G.L. Gessa. New York:Raven Press

von Hungen, K. and Roberts, S. 1973. Adenylate-cyclase receptors for adrenergic neurotransmitters in rat cerebral cortex. Eur J Biochem 36:391–401

Williams, L.T. and Lefkowitz, R.J. 1976. Alpha-adrenergic receptor indentification by [^{3}H] dihydroergocryptine binding. Science 192:791–793

Wise, C.D., Berger, B.D. and Stein, L. 1973. Evidence of alpha-noradrenergic reward receptors and serotonergic punishment receptors in the rat brain. Biol Psychiat 6:3–21

Effects of Seizures and Anticonvulsant Drugs on Cyclic Nucleotide Regulation in the CNS

James A. Ferrendelli, Robert A. Gross,
Dorothy A. Kinscherf and Eugene H. Rubin
Departments of Pharmacology and Neurology
Washington University School of Medicine
St. Louis, Missouri 63110

INTRODUCTION

It is now well known that adenosine 3′, 5′-monophosphate (cyclic AMP) is an intracellular "second messenger" which exists in many tissues and mediates actions of hormones, neurotransmitters and other "first messengers." Structurally similar to cyclic AMP, guanosine 3′,5′-monophosphate (cyclic GMP) also is found in many tissues and is assumed to be a "second messenger," but whether it mediates actions of extracellular substances has not been clearly established. Both cyclic nucleotides appear to be involved in numerous functions of nervous tissue and it is likely that both have roles in several neurological disorders (*Bloom*, 1975; *Daly*, 1977; *Nathanson*, 1977; *Phillis*, 1977). Of particular interest to us is the possible involvement of the cyclic nucleotides in seizure disorders. Several laboratories, including ours, have reported that seizures alter cyclic AMP and cyclic GMP levels in brain, and preliminary evidence indicates that cyclic nucleotides may initiate and/or modify seizure activity in mammalian CNS. In addition, there are data demonstrating that anticonvulsant drugs can influence cyclic nucleotide levels in brain *in vivo* and *in vitro*. This paper reviews some of the effects of seizures and anticonvulsant drugs on cyclic AMP and cyclic GMP regulation in mammalian brain and discusses some possible roles of cyclic nucleotides in the pathophysiology of seizures and mechanisms of action of anticonvulsant drugs.

EFFECT OF SEIZURES ON CYCLIC NUCLEOTIDE LEVELS IN BRAIN

Sattin (1971) first reported that seizure discharges altered cyclic AMP levels in brain, *in vivo*. He observed that electroconvulsive shock (ECS)

Supported in part by USPHS grants NS-09667, GM 07200, NS-13104 and the convulsant hexafluorodiethyl ether elevated cyclic AMP levels in mouse forebrain. Subsequently, *Lust*, *Goldberg*, and *Passonneau* (1976) demonstrated that ECS increased levels of cyclic AMP in mouse cerebral cortex and cerebellum and also elevated levels of cyclic GMP in cerebellum. Many others have reported that experimental seizures induced by several convulsant drugs, including homocysteine, isoniazid, picrotoxin, pentylenetetrazol and 3-mercaptopropionic acid, elevate levels of cyclic AMP and/or cyclic GMP in cerebral cortex and/or cerebellum of experimental animals (*Berti et al.*, 1976; *Folbergrova*, 1975, 1977; *Mao, Guidotti,* and *Costa,* 1974; *Rubin* and *Ferrendelli,* 1977a).

We examined the effect of pentylenetetrazol-induced seizures on cyclic AMP and cyclic GMP levels in several areas of mouse brain in order to determine the temporal relationships between seizure discharges and regional changes in CNS cyclic nucleotide levels (*Ferrendelli* and *Kinscherf*, 1977a). Mice injected with pentylenetetrazol (100 mg/kg, i.p.) exhibit a highly consistent behavioral response. At approximately 1 min after injection the animals have a myoclonic spasm involving the whole body. This is followed by a generalized clonic seizure within a few seconds. The animals continue to have intermittent myoclonic and clonic seizures for several minutes thereafter. More than 90% of the animals have a tonic seizure, usually 3–5 min after injection of pentylenetetrazol, and most of these die immediately after the tonic seizure.

Cyclic AMP and cyclic GMP concentrations were measured in cerebral cortex, cerebellum, hippocampus and striatum, and sometimes in thalamus, of mice rapidly frozen at various times after injection of 100 mg/kg pentylenetetrazol. No changes in cyclic AMP levels were observed until mice exhibited myoclonic or clonic seizure activity. At these times cyclic AMP levels rose 1- to 3-fold in all regions of brain; the largest elevation usually occurred in the hippocampus and the least in striatum. (In some experiments there was no statistically significant elevation of cyclic AMP levels in striatum.) Cyclic AMP levels remained elevated during the period of intermittent myoclonic and clonic seizures in cerebral cortex, hippocampus and striatum, but returned toward control levels in cerebellum and thalamus. Animals frozen during tonic seizure had regional levels of cyclic AMP similar to or less than those in animals frozen during clonic seizure activity.

In comparison to the changes in cyclic AMP levels, changes in cyclic GMP were more variable. Small elevations (20–50%) of cyclic GMP levels were seen in cerebellum, striatum and cerebral cortex within 30 sec after injection of pentylenetetrazol and prior to the onset of clinically apparent seizure activity. These elevations were not always statistically significant in each experiment, however. Eventually cyclic GMP levels rose 1- to 3-fold in all regions of brain with the onset of myoclonic and clonic seizures or during continued myoclonic and clonic seizure activity, and these elevations were always statistically significant ($P < 0.05$). Tonic seizures usually increased cyclic GMP levels further.

In some experiments we examined the effects of pentylenetetrazol seizures on cyclic nucleotide levels in layers of cerebellar cortex and in subdivisions of the hippocampus (*Rubin* and *Ferrendelli*, 1977b and unpublished results). Basal levels of cyclic AMP are lower in the molecular layer of mouse cerebellar cortex than in the granular layer (9 vs 14 pmoles/mg prot), whereas cyclic GMP levels are higher in the molecular layer (9 vs 5 pmoles/mg protein). In animals frozen during the first clonic seizure after injection with 100 mg/kg pentylenetetrazol, cyclic AMP levels rose 2-fold in both the molecular and granular layers of cerebellar cortex, and cyclic GMP levels rose 4-fold in the molecular layer but only 2.5-fold in the granular layer. In the hippocampus basal cyclic AMP levels are 10–13 pmoles/mg protein, and basal cyclic GMP levels are approximately 0.7 pmoles/mg protein. The concentrations of each cyclic nucleotide are similar in the major subdivisions of hippocampus—dentate gyrus, CA_3, CA_1 and subiculum. During pentylenetetrazol-induced clonic seizures the levels of both cyclic nucleotides increase throughout the hippocampus; however, the greatest increase in both is seen in CA_1.

Although most studies of seizures and CNS cyclic nucleotides have examined the effects of generalized convulsions, some information on focal seizures is available. *Walker et al.* (1973) reported that cyclic AMP levels were elevated in epileptogenic lesions (produced by freezing) in rat cerebral cortex. They also observed that cyclic AMP levels were increased in the contralateral, homologous (noninjured) cortex, but this change was less striking. *Raabe et al.* (1978) studied epileptic foci produced by penicillin injection into cat cerebral cortex. They observed that cyclic GMP levels were elevated in the epileptic focus but not in the contralateral cerebral cortex or in a distant cortical region. These investigators observed no change in cyclic AMP levels, however. In preliminary studies we have also observed that cyclic GMP, but not cyclic AMP, levels are increased in penicillin-induced epileptic foci in rat cerebral cortex (*Ferrendelli*, *Kinscherf*, and *Collins*, unpublished results).

We also have generated limited data on the effect of audiogenic seizures on cyclic nucleotide levels in mouse brain (*Ferrendelli* and *Kinscherf*, unpublished results). Mice susceptible to audiogenic seizures exhibit generalized clonic-tonic-clonic convulsions after exposure to loud noise; however, the seizure activity appears to be localized to subcortical regions and does not involve or require the cerebral cortex. Audiogenic seizures decreased cyclic GMP levels in cerebral cortex but caused an elevation in cerebellum and in the superior and inferior colliculi (the only regions of brain examined). Cyclic AMP levels increased slightly in cerebral cortex during the audiogenic seizure, but this change was much less than that in other regions of brain examined. The small elevation of cyclic AMP and the decrease of cyclic GMP levels in cerebral cortex during audiogenic seizures probably are due to cerebral anoxia, secondary to apnea during the tonic phase of the seizure.

EFFECTS OF SEIZURES ON CSF CYCLIC NUCLEOTIDES

Correlation of seizure activity with CNS cyclic nucleotide levels has also been attempted through measurement of cerebrospinal fluid (CSF) cyclic nucleotide concentrations. While this approach benefits from its application to epilepsy in humans, the meaning of cyclic nucleotide levels in CSF is not clearly defined. Although it is known that cyclic nucleotides may be extruded from brain slices (*Kakiuchi* and *Rall*, 1968), it is not known whether this is of functional importance or whether it represents a "degradation" step. In this respect it is of interest that cyclic AMP and cyclic GMP phosphodiesterase activities have been found in CSF (*Hidaka et al.*, 1975). Also, it is difficult to know if altered CSF cyclic nucleotide concentrations accurately reflect similar changes in intracellular levels and/or turnover rates, which are likely to be of greater physiological importance. This problem is compounded since the distribution of cyclic nucleotides between CSF and blood is not known. Finally, this approach does not allow the assessment of the contributions of different CNS regions to the CSF cyclic nucleotide levels. At this time, then, reports of CSF cyclic nucleotide concentrations must be examined critically, and probably should be considered no more than rough indications of the state of CNS cyclic nucleotide systems.

The finding that probenecid will increase CSF concentrations of cyclic AMP implies that the equilibrium of the nucleotide between CSF and blood may be actively controlled (*Cramer*, *Ng*, and *Chase*, 1972). This is experimentally useful in that probenecid treatment provides a rough means by which cyclic nucleotide turnover in CSF may be determined. By this means *Sebens* and *Korf* (1975) showed that enhancement of aminergic function in the CNS increased the concentration of cyclic AMP in the CSF, thus affording some evidence that CSF levels of cyclic nucleotides may be related to a functional aspect of neuronal activity in the brain.

Whereas old degenerative lesions of the CNS result in low CSF concentrations of cyclic AMP, cyclic AMP in human CSF is elevated in the presence of recent lesions or within 3 days of an epileptic seizure (*Myllylä et al.*, 1975). The same laboratory also reported supporting evidence to this finding; in rabbits that had been electroshocked cyclic AMP concentrations in CSF were greater than controls, and pretreatment of the animals with phenytoin, phenobarbital or carbamazepine partially blocked the observed increase in cyclic AMP (*Myllylä, 1976*). In contrast to these data, *Robison et al.*, (1970) reported that they could detect no difference between CSF cyclic AMP levels in epileptic patients and their control group.

Although elevated brain cyclic GMP levels are firmly associated with experimental epileptic seizures (see above), *Trabucchi et al.* (1977) did not find increased CSF concentrations of this nucleotide in epileptic patients. These and other negative findings may be the result of sampling at times when any changes in nucleotide levels or turnover have disappeared. Since cyclic nucleotides appear to be degraded in the CSF and transported into blood, these considerations are of importance. Obviously, further work is needed to clarify the relationship between CSF cyclic nucleotide concentrations and/or turnover and epileptic activity.

INFLUENCE OF DRUGS ON SEIZURE—INDUCED ELEVATIONS OF CYCLIC NUCLEOTIDES

Several drugs can modify seizure-induced elevations of cyclic AMP and/or cyclic GMP levels in brain. Pretreatment of mice with phenytoin (25 mg/kg) prevents the tonic component of ECS but does not prevent, and may even prolong, the clonic seizures (*Kupferberg et al.*, 1976). This dose of phenytoin also reduces the ECS-induced elevation of cyclic GMP in cerebellum and cerebral cortex, slightly inhibits the elevation of cyclic AMP in cerebellum and has no effect on the rise of cyclic AMP in cerebral cortex. Phenobarbital (20 mg/kg) was reported to have no effect on ECS-induced elevation of cyclic nucleotides; however, this dose did not prevent the clinical seizure (*Lust et al.*, 1976). Phenobarbital (50 mg/kg) prevented both the seizure and the elevation of cyclic AMP produced by homocysteine (*Folbergrova*, 1975).

We compared the effects of phenytoin, phenobarbital and ethosuximide on pentylenetetrazol-induced seizures and elevations of cyclic nucleotides (*Ferrendelli* and *Kinscherf*, 1977a). Phenytoin (25 mg/kg) prevented the tonic component of pentylenetetrazol seizures but prolonged and increased the severity of clonic seizures. This drug had little or no effect on pentylenetetrazol-induced elevations of either cyclic AMP or cyclic GMP. In contrast, phenobarbital (40 mg/kg) reduced seizure intensity and ethosuximide (500 mg/kg) prevented the seizure; both drugs attenuated or completely blocked elevations of both cyclic nucleotides. Thus, components generally classified as anticonvulsant drugs can block or attenuate elevations of CNS cyclic nucleotides caused by convulsant agents or stimuli. However, it appears that anticonvulsant drugs only produce this effect when they also suppress or prevent clinical seizure activity.

Some compounds which are not classified as anticonvulsant drugs also inhibit seizure-induced elevations of cyclic nucleotides. Elevations of cyclic AMP in mouse forebrain caused by ECS or by hexafluorodiethyl ether are reduced by pretreatment of the animals with methylxanthines (caffeine or theophylline) (*Sattin,* 1971). *Folbergrova* (1977) reported that propranolol partially blocked the elevation of cyclic AMP in mouse CNS caused by the convulsant agent 3-mercaptopropionic acid. Recently we examined the action of aminophylline and reserpine on changes in CNS cyclic AMP and cyclic GMP levels in mice treated with pentylenetetrazol (*Gross* and *Ferrendelli*, unpublished results). In doses of 50 mg/kg or greater, aminophylline did not alter the time of appearance or frequency of myoclonic and clonic seizures produced by pentylenetetrazol, but it caused tonic seizures to occur sooner. Aminophylline (50 mg/kg) inhibited by 50% pentylenetetrazol-induced elevations of cyclic AMP in cerebral cortex, hippocampus and striatum; a higher dose (250 mg/kg) was required to inhibit elevation in cerebellum. Neither dose of aminophylline inhibited pentylenetetrazol-induced elevations of cyclic GMP. Instead, at 250 mg/kg it elevated basal levels of cyclic GMP in all regions examined, and it augmented the effect of pentylenetetrazol.

Some animals were also pretreated with reserpine (0.1 to 10 mg/kg) 3 hr prior to pentylenetetrazol administration. At doses of 1 mg/kg and above, reserpine produced a dose-dependent decrease in the time of appearance of tonic seizures, and at these same high doses animals often exhibited increased numbers of myoclonic and clonic seizures. Reserpine pretreatment blocked, in a dose-dependent manner, cyclic AMP elevation associated with pentylenetetrazol-induced seizures in all regions of mouse brain studied, and at the highest dose studied this block was complete. Reserpine had no effect on cyclic GMP elevations produced by pentylenetetrazol in any region of brain, although it depressed basal levels in the cerebellum and cerebral cortex and elevated basal levels in the hippocampus.

MECHANISMS OF CYCLIC NUCLEOTIDE ACCUMULATION IN EPILEPTIC BRAIN

Although information concerning the relationships between seizures and CNS cyclic nucleotides is still incomplete, some conclusions can be derived from presently available data. It is clear that seizures can consistently and markedly influence cyclic nucleotide regulation in many, and perhaps all, regions of the CNS and increase levels of both cyclic AMP and cyclic GMP. This effect is particularly striking when compared to the effects of nonconvulsant drugs and stimuli. Except for anoxia or ischemia, nothing else has been reported to consistently elevate cyclic AMP throughout brain *in vivo,* and although many agents and conditions may alter cyclic GMP levels in cerebellum, their effects are frequently confined to this region of brain (*Rubin* and *Ferrendelli,* 1977a).

The fact that several different convulsant drugs and other epileptogenic stimuli increase cyclic nucleotide levels in CNS strongly indicates that the cyclic nucleotide elevations are related to the seizure discharge and are not due to some other effect of the drugs or stimuli. The observations that prevention of seizures with anticonvulsant drugs also attenuates or blocks the elevations of cyclic AMP and cyclic GMP further indicates that the changes in cyclic nucleotide levels are directly related to the seizure discharge. The temporal relationship between seizures and changes in cyclic nucleotide levels in CNS is still incompletely understood. Most of the available data indicate that cyclic AMP levels do not increase until clinical seizure activity is present, suggesting that the elevations of cyclic AMP are probably a result of the seizure discharge. It further seems that, for the most part, elevations of cyclic GMP levels are also a consequence of the seizure discharge. However, the finding that cyclic GMP levels are elevated in some regions of brain prior to the appearance of clinical seizure activity (*Berti et al.*, 1976; *Ferrendelli* and *Kinscherf*, 1977a) raises the possibility of other relationships. One possibility is that increased levels of cyclic GMP may parallel the increased neuronal activity induced by some convulsant drugs; cyclic GMP may thus be involved in the initiation or continuation of the seizure discharge (see below).

Some investigators have suggested that the elevations of cyclic AMP levels in brain seen during seizures may be a consequence of an absolute or relative cellular oxygen insufficiency (*Sattin*, 1971; *Lust et al.*, 1976). Like seizures, hypoxia or cerebral ischemia elevates levels of cyclic AMP throughout brain, and seizures, especially tonic convulsions when respiration is impaired, may cause cerebral hypoxemia. However, elevations of cyclic AMP levels occur during myoclonic and clonic seizure activity when there is no impairment of respiration (*Ferrendelli* and *Kinscherf,* 1977a). We conclude, therefore, that the elevation of cyclic AMP levels induced by seizures is not merely due to cerebral hypoxemia. However, anoxia may contribute to the effects seen during tonic seizures. Although we suggest that seizures do not elevate cyclic AMP levels indirectly by producing cerebral hypoxia, this does not preclude the idea that hypoxia and seizures elevate cyclic AMP levels by a similar mechanism, as suggested by others.

How seizure discharges cause elevation of cyclic AMP and cyclic GMP levels is an important question. Presumably, the excessive cellular depolarization characteristic of seizure discharges initiates events which eventually lead to elevation of the cyclic nucleotides. From *in vitro* studies of brain slices it is well established that agents which cause cellular depolarization (e.g., K^+, ouabain, veratridine) markedly elevate tissue levels of cyclic AMP and cyclic GMP (*Daly*, 1977; *Ferrendelli*, 1976). The following mechanism for these effects has been suggested: Cellular depolarization augments influx of calcium into intracellular spaces which leads to accumulation of cyclic GMP (*Ferrendelli*, *Rubin*, and *Kinscherf*, 1976), perhaps as a result of calcium activating guanylate cyclase. In addition, cellular depolarization increases neurotransmitter release (also a calcium-dependent process). Released neurotransmitters, particularly adenosine and/or biogenic amines such as norepinephrine, dopamine and serotonin, activate specific membrane-bound adenylate cyclases in nearby cells and thereby increase cyclic AMP levels. We suggest that seizures, via cellular depolarization, elevate levels of cyclic AMP and cyclic GMP in brain tissue *in vivo* by a mechanism similar to that proposed for the effect of cellular depolarization on cyclic nucleotides on brain tissue *in vitro*.

Much additional research will be necessary to prove this hypothesis; with regard to cyclic AMP, however, it already has some support. The observations that aminophylline, caffeine and theophylline (methylxanthines capable of blocking adenosine receptors in brain) inhibit seizure-induced elevations of cyclic AMP suggest that the increase of cyclic AMP is partially the result of release and subsequent action of adenosine on adenylate cyclase. In addition, the finding that reserpine, at doses which deplete brain norepinephrine, dopamine and serotonin, inhibits seizure-induced elevations of cyclic AMP suggests that the action of released catecholamines and/or indoleamines also is responsible for some elevation of cyclic AMP levels induced by seizures.

All data indicate that seizures elevate cyclic AMP and cyclic GMP levels by entirely different mechanisms. It appears that cyclic GMP levels in CNS are directly related to cellular depolarization and neuronal activity; increased neuronal discharges *in vitro* and *in vivo* elevate cyclic GMP levels. We assume that excessive neuronal activity occurring during seizures increases cyclic GMP levels and that the elevation of cyclic GMP occurs only in those regions where neuronal activity is augmented. Further explanation of the mechanism(s) linking cyclic GMP to neuronal activity and seizure discharges will require additional research.

EFFECT OF CYCLIC NUCLEOTIDES ON CNS FUNCTION

There is ample evidence supporting the contention that cyclic nucleotides are involved in CNS synaptic function. The enzymatic "machinery" of the cyclic nucleotide systems (adenylate and guanylate cyclases, which form cyclic AMP and cyclic GMP, respectively; phosphodiesterases, which convert the cyclic nucleotides to their respective 5′-phosphate nucleotides; and cyclic nucleotide-activated protein kinases) is primarily located in nerve terminals and synaptic membrane fractions of brain homogenates. In addition, a number of endogenous substances, which are thought to be neurotransmitters or neuromodulators in the CNS, have been shown to increase cyclic AMP levels in incubated slices of brain tissue or to activate adenylate cyclases in homogenates (*Bloom*, 1975; *Nathanson*, 1977; *Daly*, 1977). These adenylate cyclases appear to be activated via specific receptors for norepinephrine, dopamine, serotonin, adenosine and other endogenous compounds; the distribution of these cyclases is nonuniform in the CNS and is therefore suggestive of a role in the synaptic functioning of pathways which utilize specific neurotransmitter substances. In contrast, neither cyclic GMP accumulation nor guanylate cyclase activities are altered directly by neurotransmitters; cyclic GMP levels may therefore be more closely related to neuronal depolarization *per se*, perhaps through the influence of intracellular calcium concentrations (*Ferrendelli et al.*, 1976; *Ferrendelli*, 1976).

Some of the most convincing data indicating that cyclic nucleotides have a role in synaptic function, and hence modify neuronal activity, are the results of iontophoretic application of cyclic AMP and cyclic GMP onto CNS neurons. *Phillis* (1974) reported that cyclic GMP caused 51% of 32 unidentified cat cerebral cortex cells to increase their firing rate. A similar percentage of rat pyramidal tract neurons were activated by iontophoretically applied cyclic GMP (*Stone*, *Taylor*, and *Bloom*, 1975). More recently, Hoffer and his colleagues observed that iontophoretic application of cyclic GMP onto the pyramidal cells of hippocampal explants, maintained *in oculo*, produced marked cellular excitation and epileptiform discharges (*Hoffer et al.*, 1977a & b). These data are particularly intriguing and suggest that cyclic GMP may be involved in initiating and/or maintaining seizure discharges.

Bloom, Hoffer, and their colleagues have reported that iontophoretically applied cyclic AMP markedly inhibited rat Purkinje cells, hippocampal pyramidal cells, caudate neurons and corticospinal neurons (*Hoffer, Siggins,* and *Bloom,* 1971a; *Hoffer et al.*, 1971b, 1972, 1973; *Segal* and *Bloom,* 1974a & b; *Siggins, Hoffer,* and *Ungerstedt,* 1974; *Stone et al.*, 1975). Cyclic AMP also has been shown to inhibit neurons in the cat brain stem (*Anderson*, *Haas*, and *Hosli*, 1973). However, Phillis and colleagues have found cyclic AMP to be a relatively weak depressor of Purkinje cells and (unidentified) cerebral cortical neurons (*Kostopoulos, Limacher,* and *Phillis,* 1975; *Phillis,* 1977). Since they also found that other adenine nucleotides and adenosine were potent depressors of these cells, it was suggested that the activity of cyclic AMP may be due to its hydrolysis to adenosine or adenine nucleotides (for references and a summary of these data, see *Phillis,* 1977).

Regardless of the mechanism responsible for the inhibitory effect of cyclic AMP on CNS neurons, it seems apparent that this cyclic nucleotide is involved in some process associated with suppression of neuronal activity. This, plus the fact that inhibition of seizure-induced elevations of CNS cyclic AMP levels by methylxanthines or reserpine tends to precipitate tonic seizures more rapidly (see above), suggests the possibility that this nucleotide may have some role in seizure attenuation or termination. Seemingly inconsistent with this idea is the observation that dibutyryl cyclic AMP or adenosine applied to the rat cerebral cortex caused epileptiform discharges (*Walker, Lewin,* and *Moffitt,* 1975). However, the doses of these compounds were extremely high (10^{-2}M), and the observed effects may have been the result of toxicity. Also, these data are difficult to interpret because the compounds were applied to the surface of the cortex, and thus were not necessarily at "normal" sites of action. In this respect, the specificity of the iontophoretic studies is an advantage.

By analogy with peripheral systems, cyclic nucleotides are thought to exert their intracellular effects in the CNS by the specific activation of protein kinases, which are in turn responsible for the phosphorylation of other proteins. Kinases specifically activated by cyclic AMP and by cyclic GMP have been isolated in the brain, primarily in synaptosomal fractions of homogenates (*Miyamoto, Kuo,* and *Greengard,* 1969; *Miyamoto et al.,* 1973; *Uno, Ueda,* and *Greengard,* 1977; *Hofmann* and *Sold,* 1972).

Greengard and colleagues have recently characterized a bovine brain synaptic membrane phosphoprotein system regulated by cyclic AMP (*Ueda* and *Greengard,* 1977; *Uno et al.,* 1977). The membrane-bound protein kinase was found to consist of a cyclic AMP-dependent regulatory subunit and a catalytic subunit. However, the membrane-bound enzyme was quite different from cytosolic cyclic AMP protein kinases (from brain or heart) with regard to molecular weight, catalytic and immunological properties; in addition, the regulatory subunit of the membrane-bound enzyme could not inhibit the cytosolic catalytic subunits. In the presence of cyclic AMP, this unique protein kinase phosphorylated an endogenous substrate isolated from enriched synaptic membranes, which appears to be specific to nervous tissue.

The existence of a membrane-bound, cyclic AMP-dependent phosphoprotein complex is of considerable interest in view of the inhibitory effects of various iontophoretically applied neuroregulatory substances (*Bloom*, 1975; *Phillis*, 1977), their simultaneous ability to increase cyclic AMP levels in nervous tissue, and the inhibitory effects of cyclic AMP on neurons throughout the CNS. It is possible that such inhibitory effects may be mediated by the cyclic AMP-dependent phosphorylation of synaptic membrane proteins, which would then alter membrane ion permeability in order to depress neuronal activity and perhaps suppress seizure discharges.

Recent studies suggest that cyclic AMP may have functions in addition to depressing neuronal activity. Lust and his colleagues examined cyclic AMP and glucose metabolism in animals undergoing maximal electroshock seizures (*Lust* and *Passonneau*, 1976). These animals were made hypothermic (20°C) in order to minimize the freezing time of the cerebral cortex, slow the metabolic changes occurring during seizures and more easily observe the relationships between these changes and cyclic AMP accumulation. Hypothermia decreased high-energy phosphate utilization 35% and slightly lowered cerebral cortical levels of cyclic AMP. Within 60 sec after electroshock phosphorylase *b* to *a* conversion was maximal and cyclic AMP levels rose to 200% of basal levels. Although cyclic AMP levels remained elevated after phosphorylase *a* levels had returned to normal, the rise in cyclic AMP was closely paralleled by phosphorylase *b* to *a* conversion and by the conversion of glycogen synthase *a* to *b*. *Folbergrova* (1977) found similar enzyme conversions during 3-mercaptopropionic acid-induced seizures, and also correlated them to cyclic AMP changes. In addition, she observed that theophylline had no effect on the changes of either cyclic AMP or phosphorylase during seizures; propranolol only partially blocked the increases in cyclic AMP and did not prevent phosphorylase activation, but phenobarbital prevented the clinical seizure activity, cyclic AMP elevation and phosphorylase activation. The effect of propranolol suggests that only a small increase in cyclic AMP levels may initiate phosphorylase activation. This contention is supported by the data of *Lust et al.* (1976). Thus, cyclic AMP may be involved in the control of glucose and energy metabolism in brain during seizures as well as influencing neuronal activity.

At present there is relatively little data linking cyclic GMP specific protein kinases to some aspect of nervous system function or metabolism. As noted above, cyclic GMP levels in brain increase in response to cellular depolarization, probably as a result of increased free intracellular calcium concentration, and when cyclic GMP is iontophoresed onto neurons it can cause cellular depolarization. Perhaps this cyclic nucleotide acts, via a protein kinase and its substrate, as a positive feedback regulator; its elevated levels (or increased turnover) may lower depolarization threshold. Obviously, this contention is highly speculative. Nevertheless, *Schlichter, Casnellie,* and *Greengard* (1978) recently reported the existence of a cyclic GMP protein kinase substrate in the cytosol of rabbit cerebellum which is apparently unique to this tissue. Further work along these lines may lead to a better understanding of the function(s) of cyclic GMP in nervous tissue and its role in seizures.

EFFECT OF ANTICONVULSANT DRUGS ON CYCLIC NUCLEOTIDE REGULATION IN NERVOUS TISSUE

Since seizures cause such marked changes in brain cyclic nucleotide levels, one would suspect that anticonvulsant drugs may influence cyclic nucleotide regulation in brain. The effect of treatment of animals with several drugs which are generally classified as anticonvulsants, e.g., phenytoin, phenobarbital, carbamazepine, valproic acid and ethosuximide, has been reported (*Kupferberg et al.,* 1976; *Ferrendelli* and *Kinscherf,* 1977a; *Lust et al.,* 1978). In general, all of these compounds depress basal levels of cyclic GMP in cerebellum. Other regions of brain have been examined less often and it is still unclear whether or not cyclic GMP levels in noncerebellar nervous tissue are consistently altered by anticonvulsant drugs. Most laboratories have observed little or no effect of anticonvulsant drugs on cyclic AMP levels in any region of brain. Depression of cyclic GMP levels in cerebellum, *in vivo,* is not a unique effect of anticonvulsant drugs. Several agents without significant antiepileptic properties, and anoxia and hypoglycemia, also decrease cerebellar cyclic GMP levels (*Rubin* and *Ferrendelli,* 1977a). Thus, reduction of basal cerebellar cyclic GMP levels may not be related to the anticonvulsant action of drugs.

As noted above, anticonvulsant drugs may attenuate or block seizure-induced elevations of cyclic AMP and cyclic GMP levels in brain, but these effects are seen only when the drug also suppresses or prevents the seizure.

Some anticonvulsant drugs have profound effects on cyclic nucleotide regulation in brain tissue *in vitro.* Diazepam elevates cyclic AMP and cyclic GMP levels in incubated brain slices presumably by inhibiting activity of cyclic nucleotide phosphodiesterases (*Rubin* and *Ferrendelli,* 1977b). *Lewin* and *Bleck* (1977) reported that carbamazepine and phenytoin, but not phenobarbital, reduced the increase in cyclic AMP accumulation induced by ouabain in slices of rat cerebral cortex. Carbamazepine also inhibited norepinephrine- and adenosine-induced accumulation of cyclic AMP in these tissues. This effect was shared by phenobarbital, but phenytoin was ineffective in these tests.

We tested, in slices of mouse cerebral cortex, the effects of several anticonvulsant drugs on the accumulation of both cyclic AMP and cyclic GMP produced by veratridine, ouabain, K^+ and glutamate, agents which cause both cellular depolarization and marked elevations of cyclic nucleotides in tissue slices from several regions of brain of many animal species (*Kinscherf et al.*, 1976). Of the drugs tested only phenytoin, phenobarbital, carbamazepine and ethosuximide produced an effect at concentrations which are near therapeutic levels in intact animals. Three of the drugs—phenytoin, phenobarbital and carbamazepine—inhibited, in a dose-dependent manner, elevations of both cyclic AMP and cyclic GMP induced by the depolarizing agents. Ethosuximide had an inhibitory effect on depolarization-induced elevations of cyclic GMP only. Of particular interest was the observation that phenytoin and carbamazepine were more potent inhibitors of the elevations of cyclic nucleotides caused by veratridine or ouabain than those caused by K^+. Neither phenobarbital nor ethosuximide had this selective action. We have interpreted this selective effect of phenytoin and carbamazepine as indicating that the two anticonvulsant drugs are capable of blocking specific sodium channels in excitable membranes and thus preventing cellular depolarization and the subsequent elevations of cyclic nucleotides (*Ferrendelli* and *Kinscherf,* 1977b, 1978). In support of this contention is the finding that tetrodotoxin, a known blocker of specific sodium channels in excitable membranes, also selectively inhibited veratridine- and ouabain-induced elevations of cyclic nucleotides in brain slices but had no effect on those caused by K^+ or glutamate (*Ferrendelli* and *Kinscherf,* 1978). Furthermore, several laboratories have reported data indicating that phenytoin can block passive influx of sodium ions into nervous and muscle tissue (*Lipicky, Gilbert,* and *Stillman,* 1972; *Pincus,* 1972; *Perry, McKinney,* and *DeWeer,* 1978).

From the very limited amount of data now available concerning the relationships between anticonvulsant drugs and cyclic nucleotide systems, it is difficult to ascertain whether cyclic AMP and/or cyclic GMP have any role in the mechanisms of action of these drugs. In fact, most reported data suggest that anticonvulsant drugs have little or no direct action on cyclic nucleotide systems; instead, they appear to influence regulation of cyclic nucleotides indirectly by preventing cellular depolarization and seizure discharges. Nevertheless, much more information about the interactions between cyclic nucleotides and anticonvulsant drugs is needed and we believe this may be a profitable area of study.

CONCLUSIONS

On the basis of information reviewed in this paper, it is obvious that seizure discharges have a marked effect on cyclic nucleotide regulation in mammalian brain. Seizures increase levels of both cyclic AMP and cyclic GMP in several regions of brain; however, it appears that different mechanisms are responsible for the elevation of each cyclic nucleotide. At present we believe that cyclic GMP accumulates in epileptic brain as a result of increased neuronal activity and stimulus-evoked influx of calcium into neurons and perhaps other cells. In contrast, seizures appear to elevate cyclic AMP levels in CNS by augmenting release of adenosine and one or more catecholamines and/or indoleamines; these neuroregulatory agents then act at cell membranes to activate specific adenylate cyclases and increase cyclic AMP production. Anticonvulsant drugs inhibit seizure-induced elevations of cyclic nucleotides and some of these drugs also prevent accumulation of cyclic nucleotides in depolarized brain tissue *in vitro.* It appears that these actions of anticonvulsant drugs are secondary to their ability to prevent cellular depolarization and seizure discharges.

At the present time there is relatively little information concerning the functional consequence of elevated cyclic nucleotide levels in epileptic brain. However, the observations that cyclic GMP may augment neuronal activity and produce seizure discharges and that cyclic AMP may depress neuronal activity lead to the following speculative conclusions. Elevated levels, or turnover, of cyclic GMP in brain possibly may facilitate cellular depolarization, via protein kinase activation and protein phosphorylation, and thus may have some role in initiating and/or maintaining seizure discharges. On the other hand, elevated levels, or turnover, of cyclic AMP in brain, also through activation of protein kinases, may have an inhibitory influence on neuronal activity. This raises the possibility that cyclic AMP has a role in seizure attenuation or suppression. Thus, both cyclic AMP and cyclic GMP may influence cellular function and metabolism in CNS during seizures and may thereby modify behavior.

REFERENCES

Anderson, E.G., Haas, H., and Hosli, L. 1973. Comparison of effects of noradrenaline and histamine with cyclic AMP on brain stem neurones. Brain Res 49:471-475

Berti, F., Bernareggi, V., Folco, G.C., Fumagalli, R., and Paoletti, R. 1976. Prostaglandin E_2 and cyclic nucleotides in rat convulsions and tremors. In *First and Second Messengers—New Vistas,* Vol. 15/Adv. Biochem. Psychopharmacol., eds. E. Costa, E. Giacobini, and R. Paoletti, pp. 367-378. New York: Raven Press

Bloom, F.E. 1975. The role of cyclic nucleotides in central synaptic function. Rev Physiol Biochem Pharmacol 74:1-103

Cramer, H., Ng, L.K.Y., and Chase, T.N. 1972. Effect of probenecid on levels of cyclic AMP in human cerebrospinal fluid. J Neurochem 19:1601-1602

Daly, J. 1977. *Cyclic Nucleotides in the Nervous System.* New York and London: Plenum Press

Ferrendelli, J.A. 1976. Cellular Depolarization and Cyclic Nucleotide Content in Central Nervous System. In *First and Second Messengers—New Vistas,* Vol. 15/Adv.

Biochem. Psychopharm., eds. E. Costa, E. Giacobini, and R. Paoletti, pp. 303-313. New York: Raven Press

Ferrendelli, J.A., and Kinscherf, D.A. 1977a. Cyclic nucleotides in epileptic brain: Effects of pentylenetetrazol on regional cyclic AMP and cyclic GMP levels, *in vivo*. Epilepsia 18:525-531

Ferrendelli, J.A., and Kinscherf, D.A. 1977b. Phenytoin. Effects on calcium flux and cyclic nucleotides. Epilepsia 18:331-336

Ferrendelli, J.A., and Kinscherf, D.A. 1978. Similar effects of phenytoin and tetrodotoxin on cyclic nucleotide regulation in depolarized brain slices. J Pharmacol Exp Ther, in press

Ferrendelli, J.A., Rubin, E.H., and Kinscherf, D.A. 1976. Influence of divalent cations on regulation of cyclic GMP and cyclic AMP levels in brain tissue. J Neurochem 26:741-748

Folbergrova, J. 1975. Cyclic 3′,5′-adenosine monophosphate in mouse cerebral cortex during homocysteine convulsions and their prevention by sodium phenobarbital. Brain Res 92:165-169

Folbergrova, J. 1977. Changes in cyclic AMP and phosphorylase a in mouse cerebral cortex during seizures induced by 3-mercaptopropionic acid. Brain Res 135:337-346

Hidaka, H., Shibuya, M., Asano, T., and Hara, F. 1975. Cyclic nucleotide phosphodiesterase of human cerebrospinal fluid. J Neurochem 25:49-53

Hoffer, B., Seiger, Å., Freedman, R., Olson, L., and Taylor, D. 1977a. Electrophysiology and cytology of hippocampal formation transplants in the anterior chamber of the eye. II. Cholinergic mechanisms. Brain Res 119:107-132

Hoffer, B.J., Seiger, Å., Taylor, D., Olson, L., and Freedman, R. 1977b. Seizures and related epileptiform activity in hippocampus transplanted to the anterior chamber of the eye. I. Characterization of seizures, interictal spikes, and synchronous activity. Exp Neurol 54:233-250

Hoffer, B.J., Siggins, G.R., and Bloom, F.E. 1971a. Studies on norepinephrine-containing afferents to Purkinje cells of rat cerebellum. II. Sensitivity of Purkinje cells to norepinephrine and related substances administered by microiontophoresis. Brain Res 25:523-534

Hoffer, B.J., Siggins, G.R., Oliver, A.P., and Bloom, F.E. 1971b. Cyclic AMP mediation of norepinephrine inhibition in rat cerebellar cortex. A unique class of synaptic responses. Ann NY Acad Sci 185:531-549

Hoffer, B.J., Siggins, G.R., Oliver, A.P., and Bloom, F.E. 1972. Cyclic adenosine monophosphate mediated adrenergic synapses to cerebellar Purkinje cells. Adv Cyclic Nucleotide Res 1:411-423

Hoffer, B.J., Siggins, G.R., Oliver, A.P., and Bloom, F.E. 1973. Activation of the pathway from locus coeruleus to rat cerebellar Purkinje neurons: Pharmacological evidence of noradrenergic central inhibition. J Pharmacol Exp Ther 184:553-569

Hofmann, F. and Sold, G. 1972. A protein kinase activity from rat cerebellum stimulated by guanosine 3′,5′-monophosphate. Biochem Biophys Res Comm 49:1100-1107

Kakiuchi, S., and Rall, T.W. 1968. The influence of chemical agents on the accumulation of adenosine 3′,5′-phosphate in slices of rabbit cerebellum. Mol Pharmacol 4:367-378

Kinscherf, D.A., Chang, M.M., Rubin, E.H., Schneider, D.R., and Ferrendelli, J.A. 1976. Comparison of the effects of depolarizing agents and neurotransmitters on regional CNS cyclic GMP levels in various animals. J Neurochem 26:527-530

Kostopoulos, G.K., Limacher, J.J., and Phillis, J.W. 1975. Action of various adenine derivatives on cerebellar Purkinje cells. Brain Res 88:162-165

Kupferberg, H.J., Lust, W.D., Yonekawa, W., and Passonneau, J.V. 1976. Effect of phenytoin (Diphenylhydantoin) on electrically induced changes in the brain levels of cyclic nucleotides and GABA. Fed Proc 3:583

Lewin, E., and Bleck, V. 1977. Cyclic AMP accumulation in cerebral cortical slices. Effect of carbamazepine, phenobarbital and phenytoin. Epilepsia 18:237-242

Lipicky, R.J., Gilbert, D.L., and Stillman, I.M. 1972. Diphenylhydantoin inhibition of sodium conductance in squid giant axon. Proc Nat Acad Sci USA 69:1758-1760

Lust, W.D., Goldberg, N.D., and Passonneau, J.V. 1976. Cyclic nucleotides in murine brain: The temporal relationship of changes induced in adenosine 3′,5′-monophosphate and guanosine 3′,5′-monophosphate during maximal electroshock or decapitation. J Neurochem 26:5–10

Lust, W.D., Kupferberg, H.J., Yonekawa, W.D., Penry, J.K., Passonneau, J.V., and Wheaton, A.B. 1978. Changes in brain metabolites induced by convulsants or electroshock. Effects of anticonvulsant agents. Mol Pharmacol 14:347–356

Lust, W.D., and Passonneau, J.V. 1976. Cyclic nucleotides in murine brain. Effect of hypothermia on adenosine 3′,5′-monophosphate, glycogen phosphorylase, glycogen synthase and metabolites following maximal electroshock or decapitation. J Neurochem 26:11–16

Mao, C.C., Guidotti, A., and Costa, E. 1974. The regulation of cyclic guanosine monophosphate in rat cerebellum. Possible involvement of putative amino acid neurotransmitters. Brain Res 79:510–514

Miyamoto, E., Kuo, J.F., and Greengard, P. 1969. Cyclic nucleotide-dependent protein kinases. I. Purification and properties of adenosine 3′,5′-monophosphate-dependent protein kinases from bovine brain. J Biol Chem 244:6395–6402

Miyamoto, E., Petzold, G.L., Kuo, J.F., and Greengard, P. 1973. Dissociation and activation of adenosine 3′,5′-monophosphate-dependent protein kinases by cyclic nucleotides and by substrate proteins. J Biol Chem 248:179–189

Myllylä, V.V. 1976. Effect of convulsions and anticonvulsive drugs on cerebrospinal fluid cyclic AMP in rabbits. Eur Neurol 14:97–107

Myllylä, V.V., Heikkinen, E.R., Vapaatalo, H., and Hokkanen, E. 1975. Cyclic AMP concentration and enzyme activities of cerebrospinal fluid in patients with epilepsy or central nervous system damage. Eur Neurol 13:123–130

Nathanson, J.A. 1977. Cyclic nucleotides and nervous system function. Physiol Rev 5:157–256

Perry, J.G., McKinney, L., and DeWeer, P. 1978. Cellular mode of action of the antiepileptic 5,5-diphenylhydantoin. Nature 272:271–273

Phillis, J.W. 1974. Evidence for cholinergic transmission in the cerebral cortex. Adv Behav Biol 10:57–77

Phillis, J.W. 1977. The role of cyclic nucleotides in the CNS. Can J Neurol Sci 4:151–195

Pincus, J.H. 1972. Diphenylhydantoin and ion flux in lobster nerve. Arch Neurol 26:4–10

Raabe, W., Nicol, S., Gumnit, R.J., and Goldberg, N.D. 1978. Focal penicillin epilepsy increases cyclic GMP in cerebral cortex. Brain Res 144:185–188

Robison, G.A., Coppen, A.J., Whybrow, P.C., and Prange, A.J. 1970. Cyclic AMP in affective disorders. Lancet II:1028–1029

Rubin, E.H., and Ferrendelli, J.A. 1977a. Distribution and regulation of cyclic nucleotides in brain, *in vivo.* In *Alcohol and Aldehyde Metabolizing Systems,* Vol. III/ Intermediary Metabolism and Neurochemistry, eds. R.C. Thurman, J.R. Williamson, H.R. Drott, B. Chance, pp. 471–483. New York: Academic Press

Rubin, E.H., and Ferrendelli, J.A. 1977b. Distribution and regulation of cyclic nucleotides in cerebellum, *in vivo.* J Neurochem 29:43–51

Sattin, A. 1971. Increase in the content of adenosine 3′,5′-monophosphate in mouse forebrain during seizures and prevention of the increase by methylxanthines. J Neurochem 18:1087–1096

Schlichter, D.J., Casnellie, J.E., and Greengard, P. 1978. An endogenous substrate for cGMP-dependent protein kinase in mammalian cerebellum. Nature 273:61–62

Sebens, J.B., and Korf, J. 1975. Cyclic AMP in cerebrospinal fluid. Accumulation following probenecid and biogenic amines. Exp Neurol 46:333–344

Segal, M., and Bloom, F.E. 1974a. The action of norepinephrine in the rat hippocampus. I. Iontophoretic studies. Brain Res 72:79–97

Segal, M., and Bloom, F.E. 1974b. The action of norepinephrine in the rat hippocampus. II. Activation of the input pathway. Brain Res 72:99–114

Siggins, G.R., Hoffer, B.J., and Ungerstedt, U. 1974. Electrophysiological evidence for involvement of cyclic adenosine monophosphate in dopamine responses of caudate neurons. Life Sci 15:779–792

Stone, T.W., Taylor, D.A., and Bloom, F.E. 1975. Cyclic AMP and cyclic GMP may mediate opposite neuronal responses in the rat cerebral cortex. Science 187: 845–847

Trabucchi, M., Cerri, C., Spano, P.F., and Kumakura, K. 1977. Guanosine 3′,5′-monophosphate in the CSF of neurological patients. Arch Neurol 34:12–13

Ueda, T., and Greengard, P. 1977. Adenosine 3′,5′-monophosphate-regulated phosphoprotein system of neuronal membranes. I. Solubilization, purification, and some properties of an endogenous phosphoprotein. J Biol Chem 252:5155–5163

Uno, I., Ueda, T., and Greengard, P. 1977. Adenosine 3′,5′-monophosphate-regulated phosphoprotein system of neuronal membranes. II. Solubilization, purification, and some properties of an endogenous adenosine 3′,5′-monophosphate-dependent protein kinase. J Biol Chem 252:5164–5174

Walker, J.E., Lewin, E., and Moffitt, B. 1975. Production of epileptiform discharges by application of agents which increase cyclic AMP levels in rat cortex. In *Epilepsy: Proceedings of the Hans Berger Centenary Symposium,* eds. P. Harris and C. Mawdsley, pp. 30–36. New York: Churchill Livingstone

Walker, J.E., Lewin, E., Sheppard, J.R., and Cromwell, R. 1973. Enzymatic regulation of adenosine 3′,5′-monophosphate (Cyclic AMP) in the freezing epileptogenic lesion of rat brain and in homologous contralateral cortex. J Neurochem 21:79–85

Cyclic Nucleotide Levels in Brain During Ischemia and Recirculation

W. David Lust and Janet V. Passonneau
Laboratory of Neurochemistry
National Institute of Neurological and
Communicative Disorders and Stroke
National Institutes of Health
Bethesda, Maryland 20014

INTRODUCTION

The observation that cyclic AMP concentrations increase in the brain following decapitation was made by *Breckenridge* in 1964. The major impact of this finding has been that the measurement of cyclic AMP *in vivo* requires rapid fixation of the brain. However, one aspect of the observation that has been largely ignored is that ischemia itself may be a stimulus to the production of cyclic AMP. It is the latter circumstance that has focused our interest on the cyclic nucleotides during and after obstruction of the blood supply to the brain.

Complete occlusion of cerebral circulation results in a series of events which subsequently leads to a loss of brain function. The duration of ischemia required to cause irreversible brain damage is presently a subject of some controversy. Considerable evidence, both clinical and experimental, suggests that ischemia in excess of 10 minutes precludes the possibility of full functional recovery. The critical changes that lead to loss of brain function are presently not known, although various physiological, biochemical and cellular theories have been proposed. Perturbations in neurotransmitter levels, energy metabolism and ionic gradients have been variously attributed as the cause of ischemic damage. While it is well established that the levels of a number of neurotransmitters do change during ischemia (*Kogure et al.*, 1975; *Lust et al.*, 1975; *Stavinoha, Weintraub*, and *Modak*, 1973; *Welch, Chabai*, and *Buckingham*, 1975), their steady-state levels can, for a variety of reasons, be misleading. It is possible that the measurement of a transmitter-induced event might be more informative. A number of putative neurotransmitters have been shown to stimulate the production of cyclic nucleotides in brain slices (*Daly*, 1976; *Drummond*, 1973; *Ferrendelli, Chang*, and *Kinscherf*, 1974, 1975).

Further, when test agents were applied by iontophoresis to pyramidal tract neurons in the cerebral cortex, acetylcholine and cyclic GMP were excitatory to the cells and norepinephrine and cyclic AMP were inhibitory (*Stone, Taylor*, and *Bloom*, 1975). Based on these apparent neurotransmitter-cyclic nucleotide interrelationships, it seemed that the concentrations of the cyclic nucleotides *in vivo* might reflect more accurately the gross functional derangements of the various neurotransmitters in brain both during and after ischemia.

In the present study we have chosen as a model bilateral ischemia in the gerbil cerebral cortex induced by ligation of both common carotid arteries. The basis of this procedure, the anatomical anomaly of the arterial circulation, has been well established (*Levine* and *Payan*, 1966). The major advantage of this model, as opposed to decapitation, is that the subsequent impact of the ischemic insult can also be determined after the circulation has been reestablished.

MATERIALS AND METHODS

Mongolian gerbils (*Meriones unguiculatus*) obtained from the Chickline Co. (Vineland, NJ) were starved 24 hr prior to treatment. Animals were anesthetized with 35 mg/kg (i.p.) of sodium amobarbital; both common carotid arteries were exposed and were looped with surgical sutures. The gerbils were allowed to recover from the anesthesia for 2 hr, after which both arteries were occluded with Heifetz aneurysm clips for a period of 1, 5, 20, 30 or 60 min. In the recovery studies, the clips were removed after the aforementioned ischemic intervals and circulation reestablished for 1, 5, 30 or 60 min. Sham-operated animals served as controls. The animals were frozen intact with rapid stirring in liquid nitrogen and the frozen gerbils were stored at −60°C until the dissections were made. Tissue samples from the outer 1-2 mm of the cerebral cortex were removed in a cryostat maintained at −20°C. In one experiment, a cerebellar wedge and a piece of spinal cord were also dissected out.

Suspensions of the tissue were made in methanol-HCl and an aliquot for the measurement of glycogen and glucose was taken, after which the protein was precipitated with perchloric acid as described by *Nelson et al*. (1968). After centrifugation, the supernatant extract was removed, neutralized with potassium bicarbonate and the protein pellet was dissolved in 1 N NaOH. Cyclic nucleotides were measured by radioimmunoassay according to the method of *Steiner et al*. (1972). Glycogen and glucose were analyzed according to the method of *Passonneau* and *Lauderdale* (1974). All other metabolites were determined with enzymic fluorometric methods described by *Lowry* and *Passonneau* (1972). Proteins were determined by the method of *Lowry et al*. (1951).

Comment on Fixation

Rapid fixation of the brain is generally conceded as being necessary for the measurement of endogenous labile constituents. Microwave irradiation and freezing in liquid nitrogen are two such procedures that are considered to be at least acceptable, if not optimal. However, there is controversy about the relative advantages of the two methods. With rapid freezing, enzyme activities as well as metabolites can be measured, but only the superficial regions of the brain should be used due to the time-dependent freezing front from the outside of the tissue (*Ferrendelli et al.*, 1972). Conversely, enzyme activities cannot be measured in microwaved tissue, but metabolite concentrations can be measured more reliably in the deeper structures of the brain due to the uniformity of the fixation (*Medina et al.*, 1975). The situation, however, appears to be more complex in the determination of cyclic AMP. Using one of the earlier models of the microwave oven, the levels of ATP and P-creatine in the cerebral cortex from microwaved rats were significantly lower than those from rats frozen intact (*Lust, Passonneau*, and *Veech*, 1973). These results suggest that fixation by freezing was better than by microwave. However, the levels of cyclic AMP in microwaved brains were lower than those from frozen brain.

While a number of investigators have demonstrated that adenylate cyclase and phosphodiesterase are inactivated during microwave (*Jones et al.*, 1974) no one to date has examined the effects of microwave on the binding of cyclic AMP to the protein kinase. Since the cyclic AMP-binding protein interaction is quite temperature-sensitive, it is possible that during microwave irradiation the bound cyclic AMP is released and degraded to 5′-AMP before the phosphodiesterase is inactivated. This artifact would be reflected in a lower steady-state level of cyclic AMP than exists *in vivo*. In defense of rapid freezing, the levels of cyclic AMP in the cerebral cortex from a hibernating hamster (body temperature 7 °C) were approximately the same as those in the cerebral cortex from a nonhibernating hamster (body temperature 34 °C). Obviously, the time necessary to lower the temperature to a level where metabolism is essentially stopped would be less in the hibernator than in the nonhibernator. Thus, freezing the intact animals in liquid nitrogen appears to be justified both by the hamster data as well as by avoiding a potential artifact with microwave irradiation.

RESULTS

The two criteria used to establish that the brains were ischemic were 1) neurological signs, and 2) loss of energy stores. The typical neurological

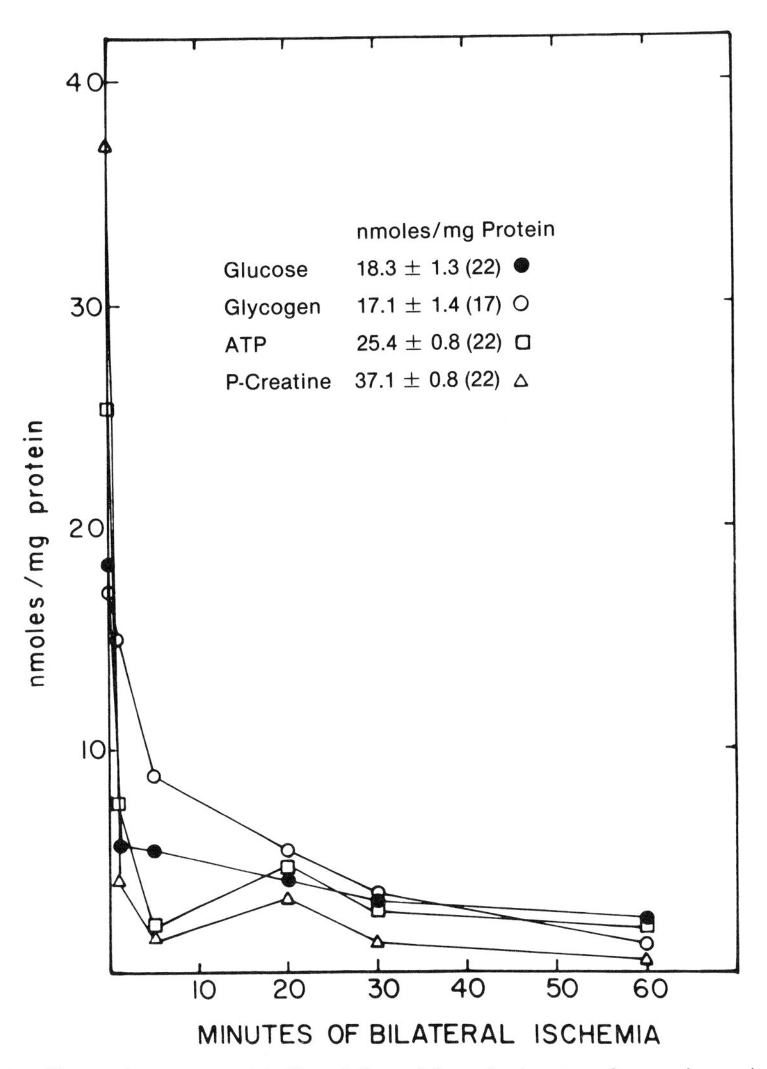

Fig. 1. Changes in energy metabolites of the gerbil cerebral cortex after varying periods of bilateral ischemia. The control values for the metabolites from the cerebral cortex of sham-operated animals are given in the center of the figure, with the corresponding symbols for each compound. The experimental animals were treated as described in the Materials and Methods section. Each value represents the mean of at least five determinations.

signs of a gerbil exhibiting bilateral stroke are ptosis of the eyelids, opisthotonic posture, sitting on the hindlimbs and vertical jumping. The second criterion is based on the finding of *Lowry et al.* (1964), who demonstrated that after decapitation the metabolism of the mouse brain

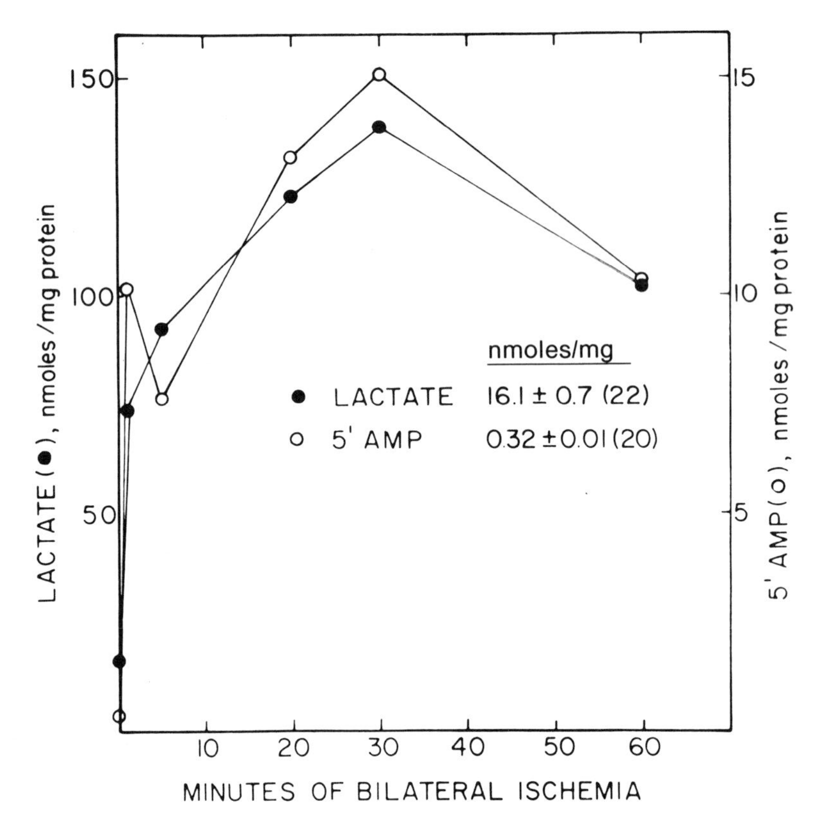

Fig. 2. Changes in lactate and 5′-AMP after bilateral ischemia. The animals were treated and the results expressed as described in Figure 1.

switched from aerobic to anaerobic glycolysis, resulting in a rapid depletion of the energy stores of the brain.

The biochemical evidence for the gerbil brain being ischemic is shown in Figures 1 and 2. The control values are in reasonably good agreement with those reported from other laboratories for brain metabolites (*Ferrendelli et al.*, 1972; *Lowry et al.*, 1964). In spite of the somewhat larger size of the gerbil, fixation of the cerebral cortex by freezing in liquid nitrogen appears to be adequate. ATP, P-creatine and glucose decrease to a minimum by 1 min of ischemia, while glycogen falls somewhat more slowly. Once these metabolites are depressed, they remain so with only minor fluctuations for up to 60 min of bilateral ischemia. The increases in both lactate and 5′-AMP also occur during the earlier stages of ischemia; thereafter, the levels tend to plateau. These results for the six metabolites are entirely consistent with those reported by *Lowry et al.* (1964) for the decapitated mouse brain. Thus, the biochemical data along with the neurological signs tend to confirm the existence of an ischemic forebrain after bilateral ligation of the common carotid arteries.

The periods of ischemia chosen were 1, 5, 20, 30 and 60 min in duration. In the 1-min group behavioral recovery was total. In the 5-min group recoverability was also quite good, but minimal neurological dysfunction did occur occasionally. In the 20-, 30- and 60-min groups overt seizures occurred both during and after the ischemic insult. The ability of the animals to survive periods of ischemia of 20 min or longer was substantially reduced: 50% of the animals died within 6 hr of recirculation after 20 and 30 min of ischemia, and all of the gerbils died within 1 hr after 60 min of bilateral ischemia. Therefore, the time intervals chosen represent a spectrum of ischemic episodes: from those that are compatible with survival to those that are not.

Once a positive ischemic insult of the cerebral cortex was established, the levels of the cyclic nucleotides were measured for up to 60 min of bilateral ischemia. As would be expected from the original work of *Breckenridge* (1964) and subsequently from a number of other laboratories (*Goldberg et al.*, 1970; *Schmidt, Schmidt*, and *Robison*, 1971), there was a statistically significant increase in cyclic AMP at all periods during the ligation (Fig. 3). The levels of cyclic AMP increased 14-fold to a peak at 1 min, declined to a level 5-fold greater than control by 5 min, and thereafter remained essentially constant for up to 1 hr of ischemia. When the gerbil data for cyclic AMP were compared to the results reported by *Steiner, Ferrendelli*, and *Kipnis* (1972) for the cerebral cortex from decapitated mice, the time course of the changes were reasonably similar. However, two differences were noted: 1) the magnitude of the cyclic AMP increase was greater in the gerbils, and 2) the cyclic AMP levels were back to control values by 20 min after decapitation. The obvious differences between the two experimental models could certainly account for these discrepancies.

In contrast to the cyclic AMP response, the levels of cyclic GMP were significantly lower than control values at all periods after bilateral ligation (Fig. 4). Generally, the reductions in the levels of cyclic GMP were greater with increasing periods of ischemia; the cyclic GMP concentration decreased to less than 20% of control after 60 min of occlusion. These results are entirely consistent with the cyclic GMP values reported by *Steiner et al.* (1972) for the cerebral cortex from the decapitated mouse. However, Kimura and coworkers (*Kimura, Thomas*, and *Murad*, 1974) have reported a 60% increase in cortical cyclic GMP approximately 30 sec after decapitation of the rat. An initial early rise in cyclic GMP undoubtedly would have been missed in our study since the earliest sampling was at 1 min. It should be noted, however, that an increase in cyclic GMP in the cerebral cortex was not observed at 15 or 30 sec after decapitation of the mouse (*Schmidt et al.*, 1971). It would appear that a

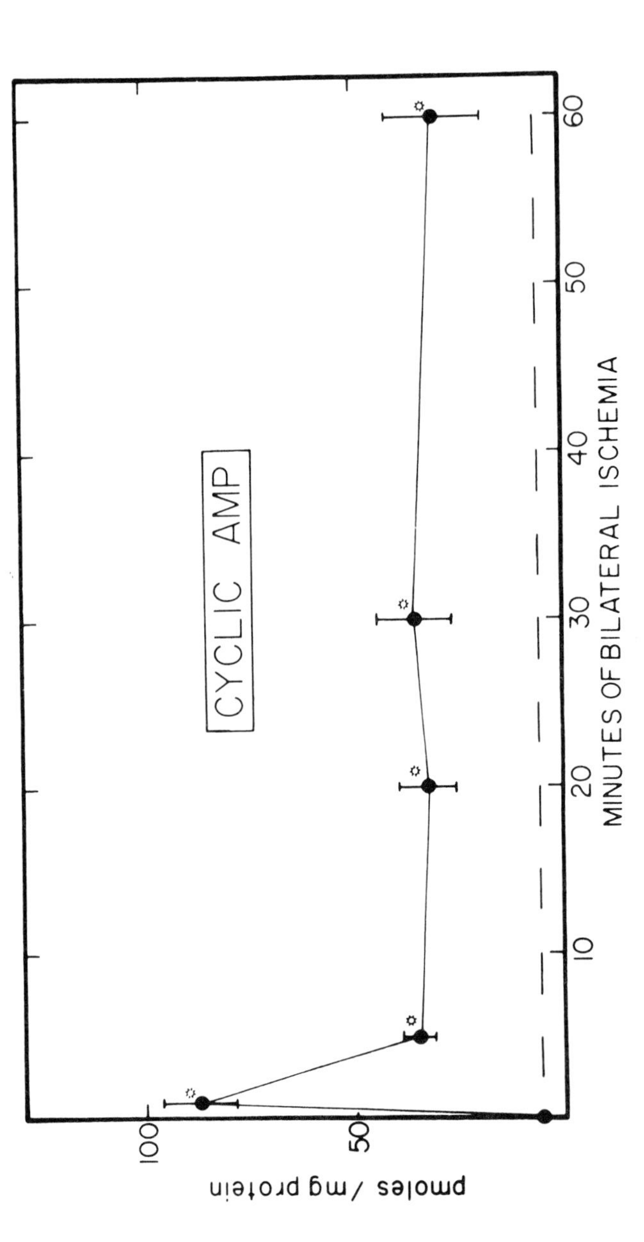

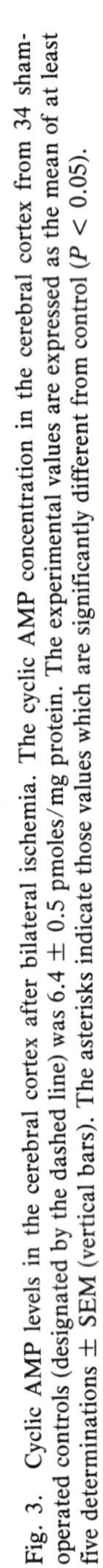

Fig. 3. Cyclic AMP levels in the cerebral cortex after bilateral ischemia. The cyclic AMP concentration in the cerebral cortex from 34 sham-operated controls (designated by the dashed line) was 6.4 ± 0.5 pmoles/mg protein. The experimental values are expressed as the mean of at least five determinations ± SEM (vertical bars). The asterisks indicate those values which are significantly different from control ($P < 0.05$).

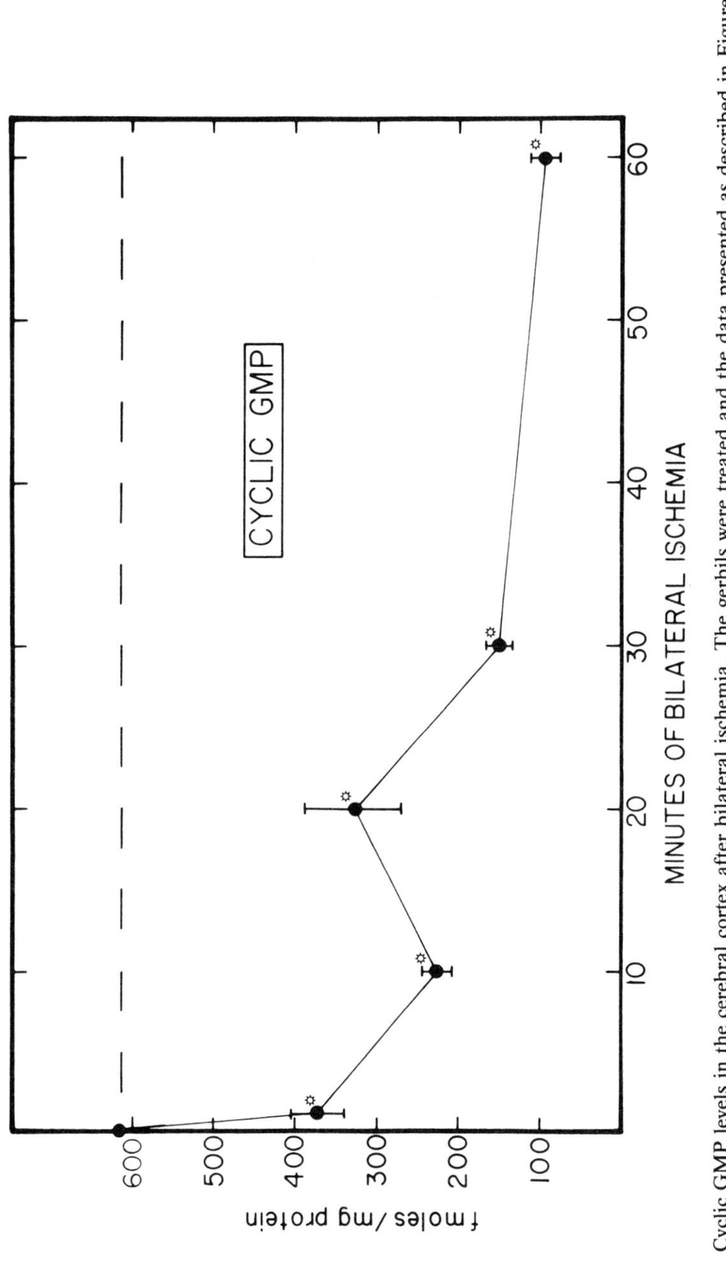

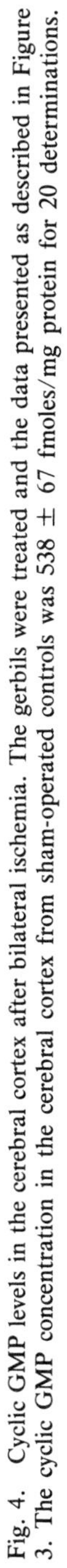
Fig. 4. Cyclic GMP levels in the cerebral cortex after bilateral ischemia. The gerbils were treated and the data presented as described in Figure 3. The cyclic GMP concentration in the cerebral cortex from sham-operated controls was 538 ± 67 fmoles/mg protein for 20 determinations.

preponderance of the results available favors a reduction in cortical cyclic GMP and an elevation in cortical cyclic AMP following ischemia or decapitation.

Could these changes in cyclic nucleotides affect the functional aspects of the brain during ischemia? Based on the neurophysiological studies by Stone and coworkers (*Stone et al.*, 1975), the possibility exists that the cyclic nucleotides may play a physiological role during ischemia or at least the initial stages thereof. It appears from their work that iontophoretically applied cyclic GMP excites pyramidal tract neurons of the cerebral cortex, whereas cyclic AMP applied similarly inhibits these neurons. Consequently, these ischemia-induced changes in cyclic nucleotides may collectively reflect and even contribute to the electrically quiescent state of the brain which is known to occur during the initial phases of ischemia. Since the electrical activity of the brain (EEG) disappears within 30 sec of ischemia (*Hossmann* and *Sato*, 1971; *Swaab* and *Boer*, 1972), the modulation of neuronal excitability by changes in cyclic nucleotide levels after 1 min of ischemia probably is not significant. However, recovery to normal levels of cyclic nucleotides may be a prerequisite for the restoration of neuronal excitability. It is for this reason that the cyclic nucleotide levels were investigated during the so-called recovery period, when circulation was reestablished.

Recovery

It is during the recirculation periods following ischemia that the extent of brain damage can truly be assessed. Survival and the general neurological condition of the gerbil are two simple measures of the degree of brain injury. If we combine the behavioral patterns with the biochemical data, certain tentative conclusions concerning the relationship of the recovery process to the metabolism of cyclic nucleotides and energy metabolites become evident.

The fluctuations in ATP during ischemia and recovery are of particular interest, not only because it serves as the major energy source for biosynthetic and transport processes within the cell but also because it is the substrate for adenylate cyclase. Normally, ATP levels in most cellular systems are reasonably stable and therefore not a factor in cyclic AMP metabolism; however, they may be of some consequence during ischemia.

The levels of ATP after all periods of ischemia were restored to a greater or lesser degree depending on the duration of the original ischemia insult (Fig. 5). This result indicates that the blood flow to the ischemic region is in part reestablished, since without a source of glucose, energy metabolism would remain depressed. While the extent of ATP recovery was severely compromised in the 60-min group, the ATP levels rapidly approached control values in the other groups. Generally, the rate of

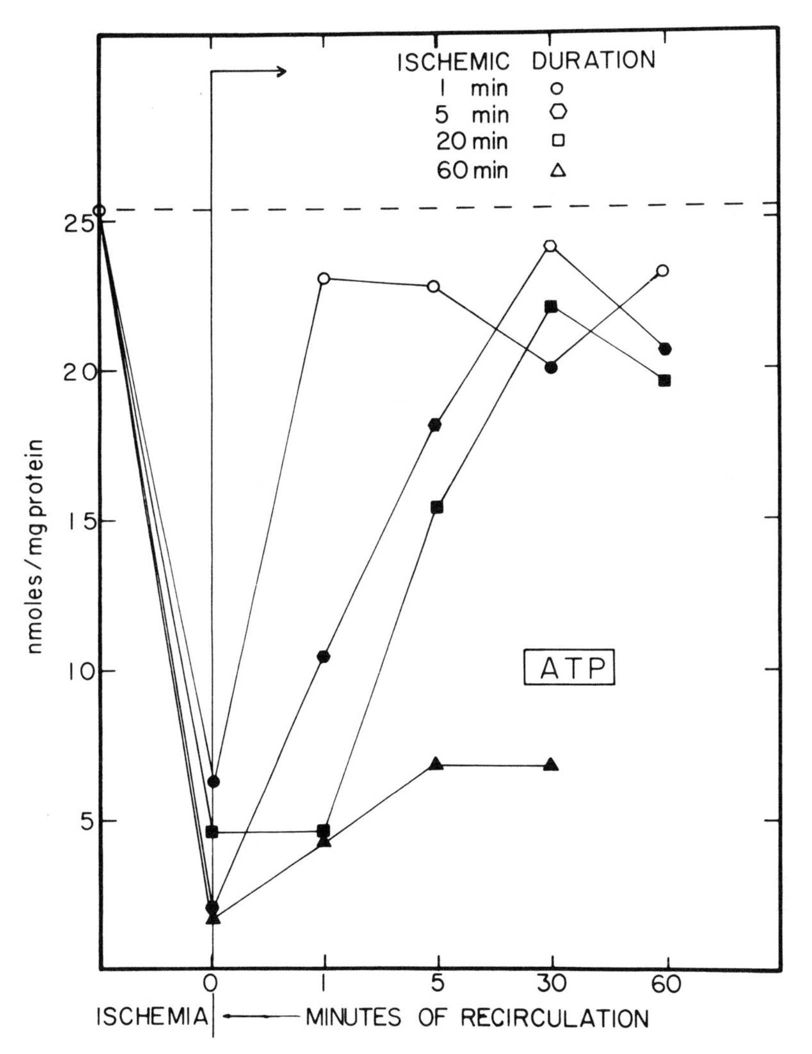

Fig. 5. ATP levels in the cerebral cortex during and after an ischemic insult. The ATP concentration in control animals (dashed line) was 25.4 nmoles/mg protein for 22 determinations. The ATP levels are shown after 1 min (circle), 5 min (hexagon), 20 min (square) and 60 min (triangle) of bilateral occlusion, and then at 1, 5, 30 and 60 min of recirculation. Note that the minutes of recirculation are not drawn to scale. Filled symbols represent values that are significantly different from those of control ($P < 0.05$). There were 6–12 gerbils in the ischemic groups (total of 36) and 5–14 in the recovery groups (total of 132).

ATP restoration decreased with increasing periods of ischemia. Although the data are not presented, other energy metabolities, including glucose and P-creatine, also exhibit a recovery pattern similar to that for ATP.

The response of the cyclic nucleotides during recirculation was somewhat more complex than the restoration of the energy stores. While the recovery of ATP, P-creatine and glucose toward control values appears to be simple and direct, this is not the case for the cyclic nucleotides. The levels of cyclic GMP were depressed during ischemia but increased to values up to 4-fold greater than control during recirculation (Fig. 6). Cyclic AMP, in contrast, continued to increase beyond the already ele-

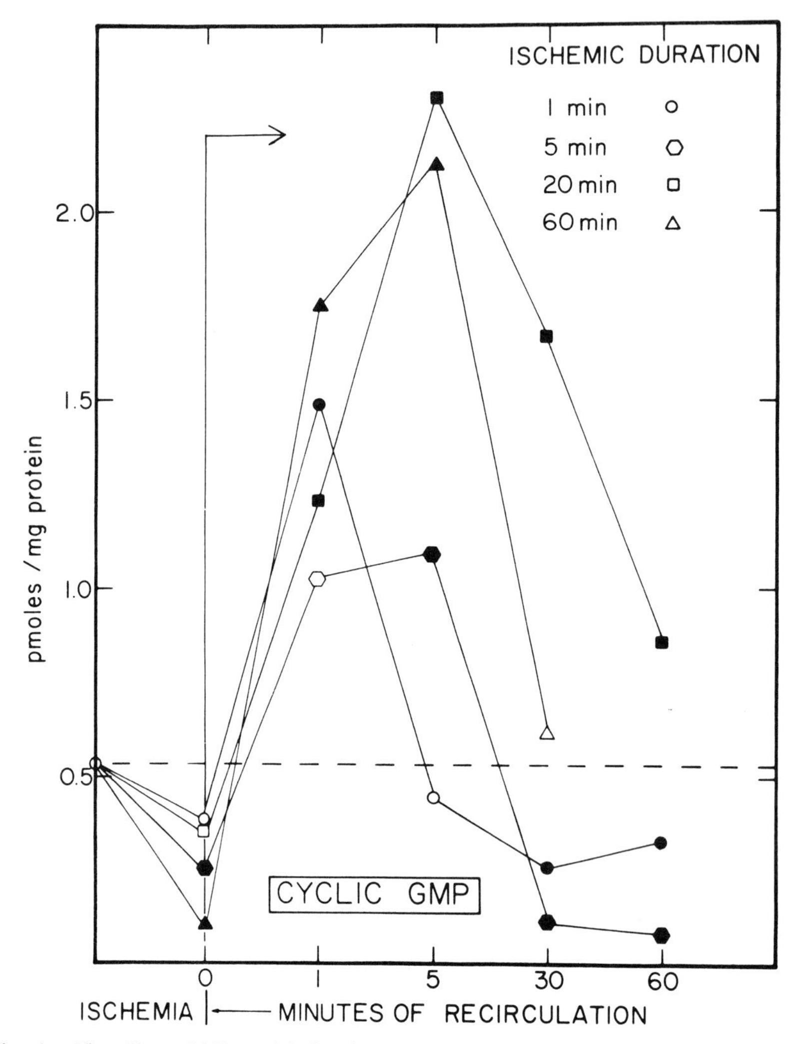

Fig. 6. The effect of bilateral ischemia and recirculation on cyclic GMP levels in the gerbil cerebral cortex. The experiments were performed and the data presented as described in Figure 5. There were 34 sham-operated controls, 6–12 animals in the ischemic groups (total of 36) and 4–14 animals in the recovery groups (total of 125).

vated levels during recirculation (Fig. 7). Thus, if the cyclic nucleotides play a role as neuromodulators, the observed fluctuations in the cyclic nucleotides during recovery may reflect the physiological condition of the ischemic cerebral cortex.

The levels of cyclic GMP increased to values greater than control during recirculation, although the relative increases were much less than those observed for cyclic AMP. If one examines the areas under the curves of Figure 6 at the true time scale, the levels increased more after

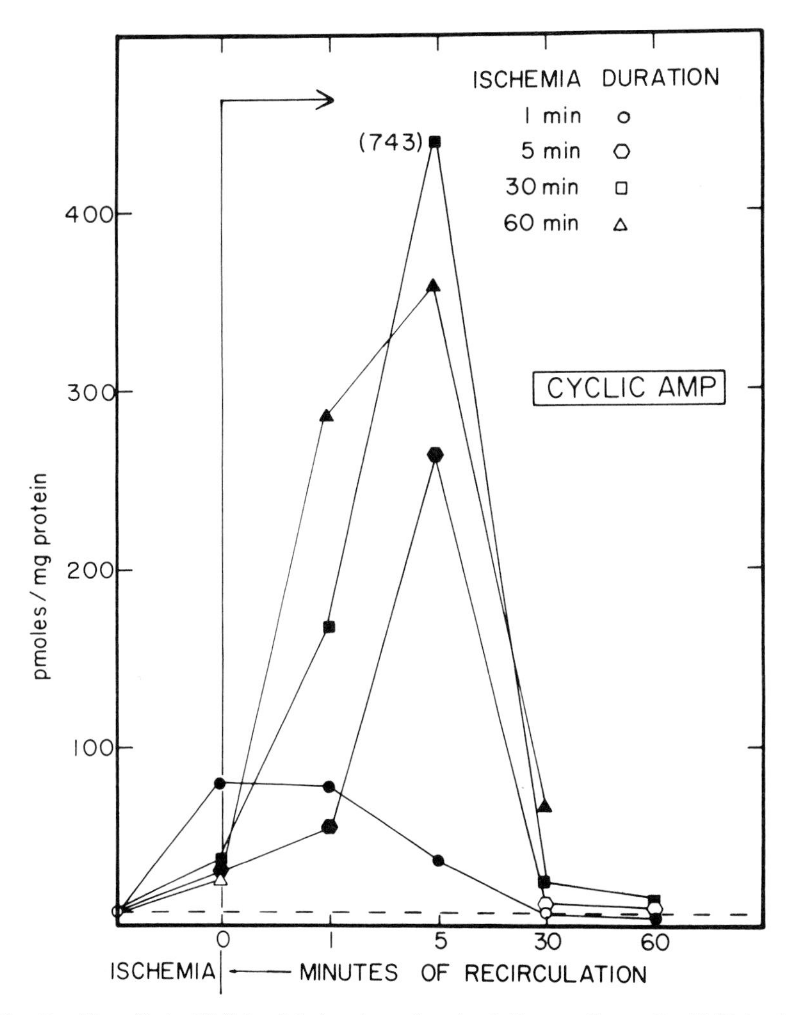

Fig. 7. The effect of bilateral ischemia and recirculation on the cyclic AMP levels in the gerbil cerebral cortex. See Figure 5 for details. The number of gerbils in each group is the same as in Figure 6.

longer periods of ischemia. The magnitude of the cyclic GMP response after varying periods when compared to that for the 1-min group was 3-fold greater in the 5-min group, 32-fold greater in the 20-min group and 14-fold greater in the 60-min group. In the 1- and 5-min ischemic groups, cyclic GMP decreased to levels significantly lower than control at 30 and 60 min of recovery. In the 20- and 60-min groups, the concentrations of cyclic GMP were still elevated.

The already elevated levels of cyclic AMP were increased during the first 5 min of recirculation in all but the 1-min ischemic group. As with cyclic GMP, the degree of the response was greater following longer periods of ischemia. Compared to that of the 1-min group, the increases were approximately 6-fold greater in the 5- and 20-min groups and 10-fold greater in the 60-min group.

What are the factors which determine the steady-state levels of cyclic nucleotides both during and after an ischemic episode? We have limited the discussion to cyclic AMP, since far more is known about its metabolism. Changes in the activities of the biosynthetic and degradative enzymes as well as the binding protein might explain the cyclic AMP changes. However, *Schwartz* and coworkers (1976) have previously looked at these enzymes and others during ischemia and recovery. While there were alterations in the activities of the particulate protein kinase and adenylate cyclase when measured *in vitro*, it seems unlikely that the observed changes could explain the elevations *in vivo*.

Another possibility is that the cyclic AMP response is a function of the differential availability of intracellular ATP and extracellular adenylate cyclase agonists (Fig. 8). Under normal circumstances (stage 1), ATP is maintained quite effectively through oxidative phosphorylation and the adenylate cyclase agonists are sequestered such that the extracellular concentrations are relatively low.

During ischemia (stage 2), cellular ATP decreases since oxidative phosphorylation is lost in the absence of oxygen and the alternative pathways for the maintenance of ATP cannot meet the energy demands of the tissue. Potassium, adenosine and norepinephrine have been clearly shown to stimulate the production of cyclic AMP in brain slices (for review, see *Daly*, 1976 and *Drummond*, 1973), and the efflux or release of these agents during ischemia has been either directly or indirectly demonstrated (*Berne, Rubio*, and *Curnish*, 1974; *Blank* and *Kirshner*, 1977; *Hossman, Sakaki*, and *Zimmerman*, 1977; *Lavyne et al.*, 1975; *Mrsulja et al.*, 1976). The extracellular agonist pool increases in size: the levels of adenosine increase as the levels of ATP decrease and those of 5′-AMP increase, potassium leaks out of the cell as the sodium-potassium pump fails, and norepinephrine is released from the tissue. Collectively, these events could account for the cyclic AMP changes dur-

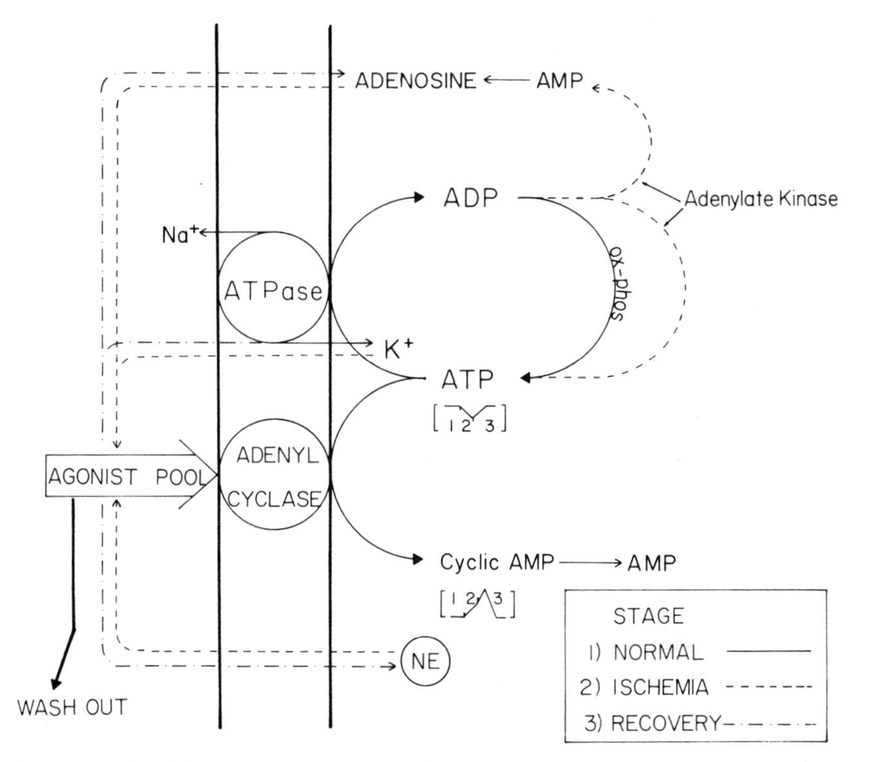

Fig. 8. Cyclic AMP metabolism during and after ischemia: a schematic representation. For detailed explanation, see text. NE = norepinephrine; ox-phos = oxidative phosphorylation.

ing ischemia. In the first minute of ischemia, sufficient agonist is released to stimulate the production of cyclic AMP (i.e., 1-min peak, Fig. 7). When the ATP is depleted, cyclic AMP accumulation is turned off (i.e., fall off of cyclic AMP after 1 min of ischemia).

A more interesting aspect of this model occurs upon the release of the occlusion (stage 3). ATP is rapidly regenerated to more than 50% of control values by 5 min of recirculation in all but the 60-min group (Fig. 5). However, the agonists are still poised at the receptor site. With the ATP now available, there is a burst of cyclic AMP production which persists until the agonists are disposed of (Fig. 8). The agonists can be removed from the extracellular site by reuptake, washout by the circulation or metabolism to an inactive compound. The levels of cyclic AMP fall to control values as the extracellular concentrations of the agonists decrease. Thus, in this relatively unique situation ATP becomes the trigger to the production of cyclic AMP. However, the size of the extracellular agonist pool may still determine the magnitude of the response.

Cyclic Nucleotides in Spinal Cord and Cerebellum of Bilaterally Ischemic Gerbils

The level of neuronal excitability decreases rapidly with the onset of ischemia, and the electrical output from an ischemic region undoubtedly would be similarly quieted. Since it is increasingly evident that the cyclic nucleotides play a role in neuronal excitability, a diminished input from an ischemic to a nonischemic region may be reflected in the steady-state levels of the cyclic nucleotides. To evaluate the effects of the ischemic cerebral hemispheres on other nonischemic regions of the brain, cyclic nucleotides were measured in the cerebellum and spinal cord, areas whose circulation is supposedly maintained after carotid artery occlusion. In fact, the high-energy phosphate metabolites, ATP and P-creatine, were at least as high in these regions of the experimental gerbils as in the sham-operated controls (data not shown). These results indicate that following bilateral occlusion of the common carotid arteries, the energy state of the cerebellum and spinal cord is preserved and therefore blood flow to these regions is not severely compromised.

The levels of cerebellar cyclic AMP were significantly increased after 1 min of occlusion and after 30 min of recirculation following 1 min of occlusion (Fig. 9). While there were other significant changes, a definite trend was not evident. The cerebellar cyclic GMP levels were not changed during and after occlusion in the 1- and 5-min groups (Fig. 10). After 20 min of ligation, the concentration of cyclic GMP was 4-fold greater than control and increased further to 10-fold greater than control after 1 and 5 min of recirculation. By 30 min of recovery, the cyclic GMP levels were less than those of control. A possible explanation for the elevated cyclic GMP after 20 min of occlusion is the presence of seizures observed in this group of gerbils. CNS depressants and anticonvulsants have been shown to decrease cerebellar levels of cyclic GMP, whereas convulsants or CNS stimulants increase the levels (*Ferrendelli, Kinscherf,* and *Kipnis,* 1972; *Lust et al.,* 1978; *Opmeer et al.,* 1976). The elevated levels of cyclic GMP could then reflect a convulsive condition common to that observed in animals after treatment with convulsants or stimulants. The exact relationship of the elevated cerebellar cyclic GMP to a seizure state is presently unclear; however, it is apparent that the elevated cyclic GMP levels and the presence of seizures are coincidental with extended periods of occlusion.

The levels of cyclic AMP in the spinal cord decreased during 1, 5 and 20 min of bilateral ligation, and decreased further after 1 min of recirculation in all three groups (Fig. 11). By 30 min of recirculation, there was some restoration of the cyclic AMP levels toward control values in all

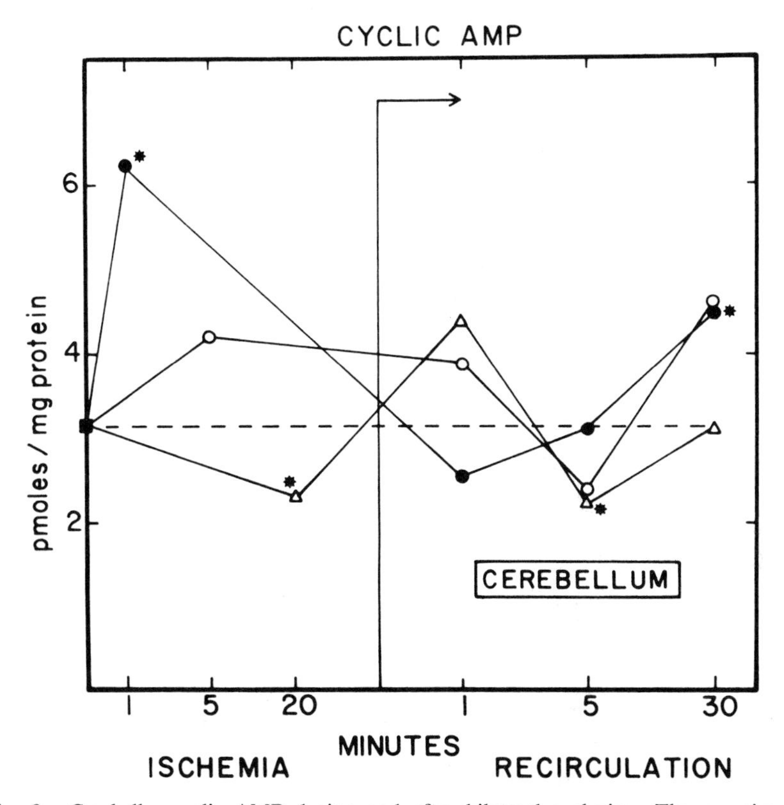

Fig. 9. Cerebellar cyclic AMP during and after bilateral occlusion. The experiments were performed as described in the Materials and Methods section. The symbols represent the cerebellar levels of cyclic AMP in sham-operated controls, ■; 1 min of occlusion, ●; 5 min of occlusion, ○; and 20 min of occlusion, △. The vertical line with an arrow separates the values during occlusion (left) and recirculation (right). The cyclic AMP concentration in sham-operated controls (dashed line) was 3.3 ± 0.3 pmoles/mg protein. Each point represents the mean of five determinations and the asterisks indicate those values significantly different from control ($P < 0.05$). A total of 65 gerbils was used in this study.

three groups. The levels of cyclic GMP were essentially unchanged in the spinal cord both during and after carotid artery occlusion (Fig. 12). An explanation for the decrease of cyclic AMP in the spinal cord is presently lacking, since information on the cyclic nucleotides in the spinal cord is scarce. However, based on the inhibitory action of cyclic AMP in other regions of the brain, the reduction in cyclic AMP may indicate an alteration in excitability.

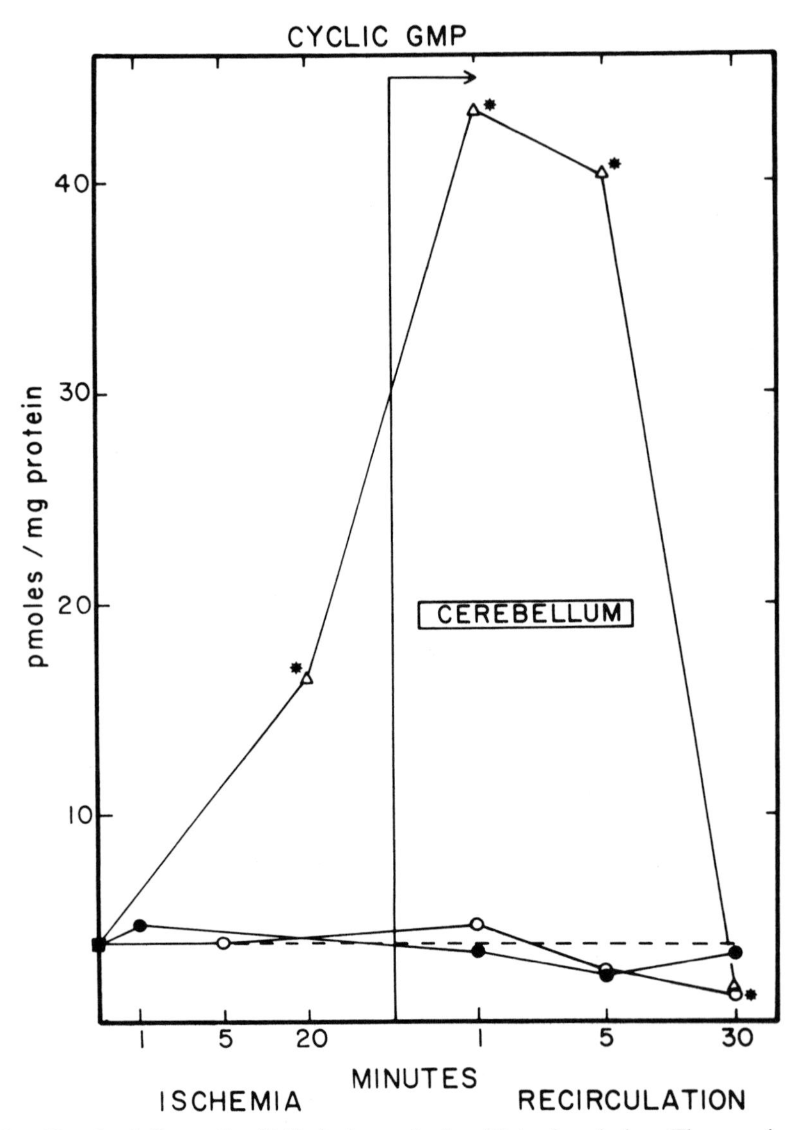

Fig. 10. Cerebellar cyclic GMP during and after bilateral occlusion. The experiment was performed and the data presented as described in Figure 9. The control level of cyclic GMP was 3.9 ± 1.1 pmoles/mg protein.

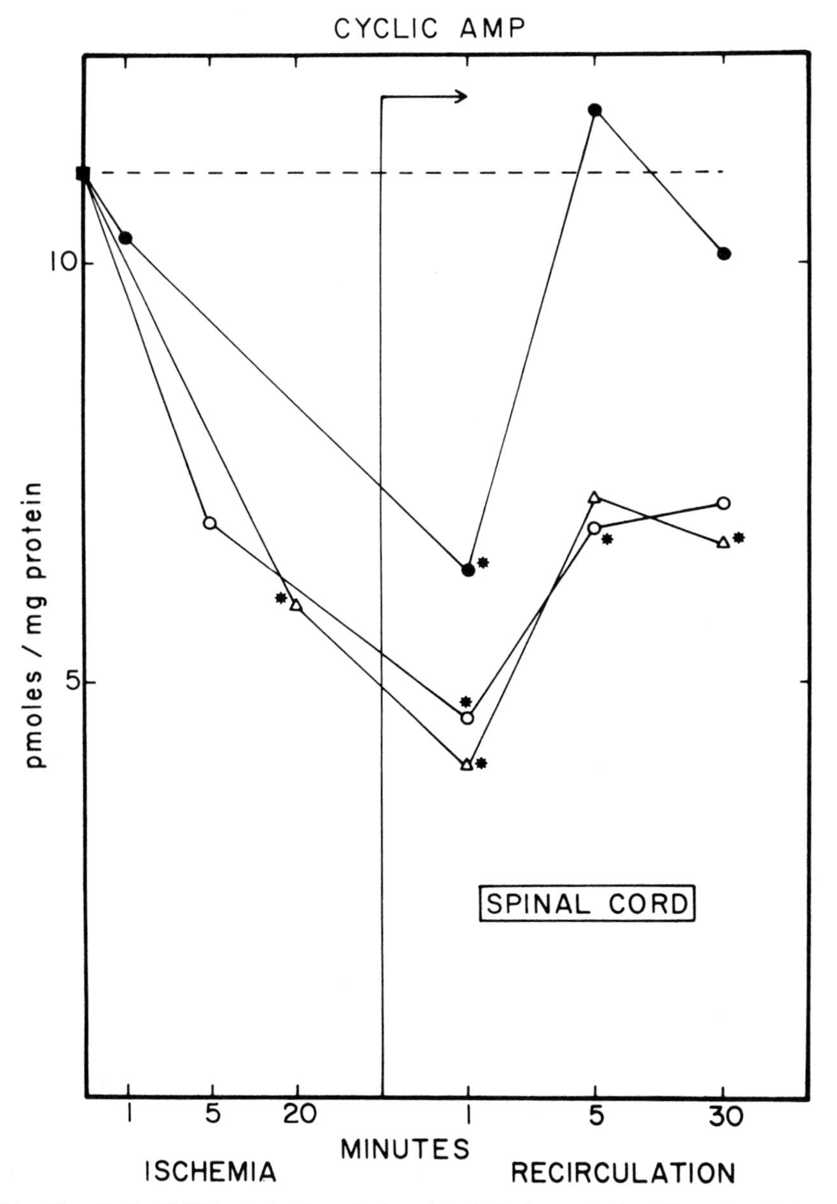

Fig. 11. Cyclic AMP levels in the spinal cord both during and after bilateral occlusion. For details, see Figure 9. The mean control value for cyclic AMP was 11.1 ± 1.4 pmoles/mg protein.

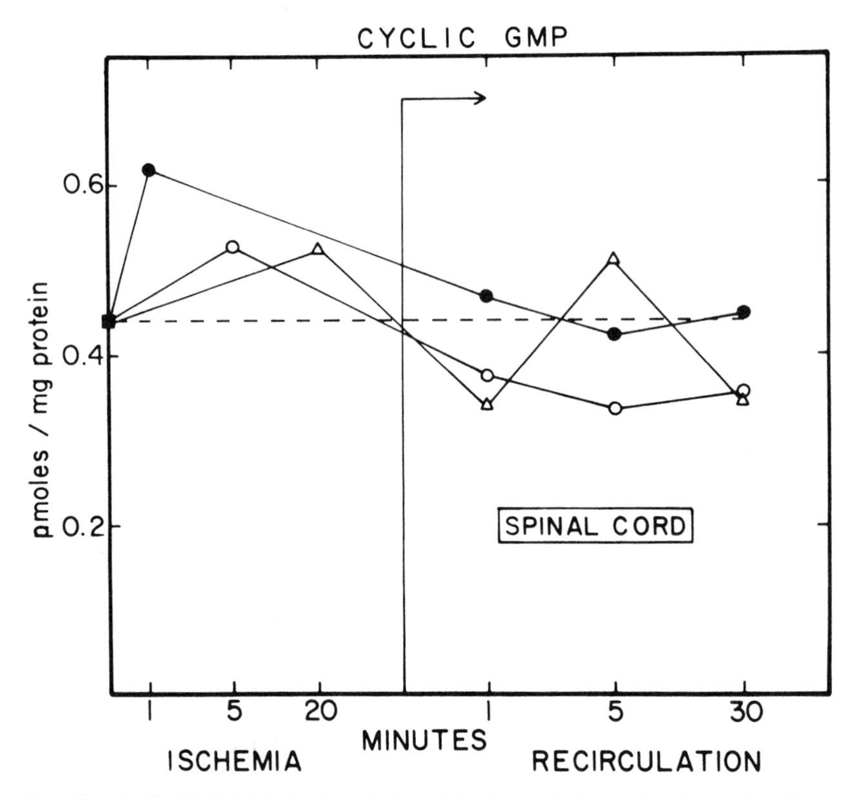

Fig. 12. Cyclic GMP levels in the spinal cord during occlusion and recirculation. For details, see Figure 9. The control value for cyclic GMP in the spinal cord was 0.51 ± 0.06 pmoles/mg protein.

Drugs[1]

Barbiturates have been demonstrated in gerbils to have a protective effect against an ischemic insult (*McGraw*, 1977). Gerbils were pretreated with 100 mg/kg (i.p.) of sodium thiamylal and after 30 min the common carotid arteries were exposed and occluded for 20 min. In some of the animals the occlusion was released for 5, 30 or 360 min of recirculation. The effect of the barbiturate on all metabolites measured, with the exception of cyclic GMP, was minimal (data not shown). The steady-state levels of cyclic GMP in the cerebral cortex of thiamylal-treated gerbils were significantly lower in all groups except at 6 hr after 20 min of occlusion (Fig. 13). The dampening of the cyclic GMP response by

[1] These studies were conducted by Dr. M. Kobayashi, who is presently at Juntendo University School of Medicine, Tokyo, Japan.

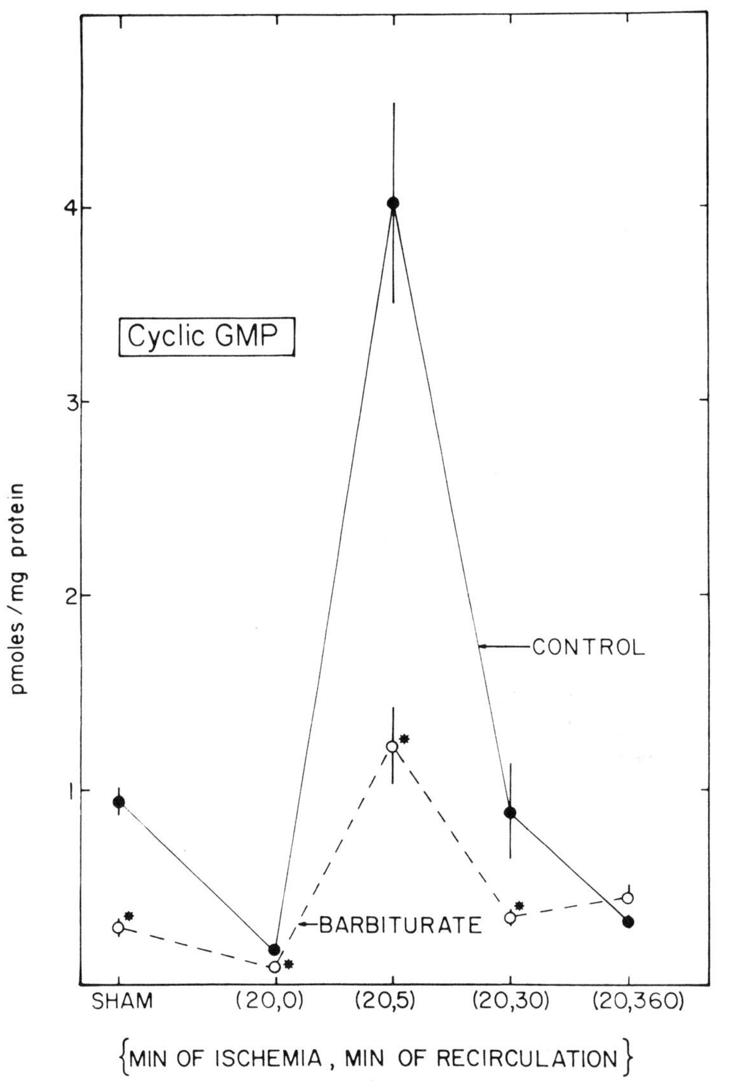

Fig. 13. Effect of thiamylal on the cyclic GMP levels in the cerebral cortex during and after 20 min of bilateral occlusion. The drug-treated gerbils were given 100 mg/kg (i.p.) of thiamylal 30 min prior to the occlusion of the common carotid arteries. After 20 min of bilateral occlusion, one group (20, 0) was frozen without recirculation and three other groups, (20, 5), (20, 30) and (20, 360), were recirculated for 5, 30 or 360 min and then frozen. The control groups received a saline injection but otherwise were treated as the drug-treated gerbils. Each point represents the mean of at least four determinations, the vertical lines the SEM and the asterisks those values significantly different from control ($P < 0.05$).

thiamylal may bear some relationship to the protective effect of the drug and thus warrants further investigation.

DISCUSSION

Ischemia

When the blood flow to the brain is blocked, the normal biochemical and physiological characteristics of the brain are profoundly affected, apparently by the loss of oxygen and glucose. The spontaneous excitability of the brain ceases within seconds of circulatory occlusion (*Hossmann* and *Sato*, 1971; *Swaab* and *Boer*, 1972) and the energy stores are essentially depleted by 1 min (*Lowry et al.*, 1964). The affected region of the brain has undergone radical changes from a metabolically active tissue to a rather quiescent one. With longer periods of ischemia, there were no further observable differences in the energy metabolities measured. Thus, the once highly integrated brain, having been submitted to a potentially lethal insult, is reduced to a simple quest for survival.

It is during this perturbed state of the brain that the cyclic nucleotides exhibit large and dramatic changes. One of the more intriguing aspects of the changes is the substantial increase in cyclic AMP in spite of a rapid decrease in its precursor, ATP. As discussed in the results, the initial burst of cyclic AMP production is probably due to the rapid release of adenylate cyclase agonists prior to the depletion of ATP. Once the ATP is exhausted at 1 min, the levels of cyclic AMP subsequently fall off toward control. In contrast, the levels of cyclic GMP decrease during ischemia, as do those of its precursor, GTP (*Kleihues, Kobayashi*, and *Hossmann*, 1974). Although the molecular events involved in the metabolism of either cyclic nucleotide during ischemia are presently unclear, the cyclic nucleotide responses indicate that there are probably two separate and distinct processes for the two cyclic nucleotides.

Is there any physiological significance to these changes during ischemia? Based on the work of *Stone et al.* (1975) alluded to previously, when iontophoretically applied, cyclic GMP increased excitability of the pyramidal tract neurons, whereas cyclic AMP depressed them. During ischemia, the elevation of cyclic AMP would depress excitability, and reduction of cyclic GMP, which is thought to be excitatory, would have a similar effect. The pattern of the cyclic nucleotide responses during ischemia strengthens the possibility that they act collectively in the transition of the brain from an active to a dormant state.

It is during the period following an ischemic episode that the fate of the brain and ultimately the animal is defined. Generally, the survival rate as well as the rate and extent of metabolite recovery decrease with increasing periods of ischemia. We have concluded previously that, on the basis of the metabolite levels during postischemia, the recovery process is more complex than a mere reversal of the ischemia-induced events. Nowhere is the demonstration of this phenomenon more clearly evident than in the fluctuations of the cyclic nucleotides during postischemia. Cyclic GMP levels increased from their depressed concentrations during ischemia to values significantly greater than control. Cyclic AMP levels already elevated during ischemia increased further to values as high as 100-fold greater than control during recirculation. Thus, a direct restoration of the cyclic nucleotides simply does not occur. Biochemically, the altered cyclic nucleotide metabolism in postischemia probably reflects the deranged conditions of the brain. Since the magnitude of the cyclic nucleotide response is proportional to the original ischemic episode, these derangements are of greater magnitude after longer periods of ischemia; therefore, the deleterious effects of the episode apparently persist for longer periods during recovery. If the same physiological rationale used for ischemia is applied to the cyclic nucleotide response during recovery, the cyclic nucleotides do not appear to be acting in concert; the elevated cyclic GMP would favor excitation and the elevated cyclic AMP inhibition. The presence of hyperexcitability, however, could possibly explain the opposing actions of the two cyclic nucleotides during recovery.

Seizures

The major concern of this paper was to evaluate the effects of ischemia on cyclic nucleotide metabolism in brain; however, seizures are an inevitable and unavoidable occurrence following longer periods of ischemia. Quite possibly, a number of the cyclic nucleotide changes could be related to the direct action of the convulsions and not to the ischemic insult. Specifically, the seizures which were prevalent in the 20-min ischemic group could account for the elevated cerebellar cyclic GMP (Fig. 10) and even for the greater cyclic GMP response in the cerebral cortex during recovery. Thus, the incidence of seizures in ischemic gerbils is an important factor to be considered in the interpretation of the data.

A number of laboratories have demonstrated that experimentally induced seizures which include at least a tonic or clonic component result in an increase in both cyclic nucleotides, whether measured in the cerebral cortex or cerebellum (*Lust et al.*, 1978; *Rubin* and *Ferrendelli*; 1977). Even in a preconvulsive state after a convulsant like bicuculline, the levels of cyclic GMP are also elevated (*Lust et al.*, 1977). This information serves to explain the elevated cerebellar cyclic GMP levels observed in the seizure-prone 20-min ischemic group (Fig. 10). Somewhat less clear is the significance of the cyclic GMP increase in the cerebral cortex of all groups during recovery. Seizures could be a factor in the increase after 20 min of ischemia, but with shorter ischemic insults seizures are not evident and yet cyclic GMP levels are significantly greater than control. Quite reasonably, the cyclic GMP response could reflect a hyperexcitable state of the cerebral cortex similar to that observed in the preconvulsive condition. The absence of overt seizures in spite of the elevated cyclic GMP in the 1- and 5-min groups could be due either to insufficient levels of the cyclic nucleotide or the presence of elevated inhibitory metabolites such as GABA (data not shown) or cyclic AMP which could offset the cyclic GMP effect. Thus, the interpretation of the cyclic GMP data favors the existence of a hyperexcitable state of the brain during recovery and, in its extreme, even convulsions. The elimination of the seizure effect in the investigation of ischemia would certainly be desirable; unfortunately, however, this condition is a natural consequence of the prolonged ischemia.

The barbiturate studies provide additional evidence that seizures are an aggravating component of the ischemic response. Barbiturates have been shown to have a protective effect against an ischemic insult. In one of our studies, the survival of the gerbils increased from 9% to 57% after 20 min of bilateral occlusion in the presence of thiamylal. Since seizures undoubtedly exacerbate the ischemic injury, it seems reasonable to assume that the drug improves the likelihood of survival by its known ability to prevent seizures. The suppression of the cyclic GMP increases by thiamylal both during and after 20 min of bilateral occlusion (Fig. 13) may similarly reflect the ability of this drug to diminish the hyperexcitable state of the brain.

CONCLUSION

In summary, the susceptibility of the brain to an ischemic insult may be reflected by the large fluctuations in the levels of the cyclic nucleotides. If the cyclic nucleotides do serve as neuromodulators, the marked derangements during ischemia when excitability of the affected region is lost probably have no physiological significance. However, during recovery, when the normal homeostatic mechanisms are once more being exerted, the perturbations of the cyclic nucleotides may indeed play a role in the restoration of brain function. Perhaps the concentrations of the cyclic nucleotides *in vivo* may be useful in determining those factors critical to the maintenance of neural viability following an ischemic episode.

REFERENCES

Berne, R.M., Rubio, R., and Curnish, R.R. 1974. Release of adenosine from ischemic brain. Circulation Res 35:262-271

Blank, W.G., and Kirshner, H.S. 1977. The kinetics of extracellular potassium changes during hypoxia and anoxia in the cat cerebral cortex. Brain Res 123:113-124

Breckenridge, B.McL. 1964. The measurement of cyclic adenylate in tissues. Proc Nat Acad Sci 52:1580-1586

Daly, J.W. 1976. The nature of receptors regulating the formation of cyclic AMP in brain tissue. Life Sci 18:1349-1358

Drummond, G.I. 1973. Metabolism and functions of cyclic AMP in nerve. In *Progress in Neurobiology,* vol. 2, eds. G.A. Kerkut, and J.W. Phillis, pp. 119-176. New York: Pergammon Press

Ferrendelli, J.A., Chang, M.M., and Kinscherf, D.A. 1974. Elevation of cyclic GMP levels in central nervous system by excitatory and inhibitory amino acids. J Neurochem 22:535-540

Ferrendelli, J.A., Gay, M.H., Sedgwick, W.G., and Chang, M.M. 1972. Quick freezing of the murine CNS. Comparison of regional cooling rates and metabolite levels when using liquid nitrogen or Freon-12. J Neurochem 19:979-987

Ferrendelli, J.A., Kinscherf, D.A., and Chang, M.M. 1975. Comparison of the effects of biogenic amines on cyclic GMP and cyclic AMP levels in mouse cerebellum *in vitro.* Brain Res 84:63-73

Ferrendelli, J.A., Kinscherf, D.A., and Kipnis, D.M. 1972. Effects of amphetamine, chlorpromazine and reserpine on cyclic GMP and cyclic AMP levels in mouse cerebellum. Biochem Biophys Res Comm 46:2114-2120

Goldberg, N.D., Lust, W.D., O'Dea, R.F., Wei, S., and O'Toole, A.G. 1970. A Role of cyclic nucleotides in brain metabolism. In *Advances in Biochemical Psychopharmacology,* Vol. 3, eds. E. Costa, and P. Greengard, pp. 67-88. New York: Raven Press

Hossmann, K.-A., Sakaki, S., and Zimmerman, V. 1977. Cation activities in reversible ischemia of the cat brain. Stroke 8:77-81

Hossmann, K.-A., and Sato, K. 1971. Effect of ischemia on the function of the sensorimotor cortex in cat. Electroenceph Clin Neurophysiol 30:535-545

Jones, D.J., Medina, M.A., Ross, D.H., and Stavinoha, W.B. 1974. Rate of inactivation of adenyl cyclase and phosphodiesterase. Determinants of brain cyclic AMP. Life Sciences 14:1577-1585

Kimura, H., Thomas, E., and Murad, F. 1974. Effects of decapitation, ether and pentobarbital on guanosine 3′,5′-phosphate and adenosine 3′,5′-phosphate levels in rat tissues. Biochim Biophys Acta 343:519-528

Kleihues, P., Kobayashi, K., and Hossmann, K.-A. 1974. Purine nucleotide metabolism in the cat brain after one hour of complete ischemia. J Neurochem 23:417-425

Kogure, K., Scheinberg, P., Matsumoto, A., Busto, R., and Reinmuth, O.M. 1975. Catecholamines in experimental brain ischemia. Arch Neurol 32:21-24

Lavyne, M.H., Moskowitz, M.A., Larin, F., Zervas, N.T., and Wurtman, R.J. 1975. Brain H^3-catecholamine metabolism in experimental ischemia. Neurology 25:483-485

Levine, S., and Payan, H. 1966. Effects of ischemia and other procedures on brain and retina of the gerbil (*Meriones unguiculatus*). Exp Neurol 16:255-262

Lowry, O.H., and Passonneau, J.V. 1972. In *A Flexible System of Enzymatic Analysis,* pp. 151-156. New York: Academic Press

Lowry, O.H., Passonneau, J.V., Hasselberger, F.X., and Schultz, D.W. 1964. Effect of ischemia on known substrates and cofactors of the glycolytic pathway in brain. J Biol Chem 239:18-30

Lowry, O.H., Rosebrough, N.J., Farr, A.L., and Randall, R.L. 1951. Protein measurement with the Folin phenol reagent. J Biol Chem 193:265-275

Lust, W.D., Kupferberg, H.J., Yonekawa, W.D., Penry, J.K., Passonneau, J.V., and Wheaton, A. 1978. Changes in brain metabolites induced by convulsants or electroshock. Effects of anticonvulsant agents. Mol Pharmacol 14:347-356

Lust, W.D., Mrsulja, B.B., Mrsulja, B.J., Passonneau, J.V., and Klatzo, I. 1975. Putative neurotransmitters and cyclic nucleotides in prolonged ischemia of the cerebral cortex. Brain Res 98:394–399

Lust, W.D., Passonneau, J.V., and Veech, R.L. 1973. Cyclic adenosine monophosphate, metabolites and phosphorylase in neural tissue. A comparison of methods of fixation. Science 181:280–282

Lust, D., Yonekawa, W., Kupferberg, H., Passonneau, J., and Penry, K. 1977. Cerebellar levels of cyclic GMP after sodium valproate and/or bicuculline.. Transactions of the American Society for Neurochemistry 8(Abs)267

McGraw, C.P. 1977. Experimental cerebral infarction, effects of pentobarbital in mongolian gerbils. Arch Neurol 34:334–336

Medina, M.A., Jones, D.J., Stavinoha, W.B., and Ross, D.H. 1975. The levels of labile intermediary metabolites in mouse brain following rapid tissue fixation with microwave irradiation. J Neurochem 24:223–227

Mrsulja, B.B., Mrsulja, B.J., Spatz, M., and Klatzo, I. 1976. Catecholamines in brain ischemia; effects of alphamethyl-p-tyrosine and pargyline. Brain Res 104:373–378

Nelson, S.R., Schulz, D.W., Passonneau, J.V., and Lowry, O.H. 1968. Control of glycogen levels in brain. J Neurochem 15:1271–1279

Opmeer, F.A., Gumulka, S.W., Dinnendahl, V., and Schonhofer, P.S. 1976. Effects of stimulatory and depressant drugs on cyclic guanosine 3′,5′-monophosphate and adenosine 3′,5′-monophosphate levels in mouse brain. Naunyn-Schmiedeberg's Arch Pharmacol 292:259–265

Passonneau, J.V., and Lauderdale, V.R. 1974. A comparison of three methods of glycogen measurement in tissues. Anal Biochem 60:405–412

Rubin, E.H., and Ferrendelli, J.A. 1977. Distribution and regulation of cyclic nucleotide levels in cerebellum, *in vivo*. J Neurochem 29:43–51

Schmidt, M.J., Schmidt, D.E., and Robison, G.A. 1971. Cyclic adenosine monophosphate in brain areas. Microwave irradiation as a means of tissue fixation. Science 173: 1142–1143

Schwartz, J.P., Mrsulja, B.B., Mrsulja, B.J., Passonneau, J.V., and Klatzo, I. 1976. Alterations of cyclic nucleotide-related enzymes and ATPase during unilateral ischemia and recirculation in gerbil cerebral cortex. J Neurochem 27:101–107

Stavinoha, W.B., Weintraub, S.T., and Modak, A.T. 1973. The use of microwave heating to inactivate cholinesterase in the rat brain prior to analysis for acetylcholine. J Neurochem 20:361–371

Steiner, A.L., Ferrendelli, J.A., and Kipnis, D.M. 1972. Radioimmunoassay for cyclic nucleotides. Effect of ischemia, changes during development and regional distribution of cyclic AMP and cyclic GMP in mouse brain. J Biol Chem 247:1121–1124

Steiner, A.L., Wehmann, R.E., Parker, C.W., and Kipnis, D.M. 1972. Radioimmunoassay for the measurement of cyclic nucleotides. In *Advances in Cyclic Nucleotide Research,* Vol. 2, eds. P. Greengard, and G.A. Robison, pp. 51–61. New York: Raven Press

Stone, T.W., Taylor, D.A., and Bloom, F.E. 1975. Cyclic AMP and cyclic GMP may mediate opposite neuronal responses in the rat cerebral cortex. Science 187:845–846

Swaab, D.F., and Boer, K. 1972. The presence of biologically labile compounds during ischemia and their relationship to the EEG in rat cerebral cortex and hypothalamus. J Neurochem 19:2843–2853

Welch, K.M.A., Chabi, E., and Buckingham, J. 1975. The effect of ischemia on catecholamine and 5-hydroxtryptamine levels in the cerebral cortex of gerbils. Transactions of the American Society for Neurochemistry. 6(Abs):157

Microwave Inactivation as a Tool for Studying the Neuropharmacology of Cyclic Nucleotides

David J. Jones and W. B. Stavinoha
Departments of Anesthesiology and Pharmacology
The University of Texas Health Science Center
San Antonio, Texas 78284

INTRODUCTION

The *in vivo* measurement of labile metabolites in the central nervous system (CNS) requires that enzyme activity be arrested as rapidly as possible following occlusion of the cerebral circulation. Such rapid enzyme inactivation insures that levels of labile compounds are measured which closely reflect endogenous concentrations immediately prior to sacrifice. In the present work we will discuss factors unique to CNS cyclic nucleotide systems which necessitate rapid and uniform enzyme inactivation for measurement of correct values. In addition, the advantages and disadvantages of utilizing various enzyme inactivation techniques will be presented, with major emphasis placed on microwave techniques.

POSTMORTEM CHANGES IN BRAIN CYCLIC NUCLEOTIDE LEVELS

The first study which established a post-decapitation rise in cyclic AMP levels in brain tissue was carried out by *Breckenridge* in 1964. The rapid conversion of phosphorylase *b* to *a* in anoxic brain (*Breckenridge* and *Norman,* 1962) and the relationships of cyclic AMP to the phosphorylase system were the bases for such studies. In these studies cyclic AMP was 80% higher in cortex 20 sec following decapitation compared to non-decapitated controls.

Subsequent studies by *Steiner, Ferrendelli,* and *Kipnis* (1972) and *Uzunov* and *Weiss* (1971) established that this postmortem increase

---

Supported in part by NIMH grant #ROI MH 25168-03, GRSB grant #507 RR 05654-06. Sincere thanks are extened to microwave engineer, A. P. Deam, for his unselfish contribution to the development and construction of the microwave exposure units. Also, the authors thank M. Webb, J. MacDougal and A. W. Guy, University of Washington at Seattle, for the use of their Thermovision camera for photographic reproduction.

in cyclic AMP content occurs at unequal rates in regional brain areas. As shown in Figure 1, the kinetics of the post-decapitation increase in cyclic AMP are different relative to the particular brain area. The rise and subsequent fall in cyclic AMP content in cerebellar tissue are much more rapid than in the cerebral cortex. Such a relationship also has been demonstrated by *Schmidt, Schmidt,* and *Robison* (1971) and *Uzunov* and *Weiss* (1971), with additional differences demonstrated to be present in other areas of brain.

Recent studies in our own laboratories have investigated postmortem changes in cyclic AMP following induction of anoxia by paralysis of respiration (*Jones* and *Stavinoha,* 1976b.) As shown in Figure 2, the rate of increase in cyclic AMP is fairly constant in all brain areas for the first 30 sec; however, levels differ significantly after 120 sec of anoxia. Interestingly, the levels and rates of change in cerebral cortex vs cerebellum during anoxia are the same. This is in contrast to the data of *Steiner et al.* (1972) and *Uzunov* and *Weiss* (1971) which showed significantly different rates of change. Moreover, in the work of *Steiner et al.*

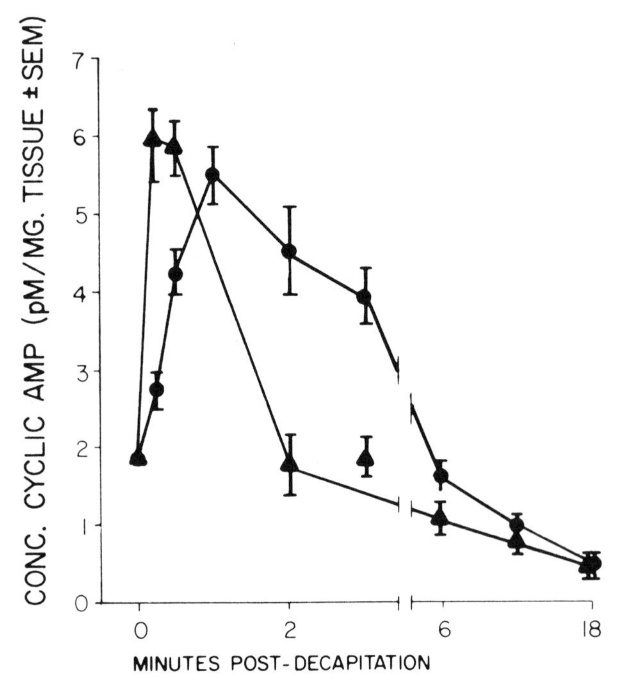

Fig. 1. Levels of cyclic AMP in mouse cerebral cortex (▲) and cerebellum (●) at various times following decapitation. Animals were decapitated and the heads frozen at the decapitated intervals in Freon (−150 °C). Controls were decapitated directly into Freon. Each value represents the mean ± SEM of 3 to 7 animals. Redrawn from *Steiner et al.* (1972), with permission of author and publisher.

(1972) levels of cyclic AMP were back to normal in the cerebellum 2 min post-decapitation. Presumably the ascending spinal shock resulting from decapitation produces disproportionate electrical and/or metabolic activity in the cerebellum, which is reflected in the rapid postmortem cyclic AMP changes. This pathway disruption is not present in the anoxia model.

The fact that postmortem changes in brain cyclic AMP are dependent upon the method of sacrifice is reflected in studies which demonstrated significantly higher levels of cyclic AMP in brain from decapitated animals vs nondecapitated controls (*Jones et al.,* 1974). The level of brain cyclic AMP from decapitated animals was 60% higher than in nondecapitated controls. In both of these groups liquid nitrogen freezing was used to inactivate enzymes and, since the freezing rates are the same in each case, it is apparent that the stress and anoxia associated with decapitation shift the concentration of cyclic AMP measured away from true endogenous levels.

Although there appears to be a relatively slower rate process controlling the postmortem alterations in cyclic GMP, both quantitative and qualitative differences in mouse regional brain areas are evident following decapitation (Figs. 2 and 3). Using respiratory arrest for production of anoxia, the regional variations in postmortem decrease in cyclic GMP are more evident (Fig. 2). In agreement with the studies of *Steiner et al.* (1972) and *Goldberg et al.* (1970), levels are unchanged at 30 sec and decreased at 120 sec following circulatory arrest.

In contrast to the above studies, *Kimura, Thomas,* and *Murad* (1974) reported that 30 sec following decapitation, levels of cyclic GMP in rat cerebellum were increased 2.7-fold and in cerebral cortex 1.6-fold. In consideration of this data, these researchers suggested that such variability in levels of cyclic GMP measured during the postmortem period appears due to unknown factors which arise during surgical removal of tissue prior to freezing. The fact that variable alterations in levels of cyclic nucleotides have been noted following occlusion of the cerebral circulation emphasizes the need for maintaining consistency from animal to animal in manipulation of the tissue prior to enzyme inactivation.

A major problem which results from the postmortem alterations in levels of CNS cyclic nucleotides involves the interpretation of data from pharmacological studies which involve intact animals. This problem occurs based on *in vitro* studies which have demonstrated that various brain regions respond differently to the effects of neurotransmitters relative to stimulating the formation of cyclic AMP (for review, see *Daly,* 1976) and cyclic GMP (*Ferrendelli, Kinscherf,* and *Chang,* 1973; *Kinscherf et al.,* 1976). In order to evalute the *in vivo* sensitivity of cyclic nucleotide systems to various pharmacological agents, the effects of which may be mediated by a neurotransmitter, one must be prepared to deal with

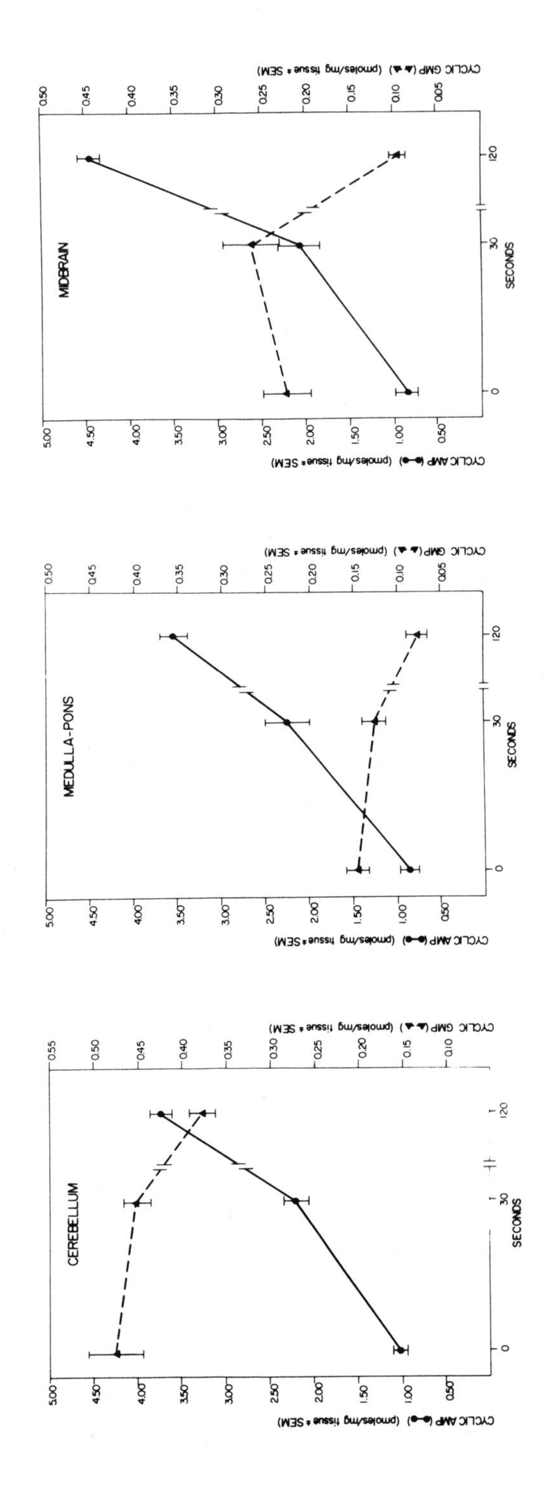
MIDBRAIN
MEDULLA-PONS
CEREBELLUM
SECONDS
0
30
120
CYCLIC AMP (●—●) (pmoles/mg tissue ± SEM)
CYCLIC GMP (▲-▲) (pmoles/mg tissue ± SEM)
5.00
4.50
4.00
3.50
3.00
2.50
2.00
1.50
1.00
0.50
0.55
0.50
0.45
0.40
0.35
0.30
0.25
0.20
0.15
0.10
0.05

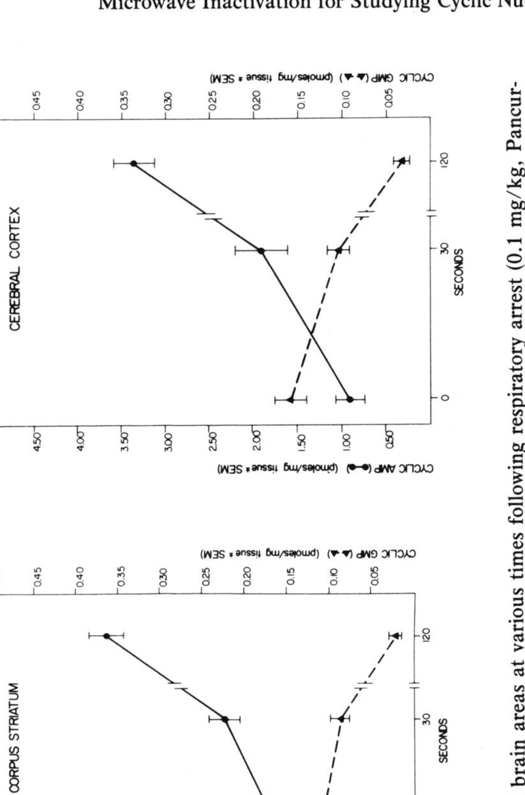

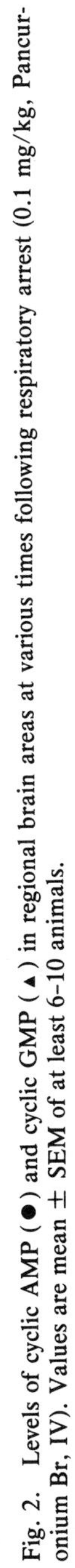

Fig. 2. Levels of cyclic AMP (●) and cyclic GMP (▲) in regional brain areas at various times following respiratory arrest (0.1 mg/kg, Pancuronium Br, IV). Values are mean ± SEM of at least 6–10 animals.

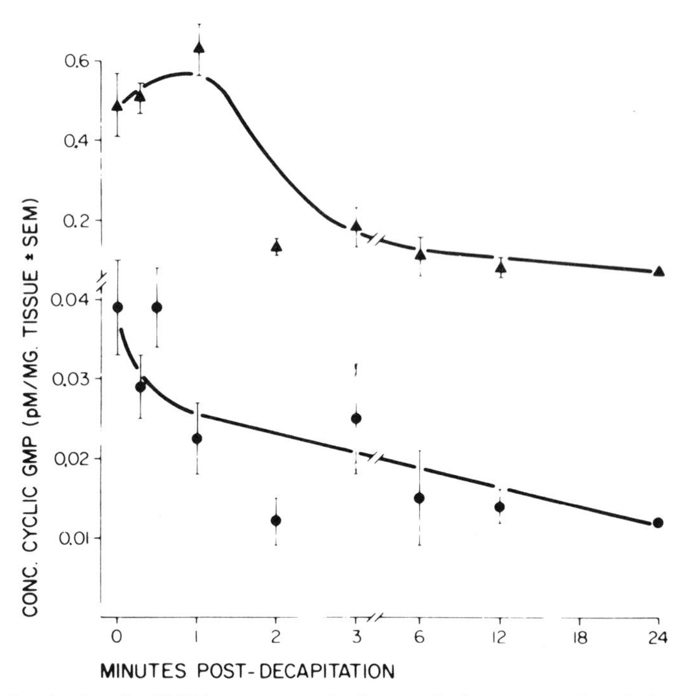

Fig. 3. Levels of cyclic GMP in mouse cerebral cortex (●) and cerebellum (▲) at various times following decapitation. See Figure 1 for details. Redrawn from *Steiner et al.* (1972), with permission of author and publisher.

the possible artifactual alteration of the measured level based on the postmortem changes induced by anoxia. Thus, the response of cyclic nucleotide systems to biologically active agents could be altered during the postmortem period such that 1) no change is detected, 2) a change is detected but not correct in value, or 3) a disproportionate change is measured relative to other responding brain areas.

If one is concerned about the differential sensitivity of the cyclic AMP system vs cyclic GMP system, the necessity for minimization of postmortem anoxia becomes more critical. Both cyclic nucleotides change by different amounts depending upon the brain area and the time of enzyme inactivation following sacrifice. Therefore, one has to consider such variables when evaluating the validity of regional brain data which attempt to characterize the pharmacological responsivity of CNS cyclic nucleotide systems.

Because of the neurochemical complexity of the CNS, it is not possible to establish a single mediator of the postmortem alterations in cyclic nucleotides. Also, it is not known whether the levels are changed due to an increase or decrease in the activity of synthetic or degradative enzymes of brain cyclic AMP and cyclic GMP. Nevertheless, it is imperative that enzymes known to be active in the cyclic nucleotide systems be inactivated as rapidly and uniformly as possible following occlusion of the cerebral circulation.

IN VIVO ENZYME INACTIVATION TECHNIQUES

Freezing

Prior to the advent of microwave inactivation techniques, the use of freezing liquids was the primary method employed for the arrest of brain enzyme activity *in situ.* While freezing does provide for the salvage of certain enzymes, there are a number of limitations inherent in the technique.

The major limitation with the rapid freezing technique is thermal diffusion, which is represented by the time required for the freezing front to pass completely through the tissue. *Takahashi* and *Aprison* (1964) and *Swaab* (1971) studied intact rat brain freezing rates in liquid nitrogen and Freon 12, respectively. With thermistors implanted in superficial cortex and deep thalamic structures, it was established that superficial structures cooled to 0 °C within 25–40 sec whereas deep structures required up to 85 sec. In these studies large variances were noted in the time requirements for freezing in each of the rat brain areas.

The same regional differential in freezing rates is also evident in the smaller volume mouse brain (*Ferrendelli et al.*, 1972; *Swaab,* 1971). In either liquid nitrogen or Freon 12, cerebral cortex required 2–8 sec and hypothalamus 14–32 sec to reach 0 °C. Previous studies by *Weiner* (1961) and *Jongkind* and *Bruntink* (1970) had validated the relationship between the smaller tissue sample, faster freezing and more representative values for labile intermediary metabolites. Moreover, studies by *Swaab* (1971) and *Ferrendelli et al.* (1972) established that levels of lactate, an indicator of anoxic changes, were lower in cerebral cortex than in slower freezing hypothalamus.

The lack of uniformity in freezing of intact brain tissue is of major concern if regional studies are to be carried out. The alterations of brain metabolism or cyclic nucleotide systems can be monitored only in superficial structures that freeze rapidly. Thus, changes in deeper structures would have to be considered to be imposed on top of the possible changes that occur during the time period between sacrifice and enzyme inactivation. As in the above data, the implication is that levels of lactate in regional brain are more representative of the sequence of enzyme inactivation rather than true endogenous levels.

An additional limitation in using quick freezing techniques is the requirement for maintaining the tissue in the frozen state prior to extraction or dissection. If during any time following freezing the brain is allowed to thaw, enzymatic activity will resume with subsequent alterations in substrate levels. Moreover, the requirement of maintaining the tissue in the frozen state makes dissection into discrete areas extremely difficult.

A number of techniques have been developed in attempts to reduce the time required for brain enzyme inactivation *in situ.* While the operative techniques of *Ponten et al.* (1973) provide for rapid freezing of superficial brain tissue, exposure of the brain can produce foci of spreading depression (*Thorn et al.,* 1958) and the use of anesthesia imposes further complications in the interpretation of metabolite data (*Stone,* 1935; *Lowry et al.,* 1964; *MacMillan* and *Siesjo,* 1973; *Michenfelder* and *Milde,* 1975). Moreover, thermal diffusion of the freezing gradient still limits the study of deeper structures.

Veech et al. (1973) introduced a technique whereby brain tissue is ejected under 25 lb/in^2 of air pressure through a hollow steel probe onto two aluminum discs previously cooled to the temperature of liquid nitrogen. The whole cortex and midbrain of the rat from the olfactory bulbs to the superior colliculi is expelled and frozen within 1 sec. The levels of intermediary metabolites and redox values of pyridine nucleotides from such tissue are much higher than those previously obtained using conventional quick freezing. *Nahorski* and *Rogers* (1973) have used this apparatus for mice and have studied brain tissue of 4- to 6-day-old chicks. In each species the time for inactivation was 1 sec.

Although rapid, the use of the "freeze-blowing" technique has several limitations. Of major concern is the loss of anatomical and structural features which subsequently precludes the study of metabolite concentrations in regional brain areas. Moreover, as with any freezing technique, one has to be concerned with possible thawing of tissue and resumption of enzyme activity. In addition, the rodent or chick is stressed by the restraining methods used to align the animal in the stereotaxic apparatus prior to sacrifice. Stress markedly alters cyclic nucleotide levels in rat (*DeLapaz, Dickman,* and *Grosser,* 1975) and mouse (*Dinnendahl,* 1975) brain such that subsequent measurements of cyclic AMP and cyclic GMP, regardless of speed of enzyme inactivation, are artifactual.

Heating

The *in vitro* denaturation of enzymes is usually achieved by placing the reaction vessel in boiling water for 3–10 min. For *in vivo* denaturation of enzymes such heating would be extremely inefficient and limited, as in freezing, by thermal diffusion. Moreover, the technique would be limited by trauma to the animal and the slow heating could produce an increase in the reaction velocity of enzyme-catalyzed reactions.

Stavinoha, Pepelko, and *Smith* (1970), realizing the apparent advantages of enzyme inactivation by heating and disadvantages of conductive heating for *in vivo* studies, used microwave irradiation to rapidly heat brain tissue prior to measuring acetylcholine levels. Microwave heating is a special type of radiation heating which depends on the propagation of electromagnetic energy into the animal brain.

The major energy conversion mechanisms involved in heating with microwaves, ionic conduction and dipole rotation, are a result of oscillating electric and magnetic fields (*McLees* and *Finch,* 1973). In ionic conduction ions are caused to migrate under the influence of the microwave electric field. Then through the collisions which occur thousands of times in every cycle of the microwave electric field, the carriers give up their induced ordered kinetic energy to random kinetic energy. In dipole rotation, when an electric field is present, dipoles such as water molecules tend to align with the field. This alignment imposes order on the system and energy is stored. When the electrical field is removed, the dipoles randomize and heat is produced. Thus, water molecules rotate in the imposed field at the frequency of the source.

The efficiency of the heat production is dependent upon the power density in the tissue, duration of exposure, frequency or wavelength of the radiation, size and dimensions of the exposed object, thermal regulatory capacity of the exposed object, thickness of tissue, composition of tissue and the heat transfer properties of the tissue (*Michaelson,* 1969). In using microwave radiation for inactivation of enzymes, these variables have been considered in the design of exposure units for this purpose in our laboratories.

The power density in the tissue is dependent upon the power output of the source and "coupling" of the electromagnetic energy to the rodent brain. The coupling efficiency is a function of position, size and method of coupling. Early studies in this laboratory used a commercially available 1.5 kW oven and involved whole-body irradiation (*Stavinoha et al.,* 1970). Eleven seconds were required for the production of 85–90 °C in rat brain. An improvement in coupling efficiency was accomplished by orientation of the rodent's head into the waveguide of this same oven (*Stavinoha, Modak,* and *Weintraub,* 1972). Temperatures for inactivation of acetylcholinesterase were reached within 6 sec (*Stavinoha, Weintraub,* and *Modak,* 1973). As might be expected with a smaller load volume, similar temperatures were reached in the mouse brain within 3 sec.

A 5.5 kW instrument was subsequently developed utilizing a different coupling geometry. In this sytem the load (animal head) is coupled to the energy source through a 3-port ferrite circulator and microwave E-H tuner. This matching technique permitted a time reduction to 2 sec for rat brain (*Jones et al.,* 1973) and 0.4 sec for mouse brain (*Medina et al.,* 1975).

The final model which is currently used in our laboratories couples 7.3 kW directly to the rodent, with a large majority going to the head. As shown in Figure 4, this unit utilizes a 4-port circulator, directional coupler and an E-H tuner to maximize the power density in the mouse brain. This power density is estimated to be 250–400 W/cm^2. The efficiency of the energy packet in delivering the power to the magnetron is shown in Figure 5, with approximately 0.5 msec required for the rise to full power.

As shown in Figure 4, the energy packet is used independently to drive a 2450 MHz microwave source and a 915 MHz source. Because of the interrelationship of the variables which contribute to microwave heating discussed above, each waveguide unit is designed in the particular configuration which most efficiently couples the source to the load. Thus, the 2450 MHz is used for the mouse and the 915 MHz source for rats and guinea pigs. The shorter wavelength 915 MHz radiation requires the use of a much larger waveguide system to couple efficiently. As shown

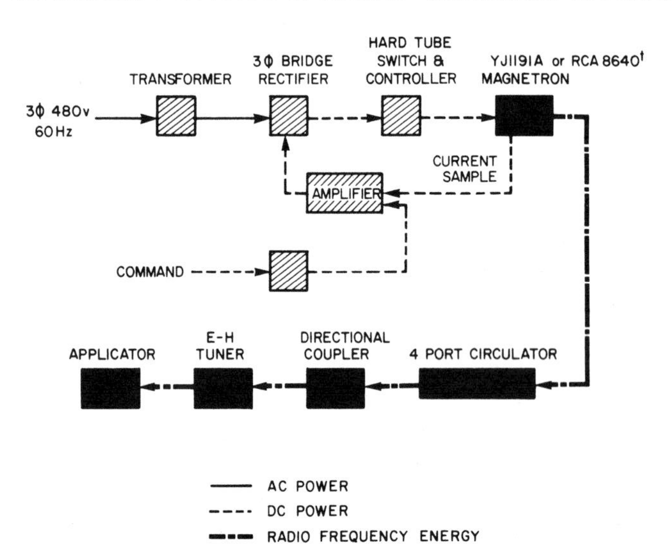

Fig. 4. Schematic of 7.3 kW, 2450 MHz microwave exposure unit. The 25 kW, 915 MHz source uses the same transformer-amplifier unit in its power packet and is included in the description.

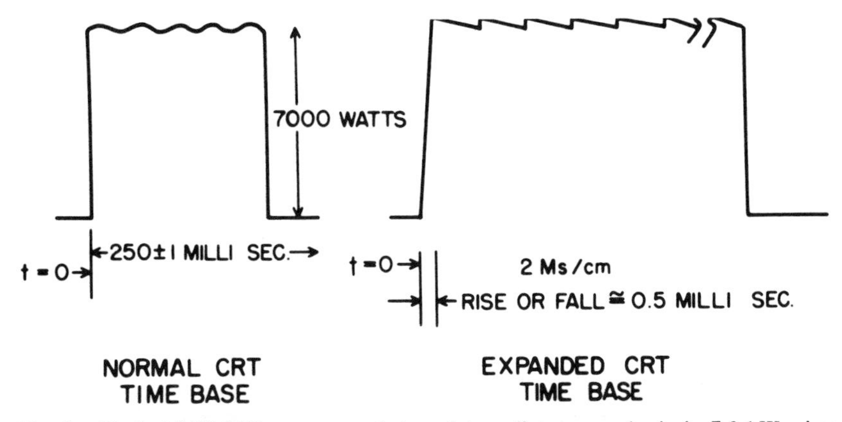

Fig. 5. Typical 2450 MHz energy packet used to radiate mouse brain in 7.3 kW microwave unit. From *Stavinoha, Frazer,* and *Modak* (1977), with permission of author and publisher.

in the photographs in Figure 6, the dimensions of the wavelength systems differ markedly.

Aside from achieving high density in the brain of mice, the internal field distribution is a major factor in ensuring the uniform inactivation of tissue in all brain areas. The patterns of the fields producing the heating are complex functions of the frequency, source configuration, tissue geometry and dielectric properties of the tissues. Reflections between interfaces separating tissues of high and low water content (e.g., brain and bone, respectively) can produce severe standing waves accompanied by "hot spots" that can be maximum in either tissue, regardless of dielectric constant or conductivity. Such unequal internal field distribution can result in brain being marginally heated in one area, while a second area "boils" and is expelled through the olfactory sinus or foramen magnum.

The field distribution can be estimated as temperature by utilizing thermographic techniques. Such methodology has been used to determine the field intensity in mouse brain following heating via the 7.3 kW instrument. The thermograph in Figure 7 demonstrates that heat production with 2450 MHz in the mouse brain occurs primarily in the brain, attesting to the effectiveness of power deposition. However, as shown in Figure 8, while the most efficient heating occurs in the central area of the brain, the anterior and posterior sections reach 82% of this value and the medullary area 58%. The maximum caloric deposition is estimated to be 250–500 cal/gm/sec at the center. The relative temperature rise contours, as viewed in a coronal section, are shown in Figure 9. It is evident that the ventral section of the brain heats much less efficiently than the dorsal section.

Fig. 6. Photographs of the power supply and attached 2450 MHz unit (A) with oscilloscope (B) for monitoring forward power. E-H tuner (C) and exposure cell for mice (D) for 2450 MHz are exteriorized from the waveguide system (not shown). 915 MHz source (E on right) uses the same power supply (A) as the 2450 MHz unit. Note the size relationship of the E-H tuner (F) for the 915 MHz unit vs the 2450 MHz unit (C). See schematic in Fig. 4 for details.

One might expect that temperatures in the anterior, posterior and ventral aspects of the brain at the end of heating would be insufficient for enzyme inactivation. In fact, experience with less efficient units indicates that the corpus striatum, situated in the anteroventral area of the brain, is the most difficult to inactivate. However, the key factor is being able to accurately overheat the central core area while the peripheral sites are heated to the required temperature for inactivation. In our laboratories, a 24 gauge needle copper-constantan thermocouple is used to measure regional mouse brain temperatures immediately following microwave exposure. In the 7.3 kW unit anterior-posterior temperatures of at least 80 °C are minimum and 250–300 msec are required.

It should be pointed out that the spatial distribution of caloric input as represented by the thermographic studies is typical only for the 7.3 kW exposure cell used presently in our laboratories. The geometry of this cell allows input of energy to the head of the animal through exchange with the magnetic field in the waveguide. Moreover, this instrument is designed specifically for mice, as is the 915 MHz source for rats and guinea pigs. Thus, conditions for optimization of brain

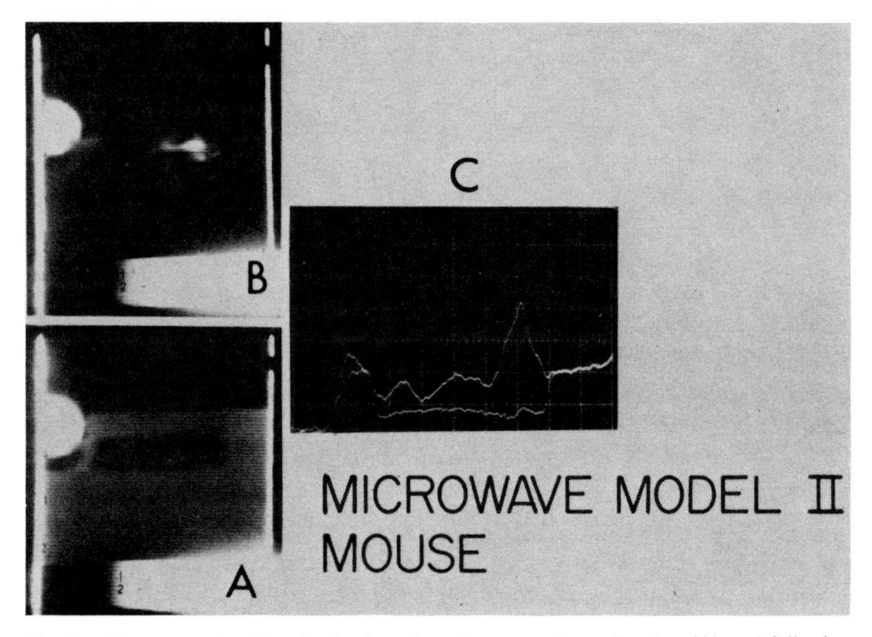

Fig. 7. Thermograph of longitudinal section of mouse prior to heating (A) and following heating (B). Lower tracing on picture on right (C) represents mouse before heating and upper tracing following heating. Each verticle line in C is 0.5 cm. Mouse nose is approximately 1 cm in from right-hand margin.

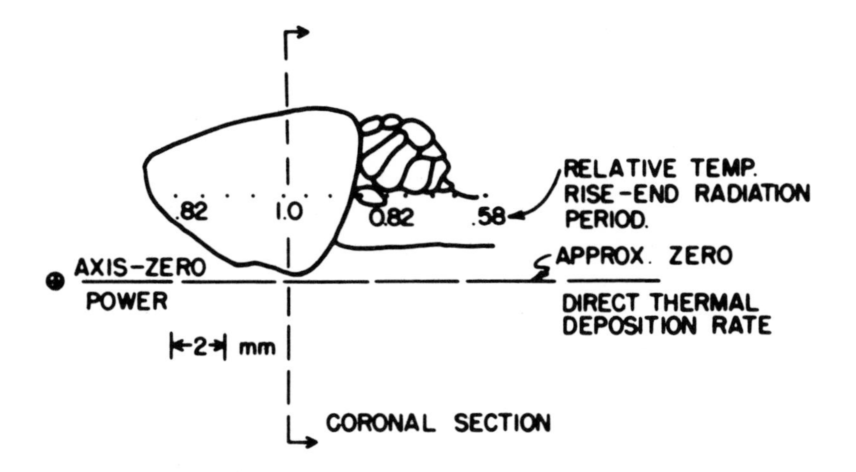

Fig. 8. Relative temperature rise at end of radiation period longitudinally along mouse brain approximately 30 sec following 2450 MHz exposure. The distribution is estimated from thermographs similar to Figure 7. From *Stavinoha, Frazer,* and *Modak* (1978), with permission of author and publisher.

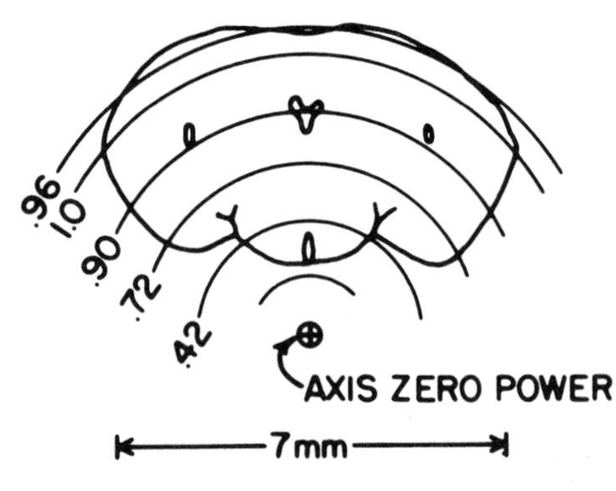

Fig. 9. Relative temperature rise contours in coronal section at end of exposure time. See Figure 8 for details. From *Stavinoha, Frazer,* and *Modak* (1978), with permission of author and publisher.

heating have been approached as nearly as possible and fruitful results are evident. One *cannot* expect such results if the speed and uniformity of heating and subsequent enzyme inactivation are not measured.

Comparison of Techniques

It is evident from the previous discussion that microwave inactivation offers a number of advantages over conventional freezing techniques. Various studies have been reported which compare this method with the rapid freezing methods. *Veech et al.* (1973) and *Lust, Passonneau,* and *Veech* (1973) compared the levels of labile intermediary metabolites and redox states in brains from "freeze-blown" rats vs brains from microwaved rats. While the brains from freeze-blown animals were inactivated within 1 sec, 14 sec were required for inactivation in the microwaved group. The microwave used in these studies was the 1.5 kW source located in our laboratories. As would be expected, compared to freeze-blown tissue the levels of metabolites and redox states in these microwaved brains were indicative of postmortem anoxia evident during 14 sec of microwave exposure.

A more valid comparison can be made from subsequent studies utilizing a 5.0 kW microwave unit which inactivated rat brain tissue within 0.6 sec (*Medina* and *Stavinoha,* 1977). Values for labile intermediary metabolites in the microwaved tissue were the same as in tissue from freeze-blown brains (*Veech et al.,* 1973). In addition to preserving the apparent *in vivo* concentration of these metabolites, microwave inactivation permitted the measurement of regional brain concentrations.

The levels of cyclic AMP in rat brain also have been compared following freeze blowing or microwave inactivation (*Lust et al.,* 1973;

Guidotti et al., 1974). The brain level of cyclic AMP measured following the use of either technique was not different. However, though cyclic AMP levels are comparable, *Lust et al.* (1973), based on levels of labile metabolites which were indicative of anoxia, indicated that freeze blowing was preferable to microwave inactivation. In these studies, the microwaved brain samples were the same as those of *Veech et al.* (1973) which were inactivated with 1.5 kW microwave irradiation for 14 sec. With the availability of more powerful units which permit measurement of comparable levels of labile metabolites, the preference of *Lust et al.* (1973) is no longer valid. This fact is evident considering the data of *Guidotti et al.* (1974), who utilized 4.5 kW, 2 sec microwave inactivation. The levels of cyclic nucleotides and labile metabolites were exactly the same in these studies, regardless of freeze blowing or microwave inactivation. Studies from our own laboratories using the 5.0 kW unit also validate this fact (*Jones et al.,* 1973, 1974).

Most studies with frozen tissue measure cyclic AMP in whole brain. As mentioned previously, microwave inactivation, through irreversible denaturation of enzymes, greatly facilitates dissection of brain tissue into regional areas. The tissue maintains its structure and the brain is easily separated along anatomical lines. If one is to dissect a frozen brain, tissue must be chipped or sliced on a cryostat. Not only is it difficult to maintain the tissue in the frozen state, it is also difficult to ensure accurate dissection of tissue into discrete regional areas.

An alternative would be the dissection of tissue prior to freezing. However, as shown in Table 1, cyclic AMP levels are markedly altered when compared to microwaved brains. Concentrations in liquid N_2 inactivated brains not only are higher than microwaved brains but also

Table 1. Comparison of Cyclic AMP Levels in Regional Brain Areas from Rats Sacrificed by 5 kW Microwave Radiation (MWR) or Decapitation with Subsequent Dissection and Freezing in Liquid N_2

	Cyclic AMP (nM/g ± SEM)	
Brain Region	MWR Inactivation	Liquid N_2 Inactivation
Cerebellum	0.68 ± 0.12	4.8 ± 0.2
Diencephalon	0.51 ± 0.02	2.2 ± 0.2
Midbrain	0.44 ± 0.04	1.4 ± 0.04
Medulla-pons	0.44 ± 0.08	1.1 ± 0.2
Corpus striatum	0.43 ± 0.04	1.4 ± 0.2
Cerebral cortex	0.59 ± 0.08	0.89 ± 0.11

Freezing data from *Ebadi, Weiss,* and *Costa* (1971), with permission from author and publisher.

the delays in freezing lead to marked regional variations in cyclic AMP content (*Ebadi, Weiss,* and *Costa,* 1971). In consideration of the marked increase in cerebellar cyclic AMP following decapitation (*Steiner et al.,* 1972; *Uzunov* and *Weiss*, 1971), it is apparent that these regional concentrations reflect more the sacrifice and inactivation procedure than true endogenous levels.

Microwave inactivation techniques are not error free with regard to regional brain studies. As pointed out earlier, the major difficulty with any microwave exposure system is producing a uniform field distribution in the rodent brain. The use of microwave exposure units which do not optimize the interaction between the source and load provides for nonuniform distribution of heat and subsequent variable inactivation of enzymes. Such a condition, as in freezing, would lead to the measurement of cyclic nucleotide levels that are representative of marked variations in regional concentrations. Apparently this is the case in the microwave studies of *Schmidt et al.* (1971), *Wellmann* and *Schwabe* (1973), and *Redos, Hunt,* and *Catravas* (1976) where significant differences are present in rat cortex vs cerebellum. The latter study utilized 1.5 kW, 3.5 sec microwave exposure. The levels of cyclic AMP in cerebellum were greater than twice the value in cortex. Such differences apparently are due to the lack of uniform enzyme inactivation. Under the conditions of microwaving used by these investigators, one might expect that enzymes in the cerebellum were only partially inactivated by the initial exposure, with subsequent heat diffusion from adjacent areas required to complete the process.

The fact that the methods of sacrifice disproportionately alter regional levels of cyclic AMP is evident from Table 2. Using microwave irradiation to sacrifice the rat, with the only variable being time for inactivation of enzymes, the regional levels of cyclic AMP are uniformly increased in

Table 2. Concentration of Cyclic AMP in Regional Areas of Rat Brain following 1.5 kW and 5 kW MWR Inactivation

	Cyclic AMP (nM/g WW ± SEM)	
Regional Area	1.5 kW Oven (8 sec)	5 kW Oven (2 sec)
Cerebellum (n = 6)	0.94 ± 0.10	0.68 ± 0.12
Medulla-pons (n = 6)	0.66 ± 0.07	0.44 ± 0.04
Midbrain (n = 6)	0.63 ± 0.06	0.44 ± 0.05
Diencephalon (n = 6)	0.74 ± 0.04	0.51 ± 0.02
Corpus striatum (n = 6)	0.59 ± 0.04	0.43 ± 0.04
Cerebral cortex (n = 6)	0.84 ± 0.08	0.59 ± 0.08

From *Jones et al.* (1974), with permission from author and publisher.

the 1.5 kW-exposure animals. Moreover, the marked regional variations in concentrations evident in the frozen brains are not present in either group of microwaved animals. Thus, with these two microwave exposure units, gradients of heating do not occur which cause uneven heating of tissue and subsequent variation in regional levels. With appropriate concern for optimizing the field distribution in the brain, profitable regional studies are permitted.

An important consideration in the utilization of microwave inactivation units is the marked lack of uniformity in performance standards. A comparison of the works of *Volicer* and *Hurter* (1977) and *Redos et al.* (1976) points out the laboratory-to-laboratory variability in optimizing heating conditions. Both studies make reference to the use of a modified microwave oven (Litton Menumaster) designed to focus microwaves at the head (*Guidotti et al.*, 1974). In the study of *Redos et al.* (1976) control cyclic AMP levels varied from 3.4 pmoles/mg protein in cerebellum to 7.2 pmoles/mg protein in cerebral cortex. Values for cerebellum and cerebral cortex in the studies of *Volicer* and *Hurter* (1977) were 1.25 and 1.06 pmoles/mg tissue, respectively. The point to be made is that the studies of *Redos et al.* (1976) indicate regional variations in endogenous cyclic AMP content, whereas those of *Volicer* and *Hurter* (1977) do not. Thus, even though the same microwave inactivation system was used in these studies (*Guidotti et al.*, 1974), the characteristics of uniform enzyme inactivation differed.

For our purposes, microwave radiation is the preferred method for the inactivation of enzymes in brain of the intact rodent. The method is relatively free from artifacts and provides one with an easily manageable tissue sample. Moreover, processing of the tissue following microwave inactivation is greatly facilitated. The addition of an appropriate amount of sodium acetate buffer (50 mM, pH 6.2), sonication, centrifugation and sampling of supernatant provide for a system without steps for loss of cyclic nucleotide or contamination by other substances.

MICROWAVE INACTIVATION AND THE MEASUREMENT OF CNS CYCLIC NUCLEOTIDES

Following the introduction in 1970 by *Stavinoha et al.* of microwave heating to produce brain enzyme inactivation, *Schmidt* and *Robison* (1971) utilized whole-body, 1.25-kW microwave exposure to inactivate adenylate cyclase and phosphodiesterase in rat brain prior to the measurement of cyclic AMP in regional areas. Although certain advantages were gained from the use of microwave inactivation, 20–30 sec were required for enzyme inactivation and levels of cyclic AMP reflected anoxic postmortem changes similar to those evident following rapid freezing (*Schmidt, Schmidt,* and *Robison,* 1971).

Subsequent studies in our own laboratories evaluated more completely the use of microwave irradiation for the study of CNS cyclic nucleotide systems (*Jones,* 1974). A series of experiments were completed which validated the utility of this technique for the inactivation of adenylate and guanylate cyclase and phosphodiesterase in rodent brain. In these studies the heat and microwave stability of cyclic AMP was validated, as well as the lack of nonenzymatic formation of cyclic AMP from ATP during microwave heating. Also confirmed was the lack of effects of microwave heating on protein or water content of brain (*Stavinoha, Weintraub,* and *Modak*, 1973; *Medina et al.,* 1975).

In studies designed to assess the effects of time for inactivation on measured levels of rat brain cyclic AMP, it was determined that the rate of brain enzyme inactivation was a major factor in the measurement of accurate, predictory *in vivo* levels of cyclic AMP (*Jones et al.,* 1974. As shown in Table 3, freezing as well as microwave data reflect this fact. Delays in freezing produce higher cyclic AMP values, and as pointed out earlier, the stress of decapitation also contributes to higher levels. Moreover, a comparison of 1.5 kW, 8 sec exposure vs 5.0 kW, 2 sec exposure indicates significantly lower levels are evident when the time for inactivation is lowered. Figure 10 represents the time vs brain temperature relationship and indicates that the above times for inactivation correlate with rat brain temperatures of 85–90 °C.

The validity of measured levels of cyclic AMP following microwave inactivation is assured by studies which demonstrate the relationship between time of heating and loss of enzyme activity. As shown in Figure 11, adenylate cyclase and phosphodiesterase are inactivated 100% within the time periods of exposure used and the temperature at this time is

Table 3. Concentration of Cyclic AMP in Rat Brain Following Various Methods of Inactivation

Fixation Methods	Time Required for Inactivation (sec)	Cyclic AMP (nM/g WW ± SEM)
Decapitation, removal of brain and freezing in liquid N_2 (n = 5)	60+	3.82 ± 0.72
Decapitation into liquid N_2 (n = 6)	30+	2.74 ± 0.17
Immersion into liquid N_2 (n = 6)	30+	1.69 ± 0.17
MWR 1.5 kW oven (n = 11)	8	0.81 ± 0.07
MWR 5 kW oven (n = 6)	2	0.54 ± 0.05

From *Jones et al.* (1974), with permission of author and publisher.

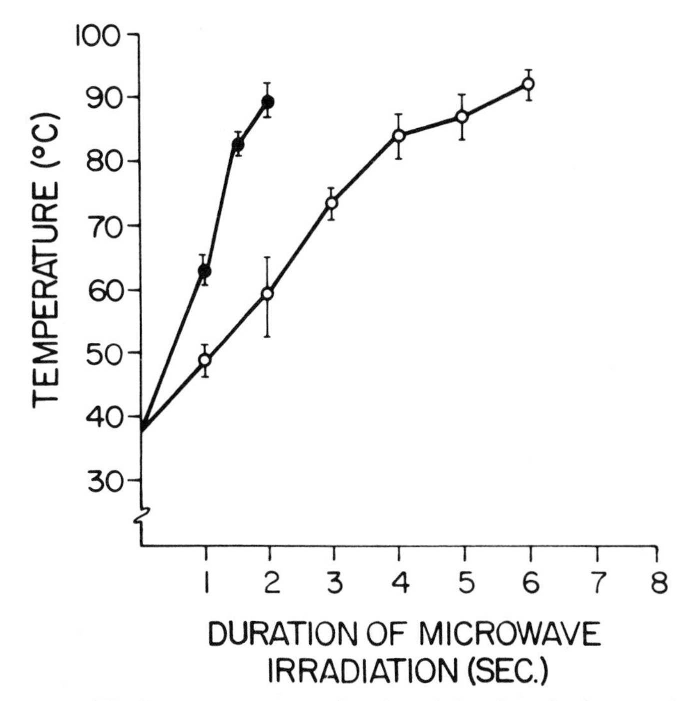

Fig. 10. Rat midbrain temperature as a function of duration of microwave heating in 1.5 kW unit (o) and 5 kW unit (●). Each value is the mean ± SEM of 8–10 determinations.

85–90°C (Fig. 10). Such data confirm the correlation between faster heating, enzyme inactivation and lowered measured levels of cyclic AMP in rat brain.

Subsequent studies measured the levels of cyclic AMP in regional rat brain areas (*Jones et al.,* 1974). Consistent with the whole brain data of Table 2, the relationship of lower measured levels of cyclic AMP with faster inactivation was also evident in regional brain tissue from 8 sec vs 2 sec microwaved groups. Moreover, the uniformity of enzyme inactivation was reflected in the lack of region-specific increases in cyclic AMP in brain areas (such as the cerebellum) following 8 sec inactivation.

The above studies were carried out utilizing rat brain tissue which in the 5 kW oven requires up to 2 sec for inactivation. If one decreases the load volume, a reduction in time of heating can be achieved. Moreover, utilizing 2450 MHz heating, the geometry of a smaller load volume, e.g., the mouse brain, permits more efficient coupling and more uniform distribution of the field intensity. Using the 7.3 kW unit, which is designed specifically for coupling the radiation to the mouse brain, temperatures of 85–90°C are produced within 250–300 msec.

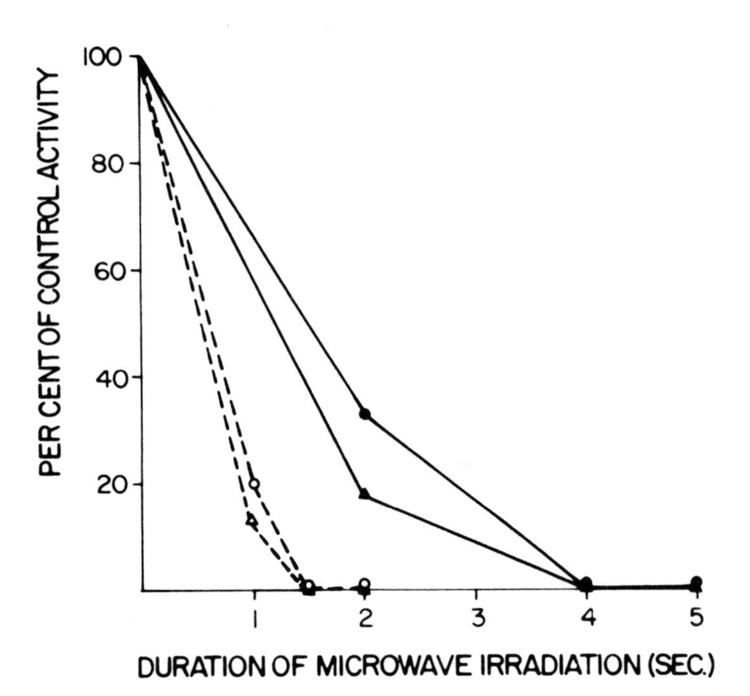

Fig. 11. Activity of rat brain adenylate cyclase as a function of duration of exposure to 1.5 kW (▲) and 5 kW (△) microwave radiation, and phosphodiesterase following 1.5 kW (●) and 5 kW (○) microwave radiation. Mean ± SEM of four rat brains for each point. From *Jones et al.* (1974), with permission of author and publisher.

As shown in Figures 12–14, 300 msec are sufficient for 100% inactivation of mouse brain adenylate cyclase, guanylate cyclase and phosphodiesterase (*Jones and Stavinoha,* 1977). More importantly, the enzymes are inactivated uniformly in dorsal, medial and ventral sections of the brain within the same time base. With such inactivation parameters, one is assured that the regional levels of cyclic nucleotides measured following 300 msec exposure are artifact free, providing an accurate base from which to study the pharmacology of CNS cyclic nucleotides (*Jones* and *Stavinoha,* 1976a; *Palmer et al.*, 1977).

Although the 7.3 kW exposure unit inactivated brain tissue within 300 msec, it was of interest to evaluate the changes in regional brain cyclic AMP and cyclic GMP occurring during slower microwave inactivation. By using the Litton 1.5 kW microwave unit which inactivated brain tissue in 4 sec, compared to the 300-msec data, a delay of 3.7 sec in inactivation time could be conveniently imposed. As shown in Table 4, there are, as expected, higher levels of cyclic AMP measured in brains from mice sacrificed in the 1.5 kW oven for 4 sec. In consideration of the regional levels from 4-sec-exposed animals and the 300 msec

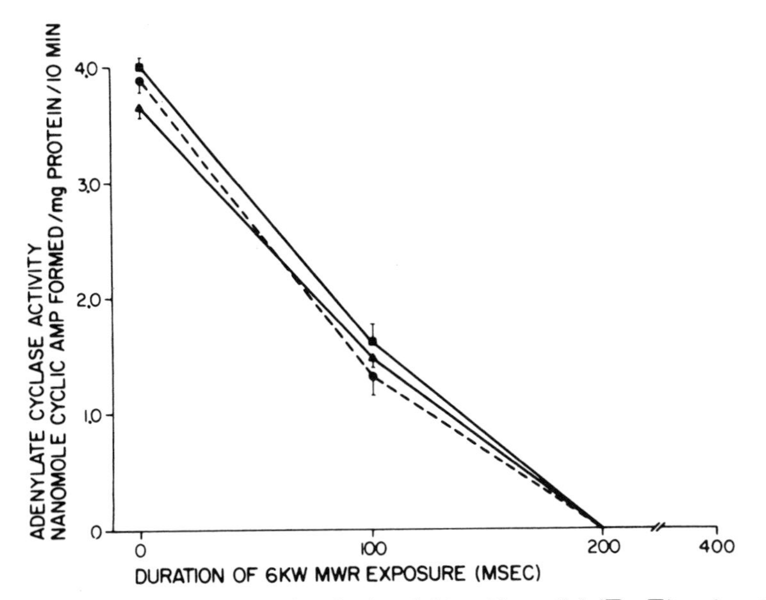

Fig. 12. Activity of adenylate cyclase in dorsal (● — ●), medial (■ — ■) and ventral (▲ — ▲) areas of mouse brain as a function of duration of 6 kW microwave exposure. From *Jones* and *Stavinoha* (1977), with permission of author and publisher.

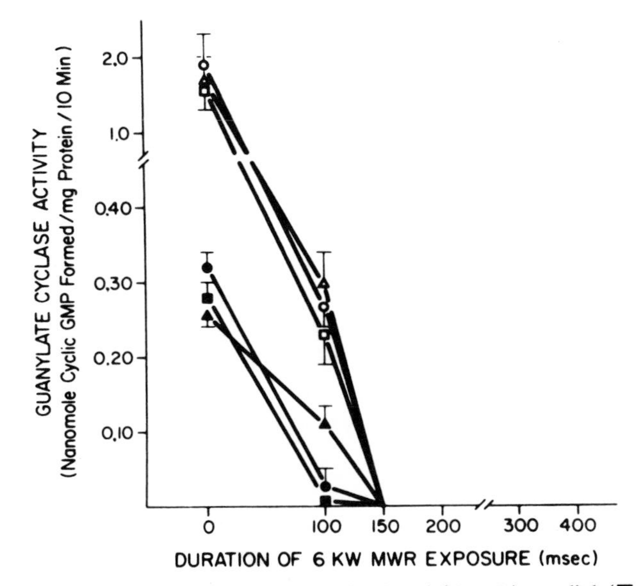

Fig. 13. Activity of soluble guanylate cyclase in dorsal (● — ●), medial (■ — ■) and ventral areas (▲ — ▲) compared to particulate (Triton X-100 dispersed) guanylate cyclase in dorsal (○ — ○), medial (□ — □) and ventral areas (△ — △) of mouse brain as a function of duration of 6 kW microwave exposure. From *Jones* and *Stavinoha* (1977), with permission of author and publisher.

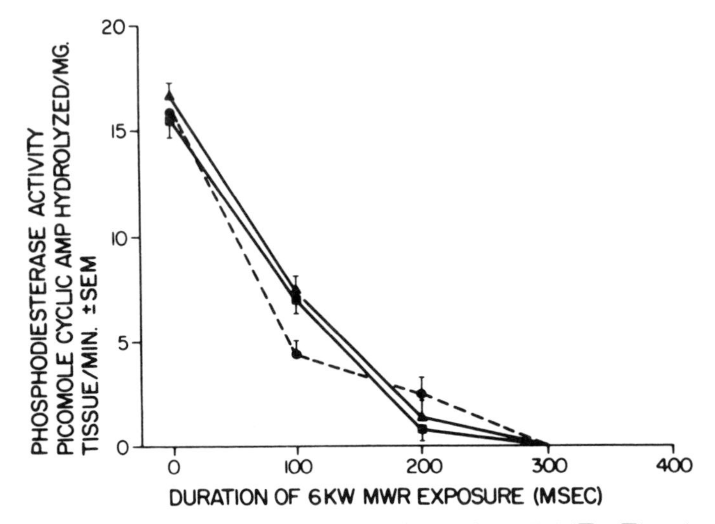

Fig. 14. Activity of phosphodiesterase in dorsal (●—●), medial (■—■) and ventral (▲—▲) areas of mouse brain as a function of duration of 6 kW microwave exposure. From *Jones* and *Stavinoha* (1977), with permission of author and publisher.

group, there does not appear to be any variation in cyclic AMP content from region to region. However, the absolute increase in each region during the delay is different, as evidenced by the change in corpus striatum which is 50% of the increase in other areas.

Also shown in Table 4 are data which establish that cyclic GMP levels are stable regardless of 4 sec or 300 msec inactivation. From the works of *Goldberg et al.* (1970) and *Steiner et al.* (1972), it was expected that the time required for enzyme inactivation would not be as great a factor for measuring *in vivo* levels of regional brain cyclic GMP. Therefore, slower microwave techniques could be efficiently used for such studies (*Katz* and *Catravas,* 1976; *Mattson, Brandt,* and *Heilbronn,* 1977; *Gumulka et al.,* 1977).

The data in Table 4 do, however, point out a problem relative to the expression of data as ratios of the levels of cyclic AMP to cyclic GMP. Many studies use this ratio as an index of the responsivity of cyclic nucleotide systems to biologically active agents. As shown in Table 5, due to the rapid postmortem increase in cyclic AMP and the lack of change in cyclic GMP, the ratio can be a direct reflection of the time required for enzyme inactivation. The ratios for brain regions in the 4-sec-inactivated mice are higher than in the 300-msec-inactivated group. This is more evident when considering cyclic AMP/cyclic GMP ratios of 35 for cerebral cortex and 70 for corpus striatum in brains from mice frozen in liquid nitrogen (*Steiner et al.,* 1972). Thus, the longer enzymes function following occlusion of the cerebral circulation,

Table 4. Comparison of Levels of Cyclic Nucleotides in Mouse Regional Brain Areas after 1.5 kW MWR Exposure for 4 sec and 6 kW MWR Exposure for 0.3 sec

	Cyclic AMP*			Cyclic GMP*	
	Time for Inactivation			Time for Inactivation	
Brain Area	4 sec	0.3 sec	Δ	4 sec	0.3 sec
Cerebral cortex	1.76 ± 0.03	0.78 ± 0.08	0.98	0.092 ± 0.014	0.118 ± 0.009
Medulla-pons	1.78 ± 0.04	0.77 ± 0.07	1.01	0.138 ± 0.019	0.129 ± 0.019
Midbrain	1.86 ± 0.04	0.84 ± 0.05	1.02	0.183 ± 0.013	0.191 ± 0.016
Diencephalon	1.98 ± 0.08	0.95 ± 0.05	1.03	0.108 ± 0.011	0.093 ± 0.018
Corpus striatum	1.62 ± 0.13	1.08 ± 0.18	0.54	0.142 ± 0.022	0.111 ± 0.017
Hippocampus	1.98 ± 0.06	0.83 ± 0.09	1.15	0.137 ± 0.009	0.148 ± 0.019
Cerebellum	1.97 ± 0.06	0.99 ± 0.11	0.98	0.584 ± 0.222	0.513 ± 0.021

*pmoles cyclic nucleotide/mg tissue ± SEM
n = 6–12 for each area
Δ absolute change in cyclic AMP levels measured in corresponding brain area after 4 sec vs 0.3 sec exposure

From *Jones* and *Stavinoha* (1977), with permission of author and publisher.

Table 5. Levels of Cyclic AMP and Cyclic GMP Measured from Same Regional Brain Tissue Sample

Brain Area	Cyclic AMP*	Cyclic GMP*	300† msec Ratio	4 sec Ratio**
Cerebral cortex	0.818 ± 0.050	0.144 ± 0.010	5.1	19.1
Medulla-pons	0.768 ± 0.034	0.131 ± 0.014	5.9	13.0
Midbrain	0.832 ± 0.035	0.171 ± 0.016	4.9	10.2
Thalamus	0.870 ± 0.043	0.136 ± 0.018	6.4	
				18.3‡
Hypothalamus	0.806 ± 0.069	0.081 ± 0.011	10.0	
Corpus striatum	1.184 ± 0.037	0.103 ± 0.008	11.5	11.4
Hippocampus	0.932 ± 0.037	0.194 ± 0.021	4.8	14.5
Cerebellum	0.717 ± 0.036	0.477 ± 0.042	1.5	3.4

n = 12 for each regional area
*pmoles cyclic nucleotide/mg tissue wet weight ± SEM
**Calculated from 4 sec data in Table 4
†Cyclic AMP/cyclic GMP ratios at 300 msec are compared to ratios from 4-sec group of Table 4
‡values for hypothalamus and thalamus combined as diencephalon in Table 4
From *Jones* and *Stavinoha* (1977), with permission of author and publisher.

the greater the disparity in ratios of cyclic AMP/cyclic GMP. One has to be aware of this type of artifact if the purpose of a study is to characterize the effects of drugs on CNS cyclic nucleotide levels.

All of this data points to the requirement of a technique to inactivate enzymes *in situ* which is both uniform and rapid. It was apparent that the 300-msec time period for mouse brain was the lower limit of our available systems. Since we could not answer the question of how fast is fast enough, we decided to answer the question of how *slow* is fast enough?

In order to answer this question validly, microwave exposure units of varying power with the same coupling efficiency were required. Previously, such systems were not available and we had to rely on exposure units in which coupling was extremely variable. However, with the addition of special timing devices and "step-down" regulators, the stepwise reduction in power output of 7.3 kW was permitted while maintaining consistency in coupling efficiency. Using such a system, "time for inactivation" was the only variable, and an accurate comparison could be made relative to the time requirements for enzyme inactivation vs levels of cyclic AMP and cyclic GMP.

It was evident from previous work that cyclic GMP levels do not change within a 4-sec period for enzyme inactivation. This is also evident from Figure 15 (right) using 2.4 kW exposure requiring 742 msec. Interestingly, cyclic AMP levels in either cerebellum or cerebral cortex do not change when 742 msec are required for enzyme inactivation (Fig. 15, left). Further reductions in power down to 456 W or one-sixteenth of the original output require 4 sec to inactivate brain enzymes. As shown in Figure 16, even at this power setting and time requirement the levels of cyclic AMP in regional brain areas are not significantly different.

The lack of change in cyclic AMP levels at 4 sec of exposure contrasts with previous 4-sec data following 1.5 kW exposure (Table 4). Although there is no accurate explanation, it might be that the exposure cell conditions in the 1.5 kW oven overly stress the animal. It does take longer to orient the animal in the 1.5 kW oven; as a result, an hypoxic environment might result. Animal alignment in the 7.3 kW oven requires less than 2 sec, whereas up to 20 sec are required in the 1.5 kW oven.

The above data point out two important facts concerning the use of microwave inactivation techniques prior to the measurement of CNS cyclic nucleotides. Previously, *Lust et al.* (1973) suggested that heating

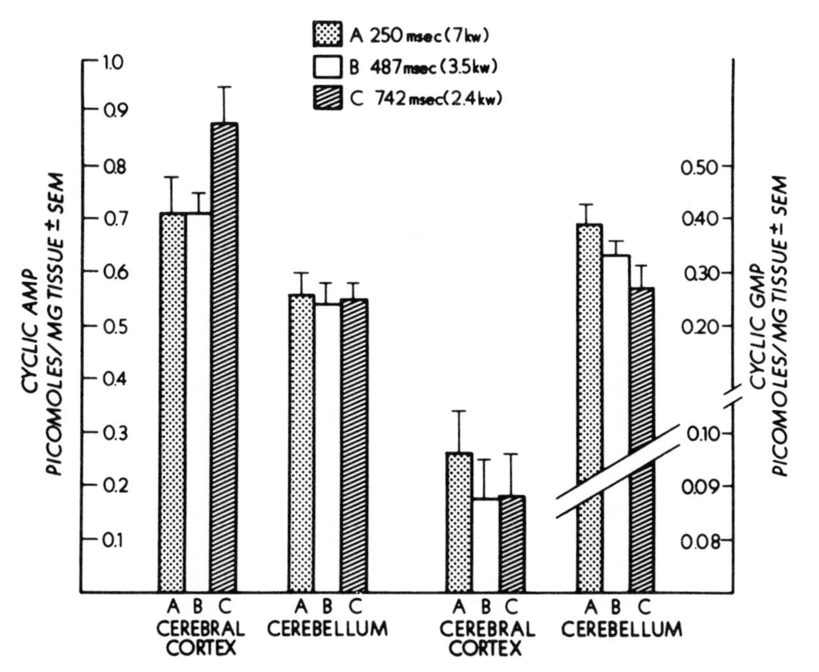

Fig. 15. Levels of cyclic AMP (left) and cyclic GMP (right) in regional brain following exposure to 2450 MHz microwave radiation at variable power outputs. Temperatures of 83–87 °C were used as the end point for heating. Each bar represents the mean ± SEM of four animals.

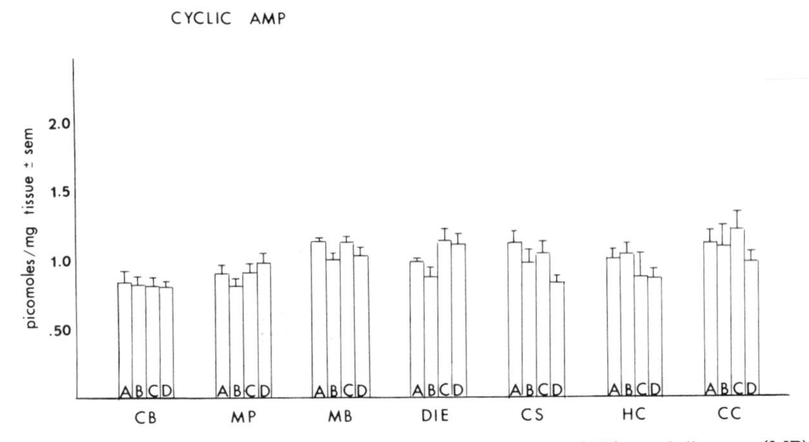

Fig. 16. Comparison of cyclic AMP levels in mouse cerebellum (CB), medulla-pons (MP), midbrain (MB), diencephalon (DIE), corpus striatum (CS), hippocampus (HC) and cerebral cortex (CC) following microwave exposure of 7 kW for 250 msec (A), 1.75 kW for 1 sec (B), 0.7 kW for 2.5 sec (C) and 0.44 kW for 4 sec (D). Each brain was validated as reaching 83–85 °C at the end of the exposure period.

during irradiation favors dissociation of cyclic AMP from protein kinase, whereby the free form is then degraded by phosphodiesterase. The result would be lowered cyclic AMP levels. The 250-msec-4-sec data shown in Figure 16 indicate that this is not true since slower heating would provide for greater dissociation and, since phosphodiesterase activity would be present for a longer time period, the levels would be still lower.

A second factor is that levels of cyclic AMP and cyclic GMP are the same regardless of whether 250 msec or 4 sec are required for inactivation. This is true provided that the energy is deposited in the brain uniformly and heating occurs efficiently without the loss factor of interphase local heating currents. The present application cell used for mice has been developed with consideration for the geometry of the mouse brain and the wave characteristics of the radiation. One cannot expect optimum heating without a system that has been developed with these factors being considered.

REFERENCES

Breckenridge, B. McL. 1964. The measurement of cyclic adenylate in tissues. Proc Nat Acad Sci USA 52:1580–1586

Breckenridge, B. McL., and Norman, J.H. 1962. Glycogen phosphorylase in brain. J Neurochem 9:383–392

Daly, J.W. 1976. The nature of receptors regulating the formation of cyclic AMP in brain tissue. Life Sci 18:1349–1358

DeLapaz, R.L., Dickman, S.R., and Grosser, B.I. 1975. Effects of stress on rat brain adenosine 3′,5′-monophosphate *in-vivo.* Brain Res 85:171–175

Dinnendahl, V. 1975. Effects of stress on mouse brain cyclic nucleotide levels *in vivo.* Brain Res 100:716–719

Ebadi, M.S., Weiss, B., and Costa, E. 1971. Distribution of cyclic adenosine monophosphate in rat brain. Arch Neurol 24:353–357

Ferrendelli, J.A., Gay, M.H., Sedgwick, W.G., and Chang, M.M. 1972. Quick freezing of murine CNS. Comparison of regional cooling rates and metabolite levels when using liquid nitrogen or freon-12. J Neurochem 19:979–987

Ferrendelli, J.A., Kinscherf, D.A., and Chang, M.M. 1973. Regulation of levels of guanosine cyclic 3′,5′-monophosphate in the central nervous system: Effects of depolarizing agents. Mol Pharmacol 9:445–454

Goldberg, N.D., Lust, W.D., O'Dea, R.F., Wei, S., and O'Toole, A.G. 1970. A role of cyclic nucleotides in brain metabolism. Adv Biochem Psychopharmacol 3:67–88

Guidotti, A., Cheney, D.L., Trabucchi, M., Doteuchi, M., Wang, C., and Hawkins, R.A. 1974. Focussed microwave radiation. A technique to minimize post mortem changes of cyclic nucleotides, DOPA and choline to preserve brain morphology. Neuropharmacology 13:1115–1122

Gumulka, S.W., Dinnendahl, V., Bartmus, D., Schonhofer, P.S., and Stock, K. 1977. On the mode of action of clonidine: Relationship between effects on behavior and cyclic nucleotide content in mouse brain. Naunyn-Schmiedeberg's Arch Pharmacol 298:7–14

Jones, D.J. 1974. Maximal Field Strength Microwave Irradiation as a Tool for Study of Brain Cyclic 3′,5′-adenosine Monophosphate. Dissertation, The University of Texas Health Science Center, San Antonio, Texas

Jones, D.J., Medina, M.A., Ross, D.H., and Stavinoha, W.B. 1973. Cyclic AMP levels in rat brain following two-second microwave inactivation. Pharmacologist 15:425A

Jones, D.J., Medina, M.A., Ross, D.H., and Stavinoha, W.B. 1974. Rate of Inactivation of adenyl cyclase and phosphodiesterase: Determinants of brain cyclic AMP. Life Sci 14:1577–1585

Jones, D.J., and Stavinoha, W.B. 1976a. Effects of morphine tolerance, dependence and withdrawal on regional mouse brain cyclic nucleotides. Pharmacologist 18:212a

Jones, D.J., and Stavinoha, W.B. 1976b. Regional brain cyclic nucleotide fluxes during tissue fixation in-situ. Trans Amer Soc Neurochem 7:116

Jones, D.J., and Stavinoha, W.B. 1977. Levels of cyclic nucleotides in mouse regional brain following 300 ms microwave inactivation. J Neurochem 28:759–763

Jongkind, J.F., and Bruntink, R. 1970. Forebrain freezing rates and substrate levels in decapitated rat heads. J Neurochem 17:1615–1617

Katz, J., and Catravas, G.N. 1976. Cerebellar cyclic GMP levels reduced by morphine and pentobarbital on a dose-and time-dependent basis. Biochem Pharmacol 25:2543–2546

Kimura, H., Thomas, E., and Murad, F. 1974. Effects of decapitation, ether and pentobarbital on guanosine 3′,5′-phosphate and adenosine 3′,5′-phosphate levels in rat tissues. Biochim Biophys Acta 343:519–528

Kinscherf, D.A., Chang, M.M., Rubin, E.H., Schneider, D.R., and Ferrendelli, J.A. 1976. Comparison of the effects of depolarizing agents and neurotransmitters on regional CNS cyclic GMP levels in various animals. J Neurochem 26:531–536

Lowry, O.H., Passonneau, J.V., Hasselberger, F.X., and Schultz, D.W. 1964. Effect of ischemia on known substrates and cofactors of the glycolytic pathway in brain. J Biol Chem 239:18–30

Lust, W.D., Passonneau, J.V., and Veech, R.L. 1973. Cyclic adenosine monophosphate, metabolites, and phosphorylase in neural tissue: A comparison of methods of fixation. Science 181:280–282

MacMillan, V., and Siesjo, B.K. 1973. The effect of phenobarbitone anaesthesia upon some organic phosphates, glycolytic metabolites and citric acid cycle-associated intermediates of the rat brain. J Neurochem 20:1669–1681

Mattsson, H., Brandt, K., and Heilbronn, E. 1977. Bicyclic phosphorus esters increase the cyclic GMP level in rat cerebellum. Nature 268:52–53

McLees, B.D., and Finch, E.D. 1973. Analysis of reported physiologic effects of microwave radiation. Adv Biol Med Phys 14:163–232

Medina, M.A., Jones, D.J., Stavinoha, W.B., and Ross, D.H. 1975. The levels of labile intermediary metabolites in mouse brain following rapid tissue fixation with microwave irradiation. J Neurochem 24:223–227

Medina, M.A., and Stavinoha, W.B. 1977. Labile intermediary metabolites in rat brain determined after tissue inactivation with microwave irradiation. Brain Res 132: 149–152

Michaelson, S.M. 1969. Biological effects of microwave exposure. In *Biological Effects and Health Implications of Microwave Radiation,* ed. S.F. Cleary, Bureau of Radiological Health, pp. 35–55 U.S. Dept H.E.W., Bethesda, MD.

Michenfelder, J.D., and Milde, J.H. 1975. Influence of anesthetics on metabolic, functional and pathological responses to regional cerebral ischemia. Stroke 6:405–410

Nahorski, S.R., and Rogers, K.J. 1973. The adenosine 3′,5′-monophosphate content of brain tissue obtained by an ultra-rapid freezing technique. Brain Res 51:332–336

Palmer, G.C., Jones, D.J., Medina, M.A., and Stavinoha, W.B. 1977. Influence of injected psychoactive drugs on cyclic AMP levels in mouse brain and lung following microwave irradiation. Neuropharmacology 16:435–443

Ponten, V., Ratcheson, R.A., Salford, L.G., and Siesjo, B.K. 1973. Optimal freezing conditions for cerebral metabolites in rats. J Neurochem 21:1127–1138

Redos, J.D., Hunt, W.A., and Catravas, G.N. 1976. Lack of alteration in regional brain cyclic adenosine 3′,5′-monophosphate levels after acute and chronic treatment with ethanol. Life Sci 18:989–992

Schmidt, M.J., and Robinson, G.A. 1971. Cyclic AMP in discrete brain regions following microwave irradiation. Pharmacologist 13:257A

Schmidt, M.J., Schmidt, D.E., and Robison, G.A. 1971. Cyclic adenosine monophosphate in brain areas. Microwave irradiation as a means of tissue fixation. Science 173:1142–1143

Stavinoha, W.B., Frazer, J., and Modak, A.T. 1978. Microwave fixation for the study of acetycholine metabolism. In *Cholinergic Mechanisms and Psychopharmacology,* ed. D.J. Jenden, pp. 169–179. New York: Plenum Press

Stavinoha, W.B., Modak, A.T., and Weintraub, S.T. 1972. Studies on 2450 MHz microwave heating of the central cholinergic system. Pharmacologist 14:221A

Stavinoha, W.B., Pepelko, B., and Smith, P.W. 1970. Microwave radiation to inactivate cholinesterase in the rat brain prior to analysis for acetylcholine. Pharmacologist 12:257A

Stavinoha, W.B., Weintraub, S.T., and Modak, A.T. 1973. The use of microwave heating to inactivate cholinesterase in the rat brain prior to analysis for acetylcholine. J Neurochem 20:361–371

Steiner, A.L., Ferrendelli, J.A., and Kipnis, D.M. 1972. Radioimmunoassay for cyclic nucleotides. III. Effect of ischemia, changes during development and regional distribution of adenosine 3′,5′-monophosphate in mouse brain. J Biol Chem 247:1121–1124

Stone, W.E. 1935. The effects of anesthetics and convulsants on the lactic acid content of the brain. Biochem J 32:1908–1916

Swaab, D.F. 1971. Pitfalls in the use of rapid freezing for stopping brain and spinal cord metabolism in rat and mouse. J Neurochem 18:2085–2092

Takahashi, R., and Aprison, M.H. 1964. Acetylcholine content of discrete areas of the brain obtained by near-freezing method. J Neurochem 11: 887–898

Thorn, W., Scholl, H., Pfeiderer, G., and Nueldener, B. 1958. Metabolic processes in the brain at normal and reduced temperatures and under anoxic and ischemic conditions. J Neurochem 2:150–165

Uzunov, P., and Weiss, B. 1971. Effects of phenothiazine tranquilizers on the cyclic 3′,5′-adenosine monophosphate system of rat brain. Neuropharmacology 10:697–708

Veech, R.L., Harris, R.L., Veloso, D., and Veech, E.H. 1973. Freeze-blowing: A technique for the study of brain *in vivo.* J. Neurochem 20:183–188

Volicer, L., and Hurter, B.P. 1977. Effects of acute and chronic ethanol administration and withdrawal on adenosine 3′,5′-monophosphate levels in rat brain. J Pharmacol Exp Ther 200:298-305
Weiner, N. 1961. The content of adenine nucleotides and creatine phosphate in brain of normal and anesthetized rats. A critical study of some factors influencing their assay. J. Neurochem 7:241-250
Wellmann, W., and Schwabe, U. 1973. Effects of prostaglandins E_1, E_2, and F_{2a} on cyclic AMP levels in brain *in vivo*. Brain Res 59:371-378

Index

AcCh, 207
Acetophenazine, 93
Acetycholine, 63, 92, 207, 230, 262
Acetylcholinesterase, 262
Acute subsensitivity in noradrenergically elicited cyclic AMP responses, 162-165
Adenine, 77, 208, 221
  incorporation into adenosine triphosphate (ATP), 77
Adenosine, 21, 22, 24, 25, 57, 75, 77, 79, 81, 83, 85, 86, 89, 160, 164, 199, 200, 202, 205, 208, 219, 220, 221, 225, 241, 242
  -activated compound, 202
  -dependent receptors, 207
  histamine-, 77
  in dopamine-sensitive assay. 24, 25
  increased synthesis of, 120
  -induced accumulation of cyclic AMP, 223
Adenosine cyclic 3′, 5′-monophosphate, *See* Cyclic AMP.
Adenosine 3′, 5′-monohosphate, 212
Adenosine triphosphate (ATP), 22-25, 27, 74, 77, 86, 113-114, 153, 231-233, 237-239, 241-243, 249, 271
  adenine incorporation, 77
Adenosine triphosphatase (ATPase), 22, 74, 83, 242
  adenosine-norepinephrine, 75
Adenosine receptor, 219
Adenylate cyclase, 1, 5, 15, 20, 22, 24, 55, 57, 59, 60, 63-67, 69, 72-79, 113-114, 152-153, 162, 165, 174-177, 179-190, 192, 193, 198, 219-220, 237, 241, 249, 270-271, 274
  activation of, 15-16, 82, 119-120, 126-127
  activity, isoproterenol-stimulated, 120
Adenylate cyclase
  acute activation of, in cyclic AMP system of rat pineal gland, 176-191
  adrenergic activation of, 59
  β-adrenergic receptors, 153
  alterations from injected neuroleptics, 78-82
  antidepressant drugs, 116-125
    imipramine, 116-119
    triiodothyronine, 116-119
  assay, 21, 22
  binding studies, 115
  catecholamine-sensitive, 20; 113, 126
  clinical studies, 125-137
    CSF cyclic AMP. 142-144
    plasma cyclic AMP. 144, 145
    platelet, 125-137
    postmortem, 137
    urinary cyclic AMP, 138-142
  cortical, 24
  -cyclic AMP, 82-87
    neuroleptic action on, 82-87
  depolarization-activated, 99
  dopamine, 62, 63
  dopamine-sensitive, 53-68, 79
    neuroleptic action, 53-68
    historical perspective, 1-3
    pharmacological characterization of the dopamine receptor, 26-43
    sources of variability, 14-26
  dopamine-stimulated, 22, 63
  during aging, 15
  during development, 15
  effect of chronic treatment with ECT on, 120
  effect of imipramine and thiiodothyronine on, 116
  electroconvulsive therapy, 119-125
  histamine-sensitive, 76-78
  hormonally sensitive, 87-92
    neuroleptic action on, 87-92
  human brain, 137
  inhibition, 67
  injected neuroleptic influence, 78-82
  leukocytes, 126, 133-137
  nigral, 64
  norepinephrine-sensitive, 67-76, 188
  phenothiazine action on, 87
  pineal gland, 182, 189, 191-193
  platelet, 129-130
  postmortem brain, 126
  receptor, 55-56
  receptor-adenylate cyclase coupling, 20
  retinal, 24
  schizophrenia, 137
  systems, 69
  transducer site, 75
Adenylate kinase, 242
Adipose tissue, lipolysis of, 91
Adrenal medulla, 207
Adrenergic neurotransmitter, 191
Adrenergic receptors, 113, 116, 125, 126, 130, 133, 145, 158
  ∝, 138, 153, 160, 162, 209
  β-, 138, 159, 162, 164, 165, 170, 175, 177, 180, 190; membrane receptor, 157, 167, 169
  sensitivity, 113, 115, 133, 134, 137, 145
Adrenergic stimulation, 190
ADTN, *See* 6, 7-Dihydroxy-tetrahydronaphthalene (ADTN).
Affective disorders, 53, 68, 73, 76, 113, 115, 126,

Affective disorders, (*Continued*)
133, 138, 139, 145, 146
histamine, 76
pathogenesis of, 73, 94, 145
urinary cAMP excretion in, 140
Agents, noradrenergic, 160
Agroclavine, 33, 35, 37
Alleles, 199
Alpha blockers, 4, 28
Alumina column, 27
Amine metabolites and receptors, 112
γ-Aminobutyric acid (GABA), 62, 72, 251
Aminophylline, 23, 217, 219
Amitriptyline, 116, 117
Amobarbital, 230
5′-AMP, 231, 233, 241
AMP, cyclic, *See* adenosine cyclic 3′.5′-monophosphate.
Amphetamine, 4, 11, 53, 79, 162, 165-169
-treated mice, 168, 169
Amygdala, 16, 54, 65, 67
Anoxia, 215, 218, 223, 225, 258-260, 267-268, 271
postmortem, 267-271
Anoxic brain, 254
Anticonvulsant drugs, 212-225, 243
effect of, on cyclic nucleotide regulation, 223, 224
Antidepressant drugs, 116-125, 162, 165, 169
imipramine, 116-119
triiodothyronine, 116-119
Antigonadotropic peptides, 194
Apnea, 215
Apomorphine, 3-6, 9-11, 15, 23, 30-32, 38-43, 42, 64, 75, 79, 93, 94
structure of, 40
Arteries, 230, 234, 247
Asthma, bronchial, 138
Astrocytoma, 97
ATP, *See* adenosine triphosphate.
ATPase, *See* adenosine triphosphatase (ATPase).
Atropine, 63
Azeperone, 61

Barbiturates, 247, 251
Batrachotoxin, 85
Behavioral genetic studies, 198
Behavioral implications of genetic determination of cyclic AMP level, 198-209
Behaviors, 199, 200, 201, 202, 204, 207-209
avoidance learning, 199, 204
avoidance performance, 201, 202, 208, 209
initial exploratory activity, 199-201, 207-209
"jump-up" performance, 208
locomotory behavior, 208
open-field situation, 209
Benperidol, 97
Benzamide, 7, 27, 28
Benzodiazepines, 28
Benztropine, 28
Beta blockers, 4, 28
Bicuculline, 250
Binding radioligand, 152, 153
Blood-brain barrier, 63
Bone, high and low water content, 264

Brain region, choice of, in dopamine-sensitive assay, 16, 17
Brain stem. 68, 81, 82, 95, 97, 161, 220
Bromocryptine, 26, 28, 35, 37, 39, 40, 43
Bronchial asthma, 134
7,8,B-triOH-CPZ, 70
Butaclamol, 6, 7, 28, 59
(−)Butaclamol, 36, 66, 93
(+)Butaclamol, 36, 66, 80, 93
(±)Butaclamol, 60, 61, 66, 77
Butyrophenones, 6, 7, 27, 28, 43, 54-55, 61-63, 76

Caffeine, 4, 217, 219
Calcium, 24, 54, 55, 57, 85, 194, 219, 222, 225
-dependent protein activator, 11, 57, 96-99
Ca^{++} regulator protein, 57
in dopamine-sensitive assay, 23, 24
release from brain mitochondria, 101
cAMP, *See* Cyclic AMP.
Capillaries, neuronal, 58
Carbamazepine, 216, 223, 224
Carbamylcholine, 92
Cardiac membranes, 180
Carotid occlusion, 244, 246, 247-251
bilateral, 246, 251
Catecholamine, 2-6, 30, 31, 34, 53-54, 68, 80, 87, 101, 116, 139, 145, 175-177, 180-182, 190-191, 198, 219, 225
binding of to receptor sites, 114
receptors, 124, 205
-sensitive enzyme system, 190
systems, 125
Catechol-O-methyl transferase (COMT), 112, 114
Catecholaminergic neuron, 114
Caudate, 17, 24, 37, 54, 59, 62, 64, 67, 68, 137
Caudate neurons, 220
Caudate nucleus, 4, 17
Cell fraction, choice of, in dopamine-sensitive assay, 20
Cell type, choice of, in dopamine-sensitive assay, 17
Central ganglia, 68
Cerebellum, 2, 17, 34, 62, 68-69, 72, 75-76, 79, 80-82, 85, 97, 161, 163, 164, 213-214, 216-218, 223, 243, 245, 250, 255, 258-259, 268-269, 270, 272, 276-279
Cerebral circulation, 229, 275
Cerebral cortex, 3, 4, 8, 11, 15-17, 19-20, 25, 54, 65, 67-68, 70-71, 74-75, 77, 79, 80, 82-83, 85-86, 95, 97, 116, 154-162, 165-166, 168, 169, 200-203, 205-209, 213-217, 220-223 230, 232-240, 243, 247, 248, 250, 251, 254-256, 258-261, 268-270, 275-279
capillary, 65
capillary enzyme, 67
cerebral homogenate, 67
cerebrum, 67
glial fraction, 67
neuronal fractions, 67
cortical adenylate cylase, 15
frontal cortex, 4, 8
glial, 65
isolated cerebral tissues, 84
Cerebral cortices, 84

Cerebral hypoxemia, 218
Cerebral ischemia, 218
Cerebral spinal fluid (CSF), 69, 112, 137, 215, 216
Character disorder, 143
Chlordiazepoxide, 28
Chlorpromazine, 11, 28, 63-71, 74-77, 79-80, 83-87, 89, 90, 92, 95-97, 166
Chlorpromazine (CPZ), 58, 60, 62
  7,8-Dioxo derivative of, 82
  high concentrations of, 77
  quaternary analogues of, 62
  -SO, 69
Chlorprothixene, 72, 81, 95, 97
α-Chlorprothixene, 60, 61
β-Chlorprothixene, 61
Choline chloride, 92
Cholinergic compounds, 63
Cholinergic compounds, *See also* choline.
Cholinomimetic agents, 92
Chromaffin granules, 207
Chromosome, 201
Chromosome 4, 205, 207
Chromosome 9, 199, 204, 209
Chronic decentralization, 186
Circadian rhythms, 174
Circulatory arrest, 258
Circulatory occlusion, 248
2-Cl-10-(3 dimethyl amino propyl)-phenothiazine, 91
2-Cl-7,8-dioxo-phenothiazine, 86
Clinical studies, 125–137
  CSF cyclic AMP, 142–144
  plasma cyclic AMP, 144, 145
  platelet, 125–137
    effect of lithium on, 130-133
    leukocyte, 133-137
  postmortem, 137
  urinary cyclic AMP, 138–142
1-Cl-promazine, 60
3-Cl-promazine, 60
4-Cl-promazine, 60
Clonidine, 75
α-Clopenthixol, 60, 61
β-Clopenthixol, 61
Clopimozide, 61
Clozapine, 6, 11, 27, 28, 60-62, 64-67, 73, 77, 78, 80, 93
CNS cyclic nucleotides, 270
CNS function, effect of cyclic nucleotides on, 220-223
Coat-color subgroups, 204, 205
Cocaine, 154, 155
Colliculus, 84
Common carotid arteries, 243, 247
Convulsants, 243
Corpus striatum, 256
Corticospinal neurons, 220
CPZ, *See also* chlorpromazine.
CPZ-CH_3Cl, 70
CPZ-MeCl, 60
CPZ-5-N-dioxide, 86
CPZ-NO, 60
CPZ-SO, 60, 62, 70, 93
Cerebrospinal fluid (CSF), 142-144, 215, 216; *see also* Cyclic AMP.
3′, 5′-Cyclic adenosine monophosphate (cAMP), *See* Cyclic *AMP*.
Cyclic AMP (cAMP), 2-5, 27, 31, 55, 67-72, 74-77, 79, 84, 87, 89, 90, 101, 112, 114, 137, 152-156, 158, 161, 163-166, 168, 174-179, 185-188, 190-194, 198-209, 212, 214-225, 229-231, 234, 235, 237, 239-244, 246, 249-251, 254, 255, 257-261, 267-270, 272, 273, 275-279
  adenosine-induced, 201, 209, 223
  basal, 100
  basal levels of, 86
  cerebrospinal fluid levels of, 79, 115, 142
  concentration in brain, 113
  -deactivating phosphodiesterases, 152
  dibutyryl, 31, 192
  effect of monoamine oxidase on response to norepinephrine, 123
  elevation of, 188
  elevation of by adenosine, 83
  excretion of in depression, decreased, 142
  fat, 79
  generating system
    effect of desipramine on, 121
    effect of electroconvulsive therapy on response to norepinephrine, 122
    effect of iprindole on, 121
  genetic determination of, 198–209
  [^{3}H]-cAMP net synthesis, 117
    norepinephrine-induced stimulation of, 117
    isoproterenol-induced stimulation of, 117
    inhibitory effect of norepinephrine, 129
    stimulatory effect of prostaglandin E_1, 128
  increased synthesis of, 120
  in vitro inhibition by PGE_1-stimulated accumulation of ^{3}H-, 132
  level in brain, 198
    genetic determination of, 198
  levels, 213
    convulsant hexafluoridiethyl ether elevated, 213
    in cerebrospinal fluid, 126
    in plasma, 126
    in urine, 126
  noradrenergic, 156, 157
  noradrenergic sensitivity, 166–169
    neuroleptic, 166
    rats, 166
  noradrenergic subsensitivity
    acute, 168
    chronic, 169
  noradrenergic supersensitivity, 158
  noradrenergically elicited responses
    chronic subsentivity in, 165–169
    chronic supersensitivity, 154–162
  norepinephrine (NE)-induced accumulation of, 202–209
    organs, 79
    kidney, 79
    liver, 79
    urinary levels of, 79
  phenothiazine action on, 86
  phosphodiesterase, 4
  postdecapitation increase in, 255
  postmortem changes, 255, 258

Cyclic AMP (cAMP), (*Continued*)
  produced by trauma, 81
  responses, 154
  "second messenger", 2
  seizure-induced elevations of, 216
  sensitivity, 153
  steady-state levels of, 99
  striatal function, 3, 4
    mediation effects of dopamine, 4
  subsensitivity, 153
  supersensitivity, 161
    chronic, 154
  synapse, 2
  synthesis
    [^{3}H]-, 135
  system of rat pineal gland, 174-194
    acute activation of adenylate cyclase, 176-191; chronic alternation, 184-190
    effect of sympathetic nerve activity on, 175, 176
    relevance, 191-193
  turnover rate of, 144
  urinary, 138-142
  urinary excretion of, 140
    effect of electroconvulsive therapy on, 141
Cyclic 3′, 5′-AMP, *See* Cyclic AMP.
Cyclic GMP, 89, 92, 193, 194, 212-220, 222-225, 230, 234, 236, 237, 239, 240, 241, 243-245, 247-251, 257-261, 273, 275-277
  basal, 100
  cerebellar, 245
  cholinergic, 100
  cholinomimetic-induced, 101
  injections of neuroleptics, 94
  lowered basal levels of, 94
  seizure-induced elevations of, 216
  system of rat pineal gland, 193, 194
Cyclic nucleotides, 69, 174, 192, 193, 212, 225, 229-251, 254-279
  accumulation in epileptic brain, 217-219
  affective illness
    pathogenesis of, 145
  controlling the levels of, 96
  effect of anticonvulsant drugs on, 212-225
  effect of, on CNS function, 220-223
  effect of seizures on, 212-216
  general role of, 138
  in bilateral ischemia, 243-247
  in cerebrospinal fluid, 215, 216
  influence of drugs on, 216, 217
  microwave inactivation and measurement of, 270-279
  neurotransmitter, interrelationships, 230
  of rat pineal gland, 175
  phosphodiesterase, 200
  postmortem changes in levels of, 254-260
  schizophrenia, pathogenesis of, 145
  seizure-induced elevations of, 216, 217
  steady-state levels of, 89
Cyclic nucleotide systems, 112
  major action of neuroleptic drugs on, 99
  summary of neuroleptic actions on
    acute amphetamine-induced cyclic AMP, 100
    adenosine cyclic AMP, 100
    chronic urinary cyclic AMP, 100
    DA-adenylate cyclase, 100
    histamine-adenylate cyclase, 100
    5-HT-adenylate cyclase, 100
    KCl cyclic AMP, 100
    NE-adenylate cyclase, 100
    ouabain cyclic AMP, 100
    phosphodiesterase calcium activator, 100
(+)3-Cyclohexylbutaclamol, 66
Cytosol, 96

DA, *See* dopamine.
D-amphetamine, 93
D-butaclamol, 11
Deactivating phosphodiesterases, 152
Decapitation, 229, 230, 233, 234, 237, 254, 258, 259, 269, 271
  stress of, 271
Decentralization, 175, 185, 191
  chronic, 186
  of superior cervical ganglia, 184
Denervation, 9, 175, 185, 188, 191
  of the pineal gland, 188
D-norepinephrine, 70
Depolarizing agents, 85
Depression, 53, 54, 116, 119, 120, 138, 144, 146
  bipolar 128, 143
  cyclic AMP, decreased excretion of, 142
  endogenous, 53
  psychotic, 139
  spreading, 261
  unipolar, 128, 134, 143
Depressive illness, 112, 113
Depressive states
  rise in cyclic AMP, 81
Desensitization, 180, 181, 183, 184, 190, 191
  adenylate cyclase, 180
    rat pineal gland, 180
Desipramine (DMI), 28, 125, 169
Desmethyl-CPZ, 60
Desmethylimipramine (DMI), 120, 125, 166-169
Desmethyl-loxapine, 61
DHA, *See* dihydroalprenolol (DHA).
Diazepam, 95, 223
Dibenzazepine, 6, 28
Dibenzazepine classes, 27
Dibenzodiazepines, 61
Dibutyryl cyclic AMP, 13, 90, 192, 221
Didesmethyl-CPZ, 60
Diencephalon, 80, 256, 268, 269, 276, 279
Di-Glycine, 19
Dihydroalprenolol, 163, 165, 167, 180
Dihydroalprenolol (DHA), 164, 168
  [^{3}H]-, 113, 124, 163, 167-168
    receptor affinity, 167
  binding, specific effect of chronic administration of antidepressants, 124
Dihydroergocryptine, 35
  [^{3}H]-, 126, 133
Dihydroergotamine, 35
Dihydroergotoxine, 35
Dihydroxyaminotetralin, 40-43
6,7-Dihydroxy-tetrahydroisoquinoline, 40
6,7-Dihydroxy-tetrahydronaphthalene (ADTN), 38, 40

3,7-DiMeO-CPZ, 70
7,8-DiMeO-CPZ, 70, 92, 95
Dimethyltryptamine, 85
DiOH-CPZ, 60, 67, 70, 86, 92
7,8-diOH-perpherazine, 69
2, 3-DiOH-promazine, 67, 71
7,8-Dioxo-CPZ, 60, 67, 70, 86
7,8-Dioxo-β-OH-CPZ, 60, 70
Diphenylbutylpiperidines, 63
Diphenylhydramine, 28
Dipole rotation, 262
Dipyridamol, 4
Directional coupler, 263
DOPAC, 7, 17, 28, 29, 30
Dopamine (DA), 1, 2, 5, 7, 9, 12, 14, 23, 26, 27, 31, 32, 33, 39, 41-43, 53, 60, 62, 63, 65-68, 71, 73, 75, 76, 78, 79, 80, 84, 87, 89, 91, 154, 160, 168, 176, 219
  adenylate cyclase, 62, 63
  agonists, 7, 10, 26
    ergot-like, 26
  alpha and beta rotomeric configurations, 41
  antagonist antipsychotics, 43
  antagonists, 7
  -apomorphine receptor, 8
  β-hydroxylase (DBH), 112, 114
  blocking agents, 1
  dopaminergic receptor system, 9
  dopaminergic system, 14
  effect of sympathectomy with, 186, 187
  enhanced basal actions, 83
  hydroxyl position, 41
  hypersensitive receptors in schizophrenia, 115
  inhibitory neurotransmitter, 31
  mediation effects on striatal function, 4
  molecule, 38
  neurons, 68
  potential configurations of compared to lergotrile, 39
  radioactive, 13
  receptor, 1, 14, 37, 38, 62, 74, 220
    adenylate cyclase, topography of, 37-48
    agonists, 8, 9, 29-37
    antagonists, 6, 7, 26-29
    classical, 14
"Dopamine2", receptor, 14
    function, 1
  homogenization-distorted adrenergic, 14
    in the pituitary, 13, 14
    pharmacological characterization, 26-43
    presynaptic, 28, 29
    sites, 7, 68, 76, 113
    supersensitivity, 99
  role as neurotransmitter, 53-56
  -sensitive adenylate cyclase, 60, 65, 66, 79
  -sensitive adenylate cyclase assay, 8, 98
  -sensitive assay, 1-44
    animal age, choice of, 14, 15
    enigmas of, 5-14
    historical perspective, 1-3
    pharmacological characterization of the dopamine receptor, 26-43
    sources of variability, 14-26
  -sensitive receptor, 67
  -stimulated adenylate cyclase, 63
    neuroleptic potency toward, 63
  theories of schizophrenia, 1
Dopaminergic, 112
Dowex, 27
Dowex 1, 22
Doxepin, 125
Droperidol, 61
Drugs, analogues, 83
Drugs, influence of, on cyclic nucleotides, 216, 217

ECS, *See* electroconvulsive shock.
EDTA, *See* ethylenediaminetetraacetic acid (EDTA).
EEG, 237
EGTA, *See* ethyleneglycol-bis-(β-aminoethyl-ether)-N-N′-tetraacetic acid.
E-H tuner, 265
Electrical stimulation, 83
Electroconvulsive shock (ECS), 165-167, 169, 212, 217
Electroconvulsive therapy (ECT), 116, 119-125, 139, 143, 144
  effect on urinary cyclic AMP excretion, 141
  effect on excretion of cyclic AMP, 141
Electrolytic lesions, 154
Electromagnetic energy, 262
Electroshock (ECS) phosphorylase, 222
Electroshock seizures (ECS), maximal, 222
Elmoclavine, 33, 35
Energy, electromagnetic, 262
Entorhinal cortex, 16
Environment hypoxic, 178, 278
Environmental lighting, 174, 184, 193
Enzyme inactivation techniques, 260-279
  comparison of freezing and heating, 267-270
  freezing, 260, 261
  heating, 261-267
(EPI), *See* epinephrine.
Epilepsy, 215, 216, 217, 220, 221, 225
  epileptiform discharges, 220, 221
Epinephrine (EPI), 2, 34, 64, 87, 89, 90, 176, 206
  -stimulated cyclic AMP levels in plasma, 145
Epinine, 4
Ergocornine, 35, 37
  -induced rotation, 37
Ergocristine, 32, 33, 34, 35, 36
Erogolines, 9, 29-43
Ergometrine, 35
Ergonovine, 33, 34, 35, 36
Ergots, 9, 32, 36, 37
  alkaloids, 13, 28, 64
  -related compounds, 35
    agonist and antagonist activities, 35
Ergotamine, 33, 34, 35, 36
Erythrocytes, 180
  frog, 180
Estrogen, 192
Ethanol, 154, 156-159, 161, 166, 167, 169
  chronic ingestion, 154
Ethosuximide, 216, 223, 224
Ethylenediaminetetraacetic acid (EDTA), 24
Ethyleneglycol-bis-(β-aminoethyl-ether)-N,N′-tetraacetic acid (EGTA), 21, 24, 27, 96, 97

Exposure units, microwave, 270

Fenfluramine, 7, 27
Festuclavine, 33, 35
Field intensity, 264, 272
  distribution of, 272
First messengers, 212
Fixation, 231
Fluoride, 159
  catalytic site, 100
  -stimulated activity, 159
Flupenthixol, 28, 72
  α-, 6, 60, 61, 64
  β-, 6, 61, 62
  α, β-, 60, 61
Fluphenazine, 6, 28, 60, 61, 62, 64-69, 72, 85, 86, 89, 93, 95
Forebrain, 213, 217, 234
Free radicals, 83
Freeze blowing, 261, 267-268
Freezing, 260, 261, 262, 269, 271
Freezing techniques, 267

GABA, *See* gamma amino butyric acid.
Galactorrhea, 29
Gamma amino butyric acid (GABA), 6
Ganglia, 175, 177, 184-187, 191, 193
  superior cervical, 175, 177, 184-187, 191, 193
    decentralization of, 184
Gene, 199, 201, 207, 208, 209
Genetic, 199-205, 207, 209
  BALB/c strain, 199
  behavioral studies, 198
  C57BL/6 strain, 207
  crosses, 202, 203
  cyclic AMP trait, 199
  determination of cyclic AMP level, 198-209
  dominant traits, 204
  linkage, 199, 207
  locus, 209
  parental strains, 209
  segregation, 202, 209
    Norepinephrine (NE)-induced accumulation of cyclic AMP, 202
    mendelian, 202
    study, 202-205
  strains, 201-205
Glia, 58, 74, 77, 162
  fractions of cerebral cortices, 82
  glial-enriched, 162
  glial-enriched fractions, 74, 77, 162
Glucagon, 89, 90, 91
Glucose, 222, 230, 232, 233, 237, 238, 239, 249
Glutamate, 223, 224
Glycogen, 58, 230, 232, 233
Glycogen synthase, 222
Glycolysis, 233
  aerobic, 233
  anaerobic, 233
GMP, 207
  action of neuroleptics on, 92-94
    in vitro studies, 92, 93
    in vivo studies, 93, 94
GMP, cyclic, *See* Cyclic GMP.
Gonadal function, 191, 192
Growth hormone (GH) 112
GTP, 249
GTP analogues, 83
24-Guage needle copper-constantan thermocouple, 265
Guanosine 3′,5′-monophosphate, 212
Guanosine cyclic 3′,5′-monophosphate, *See* Cyclic GMP,
  in dopamine-sensitive assay, 25, 26
Guanylate cyclase, 57, 93, 94, 101, 193, 219, 220, 271, 273, 274
  enzyme, 92
  inhibition of, 89
Guanyl nucleotides, 21, 25, 26, 35, 54, 62, 71, 74, 83, 100, 153, 165, 176, 177, 178, 181-184
5′-Guanylyl imidodiphosphate, 176

Haloperidol, 5-11, 13, 17, 27, 28, 30, 37, 38, 43, 59, 60-69, 73, 76, 78, 81, 82, 87, 89, 91, 93-95, 145
  ^{3}H.
    striatal binding of, 10, 63, 80
Heart, 221
Heat transfer properties, 262
Helix aspersa, 42, 44
HEPES, 19
Hexafluorodiethyl ether, 217
Hippocampal pyramidal cells, 220
Hippocampus, 54, 65, 68, 76, 84, 161, 202, 213, 214, 217, 276, 277, 279
  CA_1, 214
  CA_3, 214
  dentate gyrus, 214
  subiculum, 214
Histamine, 2, 25, 54, 75-78, 84, 85, 87, 89, 90, 160, 164, 176
  adenosine, 77
  affective disorders, 76
  as a central neurotransmitter, 76
  KCl, 77
  norepinephrine-, 75, 77
  -sensitive adenylate cyclase, 76, 78
Histones, 58
Homocysteine, 213, 216
Homogenization in dopamine-sensitive assay, 5, 6, 19, 20
Homovanillic acid (HVA) 139
Hybrid, 200
Hydroxybenzylpindolol (HYP), 113, 124, 159
6-Hydroxydopamine, 10, 30, 114, 155-161, 175, 186, 187, 188, 190, 191
5-Hydroxy indoleacetic acid (5-HIAA), 139
Hydroxyindole-O-methyltransferase, 193
13-Hydroxy lergotrile, 40
Hyperkinetic children, 142
Hyperprolactinemia, 67
Hypoglycemia, 223
Hypomanic patients, 141
Hypothalamus, 13, 16, 24, 25, 68, 69, 70, 71, 79, 82, 84, 161, 162, 260, 277
  function, 145
  lateral, 69, 70

medial, 70
Hypothermic, 222
Hypoxia, 218, 219, 278

IBMX. *See* 3-isobutyl-1-methylxanthine.
Ibotenic acid, 85
Imipramine (Imi), 28, 80, 116–119, 166, 169
in vitro effect on [^{3}H]-cAMP net synthesis, 117
-treated rats, 169
Inactivation, microwave, 267
Inbred stains, 199, 201, 202
Indole metabolism, 174
Indoleamines, 219, 225
Indoles, 193
Inhibition, 168
phosphodiesterase, 168
Inhibitors, 166
monoamine oxidase, 166
Interpeduncular regions, 54
In vivo enzyme inactivation techniques, 260–270
comparison of freezing and heating, 267–270
freezing, 260, 261
heating, 261–267.
Ionic conduction, 262
Iprindole 120, 125, 166
Ischemia, 218, 229–251
bilateral, 230, 232–236, 240, 243–247
drugs, 247–249
recovery, 237–242
Isoapomorphine, 42
Isobutylmethylxanthine (IBMX), 4, 23, 25, 27, 93, 160, 168
Isoniazid, 213
(+)Isopropylbutaclamol, 66
Isoproterenol (ISO), 2, 5, 6, 12, 25, 59, 64, 69, 71, 75, 87, 90, 117, 135, 145, 154, 156–160, 163–165, 176, 180, 181, 206; *see also* Cyclic AMP synthesis.
Isoquinoline, 38

K^+, 219, 223, 224
Kainic acid, 29
KCl, 75, 77
histamine-, 77
Kidney, 91, 144
major source of cyclic AMP in urine, 142
Kinetic energy, 262

Lactate, 233, 260
Lateral reticular nuclei, 68
L-DOPA, 4, 12, 53, 64, 80
Lergotrile, 10, 26, 28, 29–43
-induced rotation, 37
Lesions, 159, 160, 161
electrolytic, 159, 160, 161
median forebrain bundle, 159, 160, 161
Leukemia 138
Leukocyte, 133–137
human adenylate cyclase studies in, 133, 135, 137, 138
Lighting, environmental, 174, 184, 193
Limbic area, 37
Limbic forebrain, 6, 73, 75, 76, 83, 166, 168
Lisuride, 35
Lithium, 82, 130–133, 143–146
Liver, 91
Locus coeruleus, 68, 72, 156
Loxapine, 6, 28, 60, 61, 62, 66, 67
Lysergic acid diethylamide (LSD), 28, 32–36, 84–85, 93
bromo-, 35
methyl-, 35
Lung, 76

Magnetic fields, 262, 265
Magnetron, 263
Mania, 53, 54, 112, 142, 146
bipolar, 129, 134
from hypomania, 139
rise in cyclic AMP, 81
MAO. *See* monoamine oxidase (MAO).
Measurement of CNS cyclic nucleotides, 270–279
Medial forebrain bundle, 156
Median eminence, 8, 67
Medulla, 84, 264
Medulla-pons, 268, 269, 276, 277, 279
Melatonin, 175, 192, 193
in the pineal gland, 192
Memory, 170
Mendelian, 204
dominant, 200, 208
genetic segregation, 202
prediction, 204
Mental illness, 170
7-MeO-CPZ, 70, 86
3-Mercaptopropionic acid, 213, 217
3-Mercaptopropionic acid-induced seizures, 222
Mescaline, 85
Mesolimbic system, 54, 66, 68, 73
Mesoridazine, 60, 61, 92
Methadone, 144
Methdilazine, 61
Methergoline, 35
3-Methoxy-4-hydroxy-phenylethylene glycol (MHPG), 139
Methoxyindoles, 192–193
decreased synthesis, 192
5-Methoxytryptophol, 192
α-methylparatyrosine, 10, 30
Methylxanthines, 21, 25, 95, 177, 217, 219, 221
Metoclopramide, 7
Microiontophoresis, 62, 72, 73, 92, 220, 221, 222, 229, 237, 249
Microwave, 254, 260, 265, 266, 267, 270, 279
E-H tuner, 263
exposure, 265, 274, 275
exposure units, 270, 277
inactivation, 254–279
measurement of CNS cyclic nucleotides, 270–279
irradiation, 79, 89, 93, 94, 99, 262, 271, 272, 273
duration, 272, 273
mouse, 266
rats, 267
Midbrain, 65, 84, 155, 161, 162, 261, 268, 269, 276, 277, 279

Molindone, 90
Monoamine oxidase (MAO), 23, 112-116, 125
  inhibitors of, 120, 165, 166, 167
Monoamines, 53, 54
Mossy fibers, 94

N-acetylserotonin, 192, 193
N-acetyltransferase, 192, 193
NaF. *See* fluoride.
NE. *See* norepinphrine.
Nerve activity, sympathetic, 175
Nerve growth factor, 189, 191
Neurohormones, 2, 84
Neurohumoral agents, 72
  iontophoretic application, 72
Neuroleptic compounds, 69
Neuroleptic drugs, 68, 75
  clinical potency of, 113
Neuroleptics, 3, 11, 26, 27, 29, 43, 53-101, 62, 65, 68, 72, 76, 78, 89, 94, 113, 125, 143
  actions of, 94
    on central cyclic GMP systems, 92-94
    on central phosphodiesterses, 94-99
    on dopamine-sensitive adenylate cyclase systems in the CNS, 53-68; CNS in cerebral cortex, 65-67; CNS in mesolimbic-limbic structures, 65; CNS in other neural tissues, 67, 68; CNS in retina, 64, 65; CNS in striatum, 59-63; CNS in substantia nigra, 63, 64.
    on histamine-sensitive adenylate cyclase in central tissue, 76-78
    on hormonally-sensitive adenylate cyclases in peripheral organs, 87-92; adrenal cortex, 90, 91; heart, 89, 90; lung, 87-89; other peripheral organs, 91, 92; red blood cells, 90.
    on norepinephrine-sensitive adenylate cyclase systems in the CNS, 68-76; in rabbit brain, 69; in rat brain, 69-74; other species, 74, 75.
    on other central mechanisms which involve adenylate cyclase-cyclic AMP, 82-87; action of phenothiazines on basal levels of cyclic AMP, 86; adenosine, 83; catalytic site, 82, 83; depolarizing agents, 85; electrical stimulation, 83, 84; protein kinases, 86, 87; serotonin, 84, 85; transducer site, 83.
  antimuscarinic properties of, 92
  antipsychotic, 3, 62
  antipsychotic activity, 27
  diverse clinical effects of 98
  injected, 78-87
    influences of, on subsequent alterations in adenylate cyclase systems, 78-82
Neuromodulators, 220, 240, 251
Neuron, 207
  purinergic, 207
Neurons, 63, 72, 74, 77, 78, 92, 152, 162, 207, 220-225, 237, 249
  noradrenergic, 152
Neurons, dopamine, 68
  fractions of cerebral cortices, 82
  homogenates, 77
  pyramidal tract, 237, 249
Neuropharmacological control of pineal gland, 174
Neuropharmacology of cyclic nucleotides, 254-279
Neurotransmitters, 25, 53-56, 57, 113, 193, 212, 219-220, 229-230
Neurotransmitters, adrenergic, 191
  biogenic amine, 54
Neurotransmitters, central, 76
  cyclic AMP-stimulatory effects of, 25
  -cyclic nucleotide interrelationships, 230
  histamine, 76
  norepinephrine (NE), 192
Neurotubules, 58
Nialamide, 120, 166, 167
Nitrogen freezing, 258
N-methyl-DA, 64
N-methyldopamine, 90
N-methyl-5-methoxytryptamine, 85
N-methylserotonin, 85
Noradrenergic agents, 160
Noradrenergic cAMP, 156, 157
  supersensitivity, 158
    ethanol-induced, 158
Noradrenergic fibers, 68, 174
  of superior cervical ganglia, 174
Noradrenergic neurons, 152
Noradrenergic sensitivity, 166
  Cyclic AMP, 166
    neuroleptic, 166
    rats, 166
Noradrenergic stimulation, 190
Noradrenergic subsensitivity, 163-170
  acute, 164
  chronic, cyclic AMP, 165
  cyclic AMP, 166-170
    chronic, 166-169
Noradrenergic supersensitivity, 154, 155
Noradrenergic systems, 161
Noradrenergic transmission, 162, 165
Noradrenergically elicited cyclic AMP responses, 154-170
  acute subsensitivity in, 162-165
    agonist affinities, 163, 164
    changes in maximum responses, 163, 164
    molecular mechanisms, 164, 165
    pharmacological specificity, 164
    rates of appearance and disappearance, 162, 163
  chronic subsensitivity in, 165-169
    agonist affinities, 167
    anatomical and species specificity, 168
    changes in maximum responses, 167
    molecular mechanisms, 167, 168
    pharmacological specificity, 168
    rates of appearance and disappearance, 165-167
  chronic supersensitivity in, 154-162
    agonist affinities, 157, 158
    α-adrenergic *vs* B-adrenergic receptors, 158, 159
    anatomical and species specificity, 161, 162
    changes in maximum responses, 157, 158
    molecular mechanisms, 159, 160
    pharmacological specificity, 160, 161

rates of appearance and disappearance, 154-157
Norepinephrine, 2, 3, 5, 6, 11, 12, 17, 53, 54, 55, 59, 64, 65, 67-69, 71, 72, 74-78, 80, 84-86, 89, 90, 93, 112, 114, 117, 125, 137, 139, 153-155, 158, 160, 162-164, 176, 188-190, 202-209
Norepinephrine, *See also* cyclic AMP, synthesis.
adenosine-, 75, 163
at postsynaptic receptor sites 153
-histamine, 75, 77, 163
-induced accumulation of cyclic AMP, 202-207
pharmacology of, 205-207
-induced stimulation, 176
neurotransmitter, 192
receptor, 54
role as neurotransmitter, 53-56
-sensitive adenylate cyclase, 67, 74, 188-190
stimulation of adenylate cyclase by, 76
Norepinephrine (NE), 161, 164, 167, 168, 170, 175 177, 178, 180-182, 184-190, 192-194, 199, 200, 202-206, 208, 209, 219, 223, 241, 242
receptor, 220
activation, 209
Norepinephrine (NE)-sensitive adenylate cyclase, 188
development of, 188
Norfenfluramine, 7
Normetanephrine (NM) 139
Nuclei, rat, 168
accumbens, 168
amygdaloid, 168
olfactory, 168
septal, 168
5′-Nucleotidase, 57
postsynaptic, 57
Nucleotides, 176
guanyl, 176
Nucleus accumbens, 4, 8, 16, 40, 54, 65, 67, 73, 168

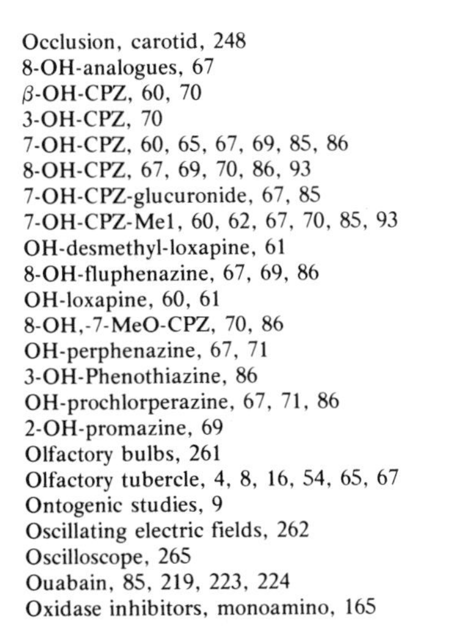

Occlusion, carotid, 248
8-OH-analogues, 67
β-OH-CPZ, 60, 70
3-OH-CPZ, 70
7-OH-CPZ, 60, 65, 67, 69, 85, 86
8-OH-CPZ, 67, 69, 70, 86, 93
7-OH-CPZ-glucuronide, 67, 85
7-OH-CPZ-Mel, 60, 62, 67, 70, 85, 93
OH-desmethyl-loxapine, 61
8-OH-fluphenazine, 67, 69, 86
OH-loxapine, 60, 61
8-OH,-7-MeO-CPZ, 70, 86
OH-perphenazine, 67, 71
3-OH-Phenothiazine, 86
OH-prochlorperazine, 67, 71, 86
2-OH-promazine, 69
Olfactory bulbs, 261
Olfactory tubercle, 4, 8, 16, 54, 65, 67
Ontogenic studies, 9
Oscillating electric fields, 262
Oscilloscope, 265
Ouabain, 85, 219, 223, 224
Oxidase inhibitors, monoamino, 165
Oxidative phosphorylation, 241, 242
Oxotremorine, 93
Ovary, 192

Pancuronium Br, 257
Papaverine, 97
Parathyroid hormone, 142
Pargyline, 23, 120, 166-168
Parkinson's disease, 1, 29, 30, 43, 44, 53, 54, 57, 80
P-creatine, 231-233, 238, 239, 243
Penfluridol, 64
Penicillin, 214
Pentylenetetrazol, 213, 214, 217
Pentylenetertrazol-induced seizures, 216
Perphenazine, 62, 67, 69, 82, 86, 93, 95
Pharmacological characterization of the dopamine receptor, 26-43
Pharmacology of norepinephrine-induced accumulation of cyclic AMP. 205-207
Phenobarbital, 216, 222, 223, 224
Phenothiazine, 6, 28, 54, 58, 64, 65, 67, 71, 75-77, 81, 82, 85-87, 125
Phenothiazine agents, 72
Phenothiazine analogues, 55
Phenothiazine derivatives, 60
Phenothiazine-like agents postsynaptic denervation supersensitivity of, 80
Phenothiazines, 7, 27, 55, 62, 69, 74, 78, 81, 82, 83, 86, 89, 90, 91, 96
inhibited basal levels of cyclic AMP, 86
monohydroxy metabolites, 76
Phenoxybenzamine, 2, 36, 134, 158
Phentolamine, 28, 67, 126, 134, 158, 177, 205, 206
(+)3-Phenylbutaclamol, 66
Phenytoin, 216, 223, 224
Phosphodiesterase, 4, 23, 37, 94-99, 114, 153, 159, 160, 164, 170, 174, 193, 200, 215, 231, 270, 275, 279
Phosphodiesterase activity, 162
AMP inhibitor, 37
cyclic AMP, 4
cyclic nucleotide, 200
enzymes, 96
inhibition of, 89, 97, 168
in rat brain, 270
neuroleptic action on, 95-98
neuroleptic-induced cellular supersensitivity, 98, 99
noncompetitive inhibition on, 95
peak I form of peak II, 97
reduction in activity, 99
specific molecular forms of, 101
Phospholipase
activation of by, 89
Phosphoprotein, 221
Phosphoprotein phosphatase, 4, 114
Phosphorylase, 58, 90, 254
Phosphorylase *a*, 222, 254
Phosphorylase *b*, 222, 254
Phosphorylation, 57, 221
oxidative, 241, 242
protein, 86
Photoreceptors, 174

Picrotoxin, 213
Pimozide, 6, 28, 36, 61, 63, 64, 65, 73, 75, 76, 97
Pindolol, 71
Pineal, 74, 193
Pineal gland, 74, 174-194
  adenylate cyclase activity, 182, 185, 189, 191
  neuropharmacological control of, 174
  rat, 174-194
    cyclic GMP system, 193, 194
    cyclic nucleotides of, 175
Pipamperone, 60, 61
Piripidel, 4
Pituitary, 67
  cells, 86
  dopamine receptor in, 13, 14
  gland, 5, 13, 14, 30, 31, 37, 38
  hormones, 112
  prolactin release, 67
Platelet, 125-137
  effect of lithium on, 130-133
  leukocyte, 133-137
  and leukocyte adenylate cyclase levels, 115
  human adenylate cyclase studies in, 137
Plasma, 112
  cyclic AMP, 115, 144-145
Pons-medulla, 161, 162
4-Port circulator, 263
3-Port ferrite circulator, 263
Post-decapitation increase in cyclic AMP, 255
Postmortem anoxia, 267
Postmortem brain, 137
Postmortem changes, 259
  anoxic
  in cyclic AMP, 255, 258
  in cyclic nucleotide levels, 254-260
Postmortem period, 17, 259
Postsynaptic membrane, 63
  -receptor complex, 63
Power density, 263
Potassium, 241
Presynaptic dopamine receptors, 28, 29
Probenecid, 143, 144, 215
Prochlorperazine, 60, 61, 62, 67, 69, 82, 86, 91, 93, 95
Prolactin, 7, 13, 14, 27, 30, 31, 37, 38, 112
Promazine, 6, 28, 60, 61, 69, 82, 83, 86, 92
Promethazine, 61, 62, 66, 67, 69, 73, 77, 78, 81, 83, 85, 91, 93, 97
Propranolol, 2, 28, 64, 67, 87, 133, 158, 164, 177, 205, 206, 217, 222
  [^{3}H]-propranolol, 160
Prostacyclin, 126
Prostaglandin, 54, 91, 160, 180
Prostaglandin E_1 $(PGE)_1$, 126, 135, 176
Prostaglandins, 25, 89, 191
Protein kinase, 4, 57, 114, 223, 225, 231, 241, 279
  phosphorylation, 100
Protein kinases, 55, 86, 220, 221
Protein phosphorylation, 225
PRT-1742, 33
Psilocybin, 85
Psychosis, 53, 54, 55, 65, 68
  paranoid, 143
  rise in cyclic AMP, 81
Psychotropic drugs, 146
PTR 17-402, 35
Pulse-labeling technique, 134
Purkinje cells, 72, 94, 220, 221
  firing rates, 73
Pyramidal cells, 207, 209, 220
Pyramidal tract neurons, 230, 237, 249
Pyridine nucleotides, 261

Quaternary-CPZ, 77

Radiofrequency lesions, 10
Radioligand binding, 152, 153
Random-bred strain, 199, 201
Rapid fixation, 231, 233
  freezing, 231
  microwave irradiation, 231
Receptor, 74, 153, 159, 162, 164, 165, 167, 169, 170, 175, 177, 180, 190, 191, 200, 205, 207, 208, 209, 219, 220
  -activated synthesis, 200
  adenosine, 219
  adenosine-dependent, 207
  adrenergic, 191
  affinity, 167
    [^{3}H]-dihydroalprenolol (DHA), 167
  α-, 205, 207, 208
  α-adrenergic, 162, 207, 209
  β-adrenergic, 159, 162, 165, 167, 170, 175, 177, 180, 190, 205
  β-adrenergic membrane, 169
  catecholamine, 205
  dopamine, 62, 68, 74, 220
  hypothetical, 83
  norepinephrine (NE) activation, 209, 220
  β-plus-α catecholamine, 207
  postsynaptic, 101
β-Receptor, 34, 38
Receptor complex, postsynaptic membrane, 63
Receptor sites, 152, 242
Recirculation, 229-251
Redox states, 267
Regulator proteins, 55
Renal artery, 43, 44
Resensitization, 184, 185
Reserpine, 11, 30, 37, 80, 114, 154, 156, 158, 159, 160, 162, 217, 219, 221
Respiratory arrest, 257, 258
Retina, 3, 16, 64
Rostral limbic nuclei, 73

S 584, 4
Schizophrenia, 53, 54, 55, 57, 65, 68, 73, 112, 126, 129, 134, 138, 146
  paranoid, 53
  pathogenesis of, 73, 145
  acute, 130, 133, 144
  subacute, 130
Second messenger, 212
Seizures, 212-225, 234, 243, 250, 251
  audiogenic, 214-215
  clonic, 213, 214, 216, 217, 218, 250
  clonic-tonic-clonic-convulsions, 214

effect on CSF cyclic nucleotides, 215, 216
effect on cyclic nucleotide levels, 212–215
epileptogenic, 214
focal, 214
generalized convulsions, 214
-induced elevations of cyclic AMP, 216
-induced elevations of cyclic GMP, 216
3-mercaptopropionic acid-induced, 222
myoclomic spasm, 213, 217, 218
pentylenetetrazol-induced, 216
tonic, 213–218, 221, 250
Sensitivity, 153
noradrenergic, 166
cyclic AMP, 166
noradrenergically induced cAMP, 153
Septal nuclei, 73
Septal region, 54
Serotonin, 23, 54, 84, 85, 100, 112, 139, 160, 168, 192, 219, 220
N-acetyltransferase, 190
Setoclavin, 33
Setoclavine, 35
Sodium, 17, 224
Sodium fluoride, 13, 24, 31, 34, 165, 176, 180, 181, 183, 185, 186, 188, 190
Sodium potassium pump, 241
Soma, 58
Species, choice of, in dopamine-sensitive assay, 15, 16
Spinal cord, 230, 243, 244, 246, 247
Spiroperidol, 6, 28, 36, 38, 60, 61, 91, 94
Spontaneous activity, 200
Spreading depression, 261
SQ 20,009, 95
Stabilization of stimulants in dopamine-sensitive assay, 23
Steroids, 192
Stimulants, 243
Stimulation, adrenergic, 190
noradrenergic, 190
norepinephrine-induced, 176
Strain, 199, 201, 202, 208
genetic crosses, 199
inbred, 199, 201, 202, 208
random-bred, 199, 201
Stress, 258, 261
Stress of decapitation, 271
Striatum, 1–5, 7, 8, 9, 11, 16, 17, 19, 24, 26–28, 31, 32–36, 37, 40, 43, 44, 54, 57, 62, 73–75, 79, 80, 83, 84, 161, 202, 213, 217, 265, 268, 269, 275–277, 279
corpus striatum, 3, 25, 36
homogenization of, 19
-plus-midbrain tissue, 208
striatal function, 3
striatal homogenate, 27, 74
striatal membranes, 7
Stroke, 232
Subcortical regions, 214
Subsensitivity, 80, 153, 162, 163, 165, 166, 170
acute, 162, 163, 165
β-adrenergic receptors, 125
chronic, 165
noradrenergic, 163–170
acute, 164
chronic, cyclic AMP, 165
cyclic AMP, 166–170
Substance P, 54
Substantia nigra, 8, 12, 16, 30, 54, 63, 64
adenylate cyclase, nigral, 64
nigral firing rates, 63
Sucrose solutions, 19
Sulpiride, 7, 27, 28, 61, 62, 66
Superior and inferior colliculi, 214
Superior cervical ganglia, 3, 174, 177, 184–186, 187, 191, 193
noradrenergic fibers of, 174
Superior cervical ganglion, 16, 175
Supersensitivity, 153, 156, 157, 160, 161, 162, 170, 186
cAMP, 161
neuroleptic-induced, 98, 99
noradrenergic, 154, 155, 158
Supersensitivity phenomenon in dopamine-sensitive assay, 9–13
Supersensitivity studies of injected neuroleptics, 78–81
Sympathetic nerve activity, 175
Synaptosomes, 11, 25
striatal, 25
synaptosomal-mitochondrial preparations, 11
Synthesis, receptor-activated, 200

Tardive dyskinesia, 78, 80
phenothiazine-induced syndrome of, 81
Telencephalon, 200–202
Temperature effects in dopamine-sensitive assay, 19
TES-EGTA, 19
TES-HC1 buffer, 26, 27
Tetrahydroxyisoquinolines, 28
Tetrodotoxin, 224
Thalamus, 67, 213, 260, 277
Theophylline, 4, 5, 23, 25, 37, 77, 79, 81, 97, 201, 202, 205–207, 217, 219, 222
Thermal diffusion, 260, 261, 262
Thermistors, 260
Thermograph, 264, 266
Thiamylal, 247–249, 251
Thiethylperazine, 61
Thioproperazine, 61
Thioridazine, 60, 61, 62, 65, 66, 67, 73, 77, 78, 89, 91, 93, 95, 97
Thiothixene, 60, 61, 69, 86
Thioxanthines, 27, 61
Thoracic ganglion, 68
Thyroid, 82, 91
Tolerance, 170
Transducer site, adenylate cyclase, 75
Transmission, noradrenergic, 162
Trauma, 81
Tricyclic antidepressant, 28, 80, 114, 116, 143–144, 165
tertiary, 93
therapy, 81
Trimeprazine, 60, 61
Trifluoperazine, 60–61, 64, 66, 71–72, 81, 82, 86, 89, 95, 97
sulfoxide, 72

Trifluphenazine, 93
Triflupromazine, 60, 61, 67, 91, 95
Triiodothyronine (T_3), 116–119
TriOH-CPZ, 60, 70
TRIS, 26
TRIS-acetate-EGTA, 19
TRIS-maleate-EGTA, 19
Tryptophan, 192
TSH (thyroid stimulating hormone), 91
Tubercle, rat, 168
  olfactory, 168
Tyramine, 23
Tyrosine hydroxylase (TH), 37, 112, 125

Urine, 112
  cyclic AMP excretion, effect of electroconvulsive therapy on, 141
  cyclic AMP in, 138
  cyclic AMP levels in, 79, 115, 142

Valproic acid, 223
Variability in dopamine-sensitive assay, 14–26
Vasopressin, 91
Veratradine, 160
Veratridine, 85, 93, 219, 223, 224

Water content, interfaces separating tissues of, 264
Waveguide, 262
  system, 265
  unit, 263
WB 4101, 153
  [^{3}H]-, 153
Withdrawal, 170

X-irradiation, 72